Securing Information and Communications Systems

Principles, Technologies, and Applications

For quite a long time, computer security was a rather narrow field of study that was populated mainly by theoretical computer scientists, electrical engineers, and applied mathematicians. With the proliferation of open systems in general, and of the Internet and the World Wide Web (WWW) in particular, this situation has changed fundamentally. Today, computer and network practitioners are equally interested in computer security, since they require technologies and solutions that can be used to secure applications related to electronic commerce. Against this background, the field of computer security has become very broad and includes many topics of interest. The aim of this series is to publish state-of-the-art, high standard technical books on topics related to computer security. Further information about the series can be found on the WWW at the following URL:

http://www.esecurity.ch/serieseditor.html

Also, if you'd like to contribute to the series by writing a book about a topic related to computer security, feel free to contact either the Commissioning Editor or the Series Editor at Artech House.

For a listing of recent titles in the *Artech House Computing Security Series,* turn to the back of this book.

Securing Information and Communications Systems

Principles, Technologies, and Applications

Steven M. Furnell, Sokratis Katsikas,
Javier Lopez, Ahmed Patel
Editors

ARTECH HOUSE
BOSTON | LONDON
artechhouse.com

Library of Congress Cataloging-in-Publication Data
A catalog record for this book is available from the U.S. Library of Congress.

British Library Cataloguing in Publication Data
A catalogue record for this book is available from the British Library.

ISBN 13: 978-1-59693-228-9

685 Canton Street
Norwood, MA 02062

10 9 8 7 6 5 4 3 2 1

Contents

Preface

The idea for this book was born within a series of courses of a pan-European collaboration that has successfully been delivered at Master's-level intensive programs since 1997 at a variety of European venues, such as Greece (Samos and Chios), Sweden (Stockholm), Finland (Oulu), Spain (Malaga), Austria (Graz), Belgium (Leuven), and the United Kingdom (Glarmorgan). The course is scheduled to be held in Regensburg, Germany, during 2008. Its title is *Intensive Program on Information and Communication Security* (IPICS), and it is based on a comprehensive IT/ICT security curriculum that was itself devised as part of an EU collaborative project under the ERASMUS program.

IPICS has a long and distinguished history. It grew from a simple idea to a very complex undertaking. It has been maintained with minimum financial support but great enthusiasm and commitment by the lecturers who gave their time and effort at free will and without any form of payment. The participating institutions also ensured that they sent very good students to take full advantage of not only IPICS courses but also learning and experiencing the culture and traditions in the country or town where the school was held. It was also fun and fostered long-term friendships.

During this period, a large number of people (expert lecturers, administrators, students, sponsors, and so on) have contributed to the evolution of this book, and we'd like to thank everyone, particularly this book's participating authors, who turned their lecture notes into text to benefit the readers. The interactions and constructive discussions between students and lecturers, and between lecturers, have certainly resulted in improvements in the delivery of course material in subsequent IPICS schools.

We'd also like to thank the commissioning editorial staff at Artech House for their support and help in this book. We thank the independent reviewers for their comments, thoughtful criticisms, and suggestions—they have been invaluable. We also thank Alexandros Tsakountakis, graduate student at the Department of Information and Communication Systems Engineering, University of the Aegean, Greece, for his invaluable support during the stage of the preparation of the final manuscript.

Last, but not least, on behalf of the whole IPICS team (expert lecturers and assistants), we thank our families and associates, who tolerated our absence from time to time when lecturing at IPICS schools and during the last few months when the book was being written and compiled under enormous time constraints and pressure.

Steven Furnell
Sokratis Katsikas
Javier Lopez
Ahmed Patel
Plymouth, UK; Piraeus, Greece; Malaga, Spain;
Kuala Lumpur, Malaysia; November 2007

CHAPTER 1

Introduction

Steven M. Furnell, Sokratis K. Katsikas, Javier Lopez, and Ahmed Patel

Over the past decade or more, the topic of IT/ICT security has worked its way from being viewed as an add-on gadget that is nice to have or nice to know about to becoming an essential consideration within the systems and applications that our society depends on. In spite of this, average knowledge of security among IT/ICT professionals and engineers is lagging much behind the evolution of the potential threats and the schemes, methods, and techniques to overcome them within the framework of international and national laws and directives. At the pace at which e-business, e-leisure, and e-social computing and electronic activities are taking place on the Internet, the whole area of security together with the need to provide privacy, trust, safety, and traceability for forensic and investigation purposes as a "total" solution requires that the professionals and newcomers to security are familiar with the fundamental aspects to deliver appropriate and valid solutions.

It is without doubt that the rapid growth of Internet-based "e-everything" (e-commerce, e-business, e-leisure, e-social, e-education, e-payment, and so forth) depends on the security, privacy, and reliability of the applications, systems, and supporting infrastructures. However, the Internet is notorious for its lack of security, and it is widely known that commercial and trustworthy applications are susceptible to failures and exposed to attacks when not properly and rigorously specified, designed, and tested. These failures and attacks can cause serious damage to the participants—commercial traders, financial institutions, government and nongovernment institutions, service providers, end-systems, and consumers. This is even more so when privacy and confidentiality is compromised and violated, and traceability is absent. In the past couple of years, it has been obvious that the area of IT/ICT security has to include powerful tools and facilities with forensically safe, verifiable, auditable, and quality-managed systems and applications upholding confidentiality and privacy in e-everything environments, which define the problems and specify the models, rules, and protocols.

It is no wonder that governments, human rights organizations, corporations, law enforcement agencies, and other organizations are realizing the power of security to combat cybercrime and interception in tracking fraudulent or unacceptable user behavior. This would include criminals, abusers, pedophiles, and a host of other malicious persons and activities.

In such situations, it becomes essential that the underlying frameworks, concepts, protocols, and standard services provide the necessary functions in their applications expected by the users and service providers. A common solution at the practical level is to use security and privacy protocols with a combination of cybercrime preventative, audit, and investigative mechanisms with good quality management and awareness. This is no mean task, but it is the only way to progress to the next stage of opening up the full e-everything services on the Internet.

It is precisely here that we address many of these problems in an intensive manner through a set of independent but well-structured and linked chapters.

The experts that contributed to this book come from a wide range of backgrounds. They have endeavored to provide up-to-date information addressing security and related emerging technologies from all of its facets—mathematical, engineering, legal, social, privacy and forensics, education, training awareness, and managerial, which includes chapters on the following:

- Basics of security concepts, services, and threats;
- Models for quality business integrated information security management;
- Principles of user authentication technologies;
- Principles of authorization and access control and their applications;
- Security of data-centric applications;
- Principles of modern cryptology and new research challenges in this field;
- Network security and its dynamics;
- Public key and privilege management infrastructures;
- Architectural and functional characteristics of smart cards and similar tokens;
- Privacy issues and privacy-enhancing technologies from legal and technical perspectives;
- Legal and technical issues relating to secure content-filtering technologies;
- A model for cybercrime investigations for forensic examination and presentation;
- Systemic-holistic approach to IT security with subjective details based on objective knowledge;
- Secure electronic voting systems using cryptology models and protocols;
- Requirements and architecture for mobile wiki systems security.

In a book of this size, it is not possible to cover every aspect of security and related subject areas in breadth and depth, but it is intensive enough to convey the "message." Given this restriction, the structure of the book is designed to make each chapter a standalone piece of work. Each chapter addresses the topics deemed necessary to convey the important aspects of the subject. The book itself starts from basic security and security management, gradually builds up to the technical chapters, and terminates with applications of security and other cognate subject areas. Each chapter has its own reference list to look up more advanced work in the topic/subject area.

The book is aimed at students taking undergraduate and graduate courses and at professionals and engineers in education, government, business, and industry. It may be used in any general security-related courses or in advanced security courses in programming, networking software and hardware engineering, software specification, software design, IT/ICT systems, applications, management, and policy. Readers, particularly professionals, engineers, and university professors, may find the book useful as general reading and as a means of updating their knowledge on particular topics such as cryptology, privacy, smart cards, cybercrime, and digital forensics. Primarily the book aims to make the reader versatile in the field of security and related subject areas without having to buy or read several books.

The book will be used at all IPICS schools in the future and updated as required.

CHAPTER 2

Security Concepts, Services, and Threats

Paul S. Dowland and Steven M. Furnell

This chapter provides a baseline introduction to the concepts of IT security, as well as a background on the need for protection from an organizational perspective. It begins by providing examples of incidents, statistics indicating the scale and nature of the problem, and a number of cases drawn from relevant surveys. These incidents are linked to the key requirements for preserving the confidentiality, integrity, and availability of systems and data. This overview leads into a more specific consideration of the main classes of abuse and abuser. The chapter introduces many of the topics and terminology covered in later sections, allowing the reader to build a solid foundation on which to base the material in later chapters.

2.1 Definitions

The need for information and communication systems security essentially arises from our dependence on information technology to support an ever-increasing range of activities in our domestic and working lives. For the individual, accessing Internet-based services such as email and the web has become a standard element of everyday life, and going online is often the first route in a range of situations, from simply seeking information to tasks such as shopping and banking. For the organization, the use of technology is equally well established, with services such as email being part of the lifeblood of modern business communication, and online transactions forming an increasing part of commercial activity. As a consequence, there is a significant requirement for protection, which leads us to consider the security of the information and communication systems that we depend on.

But what exactly is information security anyway? Despite what other books (and indeed other chapters in this one!) may try to tell you, there is no single short answer. To illustrate the point, we can consider a series of brief definitions, taken from a range of respectable sources. Each one says something different . . . and all of them are correct!

- "Information security is the protection of information from a wide range of threats in order to ensure business continuity, minimize business risk, and maximize return on investments and business opportunities" ISO/IEC 17799, Code of Practice for Information Security Management, 2005 [1];
- "Computer security is the protection of a company's assets by ensuring the safe, uninterrupted operation of the system and the safeguarding of its

computer, programs and data files," Prof. Harold J Highland, State University of New York;
- "The protection of information assets through the use of technology, processes, and training" Microsoft Security Glossary [2];
- "The ability of a system to manage, protect, and distribute sensitive information," Software Engineering Institute (SEI), Carnegie Mellon University [3].

As we can see, each definition serves to highlight a different aspect. The definition from the ISO standard has a clear emphasis on the role of security to safeguard the business, recognizing that there could be significant impediments without it. The definition from Highland can be seen as a further development of this theme, in the sense that it still takes the company perspective (emphasizing that, at the core, there will be assets of value that require protection) but adds an insight into what needs to be considered in order to achieve this. The Microsoft definition also refers to assets, but specifically in the context of information, and usefully highlights the variety of means that need to be used. By contrast, the definition from SEI can actually be considered to be quite a restricted view, in that it places emphasis specifically on sensitive information. Although this is certainly a requirement in many contexts, it can easily be argued that *all* information requires some level of protection—regardless of whether or not it is sensitive. For example, even mundane information on a public web site requires protection to ensure that it remains free from unauthorized modification and remains accessible when needed. However, the SEI definition does link to the view that many people have when the issue of security is raised—namely, that it is all about protecting secrets and critical data—which can lead them to overlook the requirement to protect other things.

Since 1994, the U.K. Department of Trade and Industry (DTI) has undertaken an annual survey of small to medium-sized organizations to evaluate the level and impact of computer crime and abuse. In the most recent survey [4] (2006), it was reported that the cost of security incidents has risen by approximately 50 percent over the previous year, with 62 percent of U.K. companies experiencing some form of security incident. This figure is somewhat different in the United States, where the recent CSI/FBI survey [5] showed a decrease in both the overall level and cost of reported incidents (although the report does note that respondents did indicate that incidents are still not reported due to the potential adverse publicity that the organizations concerned may receive). The CSI/FBI survey specifically asks about the reporting of incidents to either law enforcement or legal counsel, with the most recent survey indicating that 30 percent of organizations that identified an incident of computer intrusion did not report it, with the majority simply patching the holes utilized in the attack (more than 48 percent stated that the potential for negative publicity was the primary reason for not reporting incidents).

Before considering the issues surrounding computer security, it is important to first define a series of keywords that form a vital part of the understanding and inter-relationship between threats and vulnerabilities.

- *Threat*: a circumstance that has the potential to cause loss or harm. Threats may be accidental (e.g., fire, power failure, operator error), environmental (e.g., flood, lightning), or deliberate (e.g., hacking, malicious software, eavesdropping).

- *Vulnerability*: influences the likelihood of a threat to become a reality and relates to a weakness in the system that might be exploited to cause loss or harm. Whereas threats are generic, the vulnerability to them is environment-specific, as it depends on factors such as the systems in use and the countermeasures in place to protect them.
- *Asset*: everything and everybody forming part of an information system, including hardware, software, and people.
- *Risk*: the threats and vulnerabilities relating to a particular asset. Risk is addressed by reducing the vulnerability of our assets to relevant threats. This is achieved through the use of controls/countermeasures, which are mechanisms or procedures used to reduce one or more elements of risk.
- *Impact*: results from the occurrence of a security breach and represents the effect of a failure to preserve confidentiality, integrity, and/or availability. Impacts can be measured in terms of disclosure, denial, destruction, or modification of systems and data.
- *Consequence*: an adverse outcome that results from an impact. Consequences effectively represent the harm that an incident has done to a system or those connected with it. Indicative consequences could include business disruption, danger to personal safety, disclosure of commercial information, infringement of personal privacy, failure to meet legal obligations, financial loss, and loss of reputation or goodwill.

Having introduced the terms, Figure 2.1 illustrates how they all fit together, indicating the point at which the risk becomes a reality following a security breach.

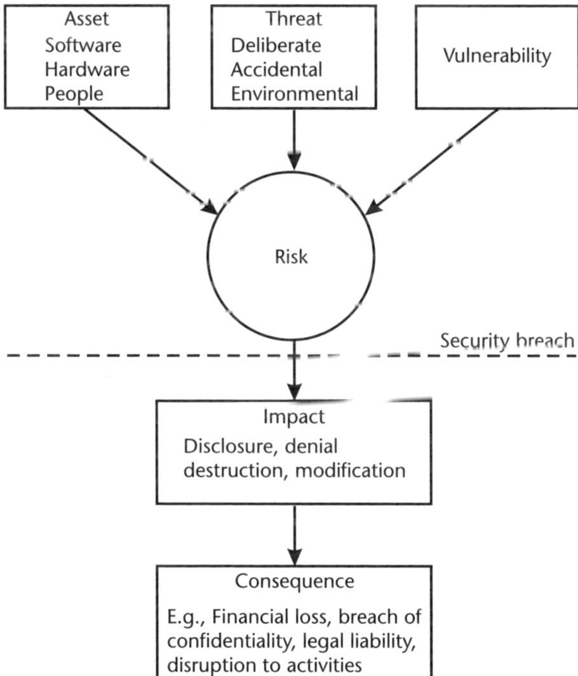

Figure 2.1 Factors relating to IT risk.

2.2 Threats and Vulnerabilities

2.2.1 Threat Types

At a high level, threats may be grouped into three main categories, resulting from a variety of accidental or deliberate acts against which systems must be protected:

- Physical threats (e.g., fire, flood, building or power failure);
- Equipment threats (e.g., CPU, network, or storage failure);
- Human threats (e.g., design or operator errors, misuse of resources, various types of malicious damage).

However, the extent to which a threat actually represents a problem will depend on the vulnerabilities of the organization or system concerned.

2.2.2 Vulnerabilities

Vulnerabilities can take many forms, and some typical examples are listed here, along with the ways in which they could be exploited. It may be noted that these points reinforce the earlier observation that computer security relates to more than just the protection of the computer itself.

1. Environment and infrastructure vulnerabilities:
   - Lack of physical protection of the building, doors, and windows (could be exploited by theft);
   - Inadequate access control to rooms (could be exploited by unauthorized access);
   - Unstable power grid (could result in power failure).
2. Hardware vulnerabilities:
   - Susceptibility to temperature variation (could be caused by overheating);
   - Insufficient maintenance of storage media (could result in media failure);
   - Lack of efficient configuration change control (could be exploited by operational staff).
3. Software vulnerabilities
   - Complicated user interface (could result in user error);
   - Lack of authentication (could be exploited by masquerading);
   - Lack of audit trail (could be exploited by unauthorized software use).
4. Communications vulnerabilities:
   - Unprotected communication lines (could be exploited by eavesdropping);
   - Unprotected sensitive traffic (could be exploited by eavesdropping);
   - Unprotected public network connections (could be exploited by unauthorized users).
5. Personnel vulnerabilities:
   - Unsupervised work by outside staff (could be exploited by theft);
   - Insufficient security training (could result in user error);

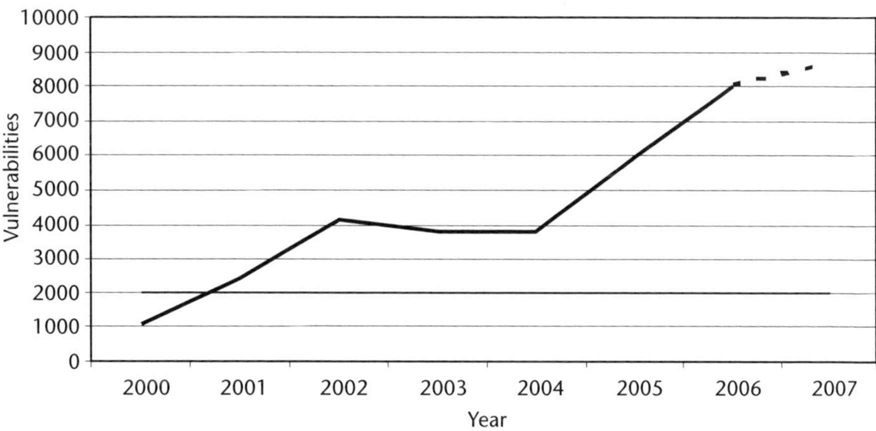

Figure 2.2 CERT/CC vulnerability statistics from 2000–2007 [6].

- Lack of monitoring mechanisms (could be exploited by use of systems in an unauthorized way);
- All of the things that could potentially exploit the vulnerabilities are regarded as threats to the system.

It should be noted that while software vulnerabilities are usually the most commonly encountered form, they are also often the easiest to fix. In most cases, vulnerabilities are fixed by the simple installation of a patch from the operating system (or application) vendor. However, this task is made increasingly difficult when organizations operate a range of operating systems and vendor applications. The United States Computer Emergency Response Team Coordination Center (CERT/CC) monitors vulnerabilities reported worldwide and produces annual reports. Figure 2.2 clearly shows the scale of the problem indicating a substantial increase over the period 2000–2007 (Note: the figure shown for 2007 is an estimate based on the results available from the first quarter of 2007).

2.2.3 Attacks and Misuse

One of the most commonly recognized reasons (although far from the only reason) for requiring security is to protect our systems against those people who would intentionally attack them. Table 2.1 presents an overview of the traditional classes of human abuser, based on the classification proposed by Anderson [7].

When considering computer-related threats, one of the most widely recognized terms is *hacker*. However, the precise meaning of this name is the subject of some debate. Whereas, in the early days of computing, it was used as a recognition of someone's technical skills, the name has recently become more closely associated with individuals attempting unauthorized access to systems and data. However, using such a term as a catchall omits the consideration of what people are actually doing when such access is gained, and here we begin to encounter other labels based on the hacker's actions and motives. For example, the terms *black hat* and *white hat* are often used to provide a high-level distinction, with the former referring to those

Table 2.1 Categories of System Abuser

Abuser Type	*Description*
External penetrators	Outsiders attempting to gain or gaining unauthorized access to the system (e.g., a hacker trying to download the password file(s) from a server or a rival company trying to access the sales database).
Internal penetrators	Authorized users of the system who access data, resources, or programs to which they are not entitled. Subcategorized into the following: Masqueraders: Users who operate under the identity of another user (e.g., someone using another user's PC while they are absent from their terminal, or someone using another's user name/password). *Clandestine users*: Users who evade access controls and auditing (e.g., someone disabling security features/auditing).
Misfeasors	Users who are authorized to use the system and resources accessed, but misuse their privileges (e.g., someone in the payroll department accessing a colleague's records or misappropriating funds).

intruding into systems in an unauthorized, and frequently malicious, manner (for which another oft-used label is *cracker*), while the latter refers to "ethical" hackers, working for the benefit of system security. A further term, *grey hat*, is used to refer to individuals whose activities and motives are somewhat unclear or may be prone to change.

In order to further illustrate the lack of a clear-cut black and white, good and bad distinction, here are some names that are frequently ascribed to members of the hacker community (it should be noted that even this still does not provide an exhaustive list) [8]:

- *Cyberterrorists*: terrorists who employ hacker-type techniques to threaten or attack against systems, networks, and/or data. As with other forms of terrorism, cyberterrorist activities are conducted in the name of a particular political or social agenda. The underlying objective will typically be to intimidate or coerce another party (e.g., a government).
- *Cyber warriors*: persons employing hacking techniques in order to attack computer systems that support vital infrastructure, such as emergency services, financial transactions, transportation, and communications. This essentially relates to the application of hacking in military and warfare contexts.
- *Hacktivists*: hackers who break into computer systems in order to promote or further an activist agenda. Incidents such as the defacement of web sites are very often linked to these individuals.
- *Malware writers*: not strictly a classification of hacker—but often considered alongside them—these individuals are responsible for creating malware programs such as viruses, worms, and Trojan horses.
- *Phreakers*: individuals who specifically focus on hacking telephone networks and related technologies. Their objectives may range from simple exploration of the infrastructure to actually manipulating elements of it (e.g., to enable free phone calls to be made).
- *Samurai*: individuals who are hired to conduct legal cracking jobs, penetrating corporate computer systems for legitimate reasons. Such hackers may also be termed *sneakers*.

- *Script kiddies*: individuals with fairly limited hacking skills who rely on scripts and programs written by other, more competent hackers. Hackers of this type typically cause mischief and malicious damage and are generally viewed with scorn by more accomplished members of the hacking community. Such individuals may also be referred to as *packet monkeys*.
- *Warez d00dz*: a subclass of crackers who obtain and distribute illegal copies of copyrighted software (after first breaking any copy protection mechanisms if appropriate). The spelling used is representative of a common form of hacker slang—in this case the two words, when written properly, are "wares dudes." More commonly, these individuals are known as *software pirates*.

2.2.4 Impacts and Consequences of Security Breaches

Following a security breach, the impacts can be measured in terms of four critical factors (linked to the security objectives outlined later in this chapter):

- *Disclosure*: This most closely relates to a breach of confidentiality, in which private or sensitive information is disclosed to a third party. The severity of the disclosure is likely to be higher in which the disclosure is to an individual outside of the organization.
- *Denial*: The denial of access to data or systems is linked to the concept of availability and could have far-reaching consequences (see examples later in this section).
- *Destruction*: This is also linked to the issue of availability. The destruction of data will lead to the same consequences as denial until the point at which the data is recovered; however, if the data is irrecoverable, then this is likely to lead to greater problems.
- *Modification of systems and data*: This is the most difficult impact to evaluate, as it relates to the issue of integrity. If the modification is detected, the recovery can be achieved from suitable backups; however, if it goes undetected for a period of time, there is considerable opportunity for harm.

As indicated earlier, all of these impacts can be caused through malicious actions (e.g., a hacker) or through accidental means. For example, simply tripping over a network cable can have a significant impact on an end user, or, at the other extreme, a fire in the server room can halt an entire organization's IT infrastructure.

The consequences of a security breach can be widespread and are not always immediately obvious. While some consequences can be directly related to financial loss, an organization could face more subtle implications.

- *Embarrassment*: Large organizations are often reluctant to publicly acknowledge security breaches (especially in the banking/insurance sectors), as there is likely to be a negative impact on the reputation or perception of the company.
- *Breach of commercial confidentiality*: An example would be customer data in an organization being released to a competitor.

- *Breach of personal privacy*: Many organizations have large volumes of information about individuals in which there is an expectation of privacy (e.g., insurance companies).
- *Legal liability*: If the breach were to lead to personal information being disclosed to third parties, or there was evidence that there was inadequate control over the protection of the resources, the organization could face legal proceedings.
- *Disruption to activities*: Some level of disruption is almost always unavoidable, but in particularly serious circumstances, an organization may be unable to continue business for the duration of the incident (and follow-up rectification) and may even cease to trade completely.
- *Financial loss*: In almost every case there is some form of financial loss incurred as a consequence of an incident. Losses can include direct loss of business, loss of confidence resulting in reduced business, and costs incurred in investigating and rectifying the problem.
- *Threat to personal safety*: It is possible that a failure in availability (or integrity) could result in a failure to access information that could jeopardize personal safety—for example, in a health care environment, failure to access pertinent information could result in patient care being denied or incorrectly prescribed.

To illustrate the range of impacts and consequences, a number of example scenarios are presented next:

- A construction worker accidentally cuts through the network link between a hospital building and the IT data center, preventing access to patient records. In an ideal world, the network would be fault-tolerant and the data center would be dual-homed to ensure continued access to the data. However, a worst-case scenario could result in patients being denied treatment, or, worse, given the wrong treatment as previous drug treatments and allergy records would be unavailable.
- A hacker compromises an online banking server and steals 10,000 credit card details. While there is clearly a risk to personal privacy and financial loss, it is likely that the bank would also suffer considerable embarrassment—particularly as online banking depends on a high level of trust between the customer and the bank. If the bank were found to be negligent, then there could also be issues of legal liability.
- A denial of service (DoS) attack occurs against an online auction site. Apart from the obvious inconvenience to buyers and sellers alike, an attack of this nature would be likely to cause damage to the reputation of the auction site. As an auction site derives income from commission on sales, a DoS attack would also have financial consequences for both the auction site and the sellers using the site.

2.3 Security Services and Safeguards

The breadth of the issues that need to be considered in order to achieve comprehensive protection can be illustrated by examining the 11 top-level clauses expressed within the international code of practice for information security management:

- *Security policy*: outlines the need for an organization-wide information security policy, which is documented and available to all staff. Specific guidelines cover the definition of the policy and procedures for allocating responsibilities in its implementation. Issues regarding policy dissemination, monitoring, and review are also considered.
- *Organization of information security*: covers the establishment of a management structure for maintaining information security, including the setup of the forum, its functions, and the allocation of responsibilities. It also covers the establishment of an authorization procedure for hardware and software purchases, access to corporate data by third parties, and the measures required to prevent and detect unauthorized access.
- *Asset management*: relates to the protection of organizational assets and deals with the establishment of an asset inventory for hardware, software, and information, and advises on the classification, labeling, and handling of assets.
- *Human resources security*: covers the risks posed by deliberate and accidental human actions, including user error, fraud, and theft. It addresses how to include security responsibilities as part of the job description, the screening of potential staff, security training and awareness, and the establishment of a framework for reporting security incidents and suspected weaknesses.
- *Physical and environmental security*: covers the need to establish secure areas with physical entry controls, the physical placement and protection of equipment (e.g., to minimize risks such as fire, theft, and power failure), equipment maintenance, and the security of physical assets when they need to be taken off site or disposed of.
- *Communications and operations management*: highlights a wide variety of operational and housekeeping issues, including protection against viruses and other malicious software, change control, data backup, security of system documentation, disposal of media, the protection and authentication of data during transfers and in transit, and email and web security requirements.
- *Access control*: covers access to information, and the formulation of policies and rules to control it appropriately. This includes user authentication and privileges, network access control (including external and remote access), use of controls in operating systems and applications, and methods for monitoring system access and use. The recommendations here also cover situations involving users working offsite.
- *Information systems acquisitions, development, and management*: identifies security considerations pertaining to the acquisition of new systems, and the modification and maintenance of existing ones. Issues covered include application security (e.g., validation of input data), use of cryptographic controls (such as encryption and digital signatures) to protect particularly sensitive data, the security of system files, and security during software development activities (e.g., change control and guidelines on outsourcing).
- *Information security incident management*: highlights the need to establish mechanisms to enable the reporting of security events, as well as observed or suspected security weaknesses. Coverage is also devoted to the management

of security incidents (e.g., responsibilities, procedures, collection of evidence), and measures might be taken to improve controls as a result.

- *Business continuity management*: identifies how a comprehensive business continuity plan should be designed, implemented, tested, and maintained for use in the event of major failures and disasters (e.g., fire, flood, hardware failure). The intention is to ensure that critical organizational processes can continue with minimal disruption, with appropriate fallback, containment, and recovery procedures in place.
- *Compliance*: addresses the various legislative and regulatory requirements with which an organization must comply. Legislative requirements will include compliance with laws relating to copyright, data protection, computer misuse, and safeguarding of business records. In addition, organizations also need to ensure that their own internal policies and procedures are being complied with. The guidelines highlight the various aspects of compliance and advise how to monitor them via audits.

2.3.1 Identifying Assets and Risks

If security is to be incorporated effectively, a means is needed to determine which assets require protection and to what level. This can be achieved by assessing the level of risk in the system and then by applying security controls and countermeasures accordingly.

2.3.2 Security Objectives

The goal of computer security is commonly recognized as maintaining three top-level principles or characteristics: confidentiality, integrity, and availability (CIA).

Confidentiality refers to the prevention of unauthorized information disclosure and is normally the aspect that people most closely associate with the concept of security. In the majority of organizations, it is desirable to restrict access and allow only authorized users (those possessing a legitimate "need to know") to see specific systems and data. The type of access is read-type access: reading, viewing, printing, or even just knowing the existence of an object. The seriousness of an unauthorized disclosure may often be dictated by whether it occurs to a member of the same organization or a total outsider (with the consequences from the latter being potentially more severe).

Integrity relates to the prevention of unauthorized modification of information. Users must be able to trust their systems and be confident that the same information can be retrieved as was originally entered. Furthermore, assets must only be modified by authorized parties in authorized ways. In this context, modification includes writing, changing, changing status, deleting, and creating. The integrity of a system or its data may be compromised as a result of accidental error (e.g., during data entry) or malicious activity (e.g., virus infection).

Availability relates to the need for data and systems to be accessible and usable (by authorized parties) whenever and wherever they are required. This necessitates both the prevention of unauthorized withholding of information or resources, as well as adequate safeguards against system failure. The seriousness of any denial of service will, in most cases, increase depending on the period of unavailability.

There are many variations on this theme, particularly when you look at the services that are required to implement security, but the majority of published materials on the topic introduce things from the CIA standpoint. Indeed, any practical security measures can be related back to one or more of these top-level principles.

In addition to the classic C-I-A triplet, a fourth objective of *accountability* is often included. Accountability is vitally important to allow actions and intrusions to be tracked. Without some form of accountability, it is impossible to directly attribute an action to an individual or to be able to prove that an individual did not perform a specific action (i.e., through authentication we should be able to hold individuals accountable for their actions, or, alternatively, to be able to defend individuals or organizations). Accountability is usually achieved through historical logs; however, this only allows action to be taken after the event. Therefore, many systems utilize some form of interactive monitoring (e.g., intrusion detection systems) to audit (and respond to) user actions in real time.

2.3.3 Perspectives on Protection

Although it is often tempting to regard IT security as a technological problem, requiring similarly technology-related solutions, an important realization is that it is not just about installing controls on the computers themselves. Although this is clearly part of the puzzle, there must also be security in the environment *surrounding* them, and several other perspectives also need to be considered, as illustrated in Figure 2.3 [9]. A comprehensive security solution will involve recommendations and countermeasures in each of these areas.

These elements are, of course, easiest to identify within a business setting rather than on an individual basis, and some of the associated issues are highlighted here:

- Technical (e.g., system-based safeguards, such as authentication, access control, antivirus protection, and data encryption);

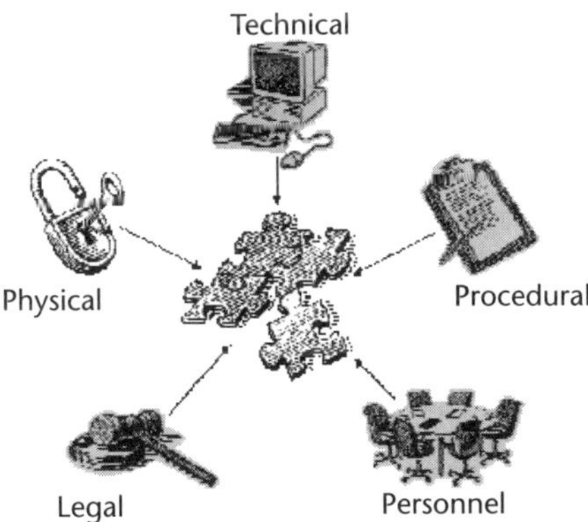

Figure 2.3 Elements of the security puzzle.

- Physical (e.g., issues such as physical access to systems, protection against theft, and safeguards against fire, flood, and other environmental incidents);
- Personnel (e.g., appropriate controls relating to the people that use systems, such as recruitment procedures including reference checks, training and awareness programs, and termination procedures);
- Procedural (e.g., the need for a security policy and for conducting risk assessment, as well as issues such as disaster planning and recovery);
- Legal (e.g., the need to comply with relevant legislation, such as data protection laws, as well as the need for awareness of laws that might become relevant as a result of a security breach, such as computer crime and misuse).

The requirements listed under each heading are far from exhaustive, but they give a clear indication of the fact that securing a system does not stop at the technological measures. Indeed, it should be clear that security is a multifaceted problem, and a range of expertise is typically needed to ensure appropriate solutions.

Another important realization is that security cannot be achieved via a one-off activity. Instigating security measures is just one part of an overall management process and should be supported by a series of policies. A view of this situation is illustrated in Figure 2.4, and discussed in the paragraphs that follow.

As indicated at the top of the diagram, any attempt to address security in an organizational setting ought to be informed and guided by appropriate policy. At the highest level, the value of the policy comes as a means of highlighting the importance of information security to the mission and goals of an organization. The scope of the policies should include both human and technical considerations. On the people side, it needs to clearly define the bounds of permitted behavior and what people need to do in order to play their part in keeping the system secure. On the technical side, the policy must provide a backdrop for a range of more detailed guidelines and procedures, which need to be applied on systems themselves. Rather than make reference to specific technologies and measures, the high level policy would typically be framed in more general terms—making it independent from

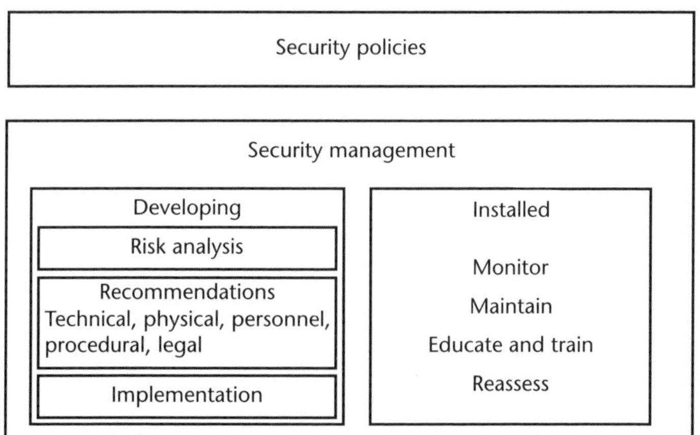

Figure 2.4 Managing security.

advances in individual technologies. A good security policy can be summarized as follows [10]:

- Is short and backed from the top of the organization;
- Recognizes that information is critical and must be protected;
- Emphasizes the importance of security awareness and training;
- Emphasizes compliance with legal and regulatory requirements;
- Emphasizes relations with third parties;
- States roles and responsibilities for information security;
- Outlines standards and procedures;
- States the consequences of violations and noncompliance.

In addition to a high-level policy spanning the whole organization, further policies ought to be developed at lower levels in order to provide more specific guidance (e.g., addressing particular departments, roles, and systems).

These policies then support the wider process of security management, which (as the figure shows) can be considered to fall into two phases: the initial development and the ongoing management of the installed solution. The starting point in the development phase is to determine the appropriate level of countermeasures for the assets requiring protection. However, this is rather difficult to achieve if an ad hoc approach is taken, and there is potential to devote too much attention (in terms of time, effort, and expense) to protect one system while inadvertently neglecting something else. As a result, organizations require a means by which they can select an appropriate level of protection for their assets, and address the problem in a consistent and structured manner. The solution is provided by risk analysis, which aims to identify the potential risks to a system and then suggest appropriate countermeasures. As previously illustrated in Figure 2.1, risk is typically measured in terms of the threats and vulnerabilities that relate to a particular asset. Outcomes are then assessed in terms of the impacts and consequences that would occur in the event of a security breach.

The risk analysis will lead to a series of countermeasure recommendations, addressing the various security perspectives identified in the earlier discussion. Indeed, a proper risk analysis will not only yield recommendations on the type of protection, but also on the strength of mechanism that is required in a particular context. This is clearly preferable to an ad hoc approach and provides a means by which security purchases can be more clearly justified on the basis of real evidence from within the organization, rather than just using scare tactics.

It is worth noting that countermeasures will rarely eliminate risk altogether. For example, if we consider the threat of virus infection, then we could clearly reduce the risk by using antivirus software. However, this will never guarantee that our system is totally immune to viruses. New strains appear all the time, and so our vulnerability will also depend on how frequently we update the virus signatures that the scanner uses to detect them, and indeed how quickly the vendor of the antivirus package actually releases new signatures in response to new strains. If our scanner is updated every day at 2 am, but a rampant new strain of virus appears at 2:30 am, then the system will not be protected for almost 24 hours—despite our best efforts with the countermeasure. This does not mean that the countermeasure is pointless—it is still providing protection against all the other strains that might hit us—but it shows that the risk has only been reduced, not removed. The idea here is that the residual level of risk is acceptable or manageable from the organization's perspective.

For completeness, it is also worth noting that some countermeasure options may not intend to reduce or remove risk—they may instead transfer it to another party. The classic example here would be the option of taking out an insurance policy against a particular type of incident. However, this option is clearly not applicable to a significant proportion of IT security threats, as it is only really relevant to guard against financial losses. Most security incidents will have unacceptable consequences beyond this. For example, even a threat such as theft of computers, which can certainly be insured against, could have more profound implications than the replacement cost of the systems. Although the insurance would enable new PCs to be purchased, what about the data that was on the old ones? If regular backups had not been taken, then we may be left with nothing, so disruption to normal activity would continue. In addition, without access-related countermeasures on the stolen machines, the original data could now have fallen into the wrong hands, which may lead to a breach of confidentiality and then legal liability. The point here is that a mixture of countermeasures will typically be required, rather than expecting to rely on one particular type.

Having got the countermeasure recommendations, the task then is to implement them—and appreciate their implications. For many of the recommendations, the implementation process will first involve an element of fact finding to determine what products are available to meet the security requirement. There may also be a requirement to consider constraints. The most obvious of these may be the available budget, but other more subtle considerations may also include such factors as the practicality of a particular countermeasure within the target environment, in terms of its compatibility with the organizational culture or the degree to which the culture might be changed as a result of the countermeasure. For example, one countermeasure that a company could consider against Internet-related threats is a firewall, which would enable it to impose restrictions on the type of traffic that can enter or leave the internal network. However, if employees have previously enjoyed unrestricted Internet access, then imposing restrictions that are too Draconian could have adverse effects on their morale and productivity. It might also change their perception of the company, and an organizational culture that had previously been viewed as open and liberal could quickly be regarded as the opposite. In some cases, of course, adverse implications may be unavoidable in the pursuit of security, but if they are at least considered (and then explained to those affected) it reduces the chances of a cavalier attitude being perceived. If the situation is handled appropriately, then the actual implementation of countermeasures can occur successfully. All of the discussion so far has related to the left-hand side of the security management process depicted in Figure 2.3. Unfortunately, security cannot just be installed and then forgotten about, and the success of all previous efforts will still come down to the ongoing commitment that the organization is prepared to make. For example, simply buying and installing the software will typically achieve very little unless someone takes the time to configure and use it properly. Unless the software is completely automated and autonomous (and few current security packages are), it will still require something in terms of administrator involvement, sometimes on a regular and frequent basis. Without it, the sense of security could be entirely false. A good example can again be cited with firewall software—the *de facto* approach used to protect a system or network from attack on the Internet. When used effectively, a firewall can limit vulnerabilities by controlling the traffic that is permitted to enter and leave the network. However, a study from 3i in January 2002 reported that around 80 percent of firewalls are ineffectively installed and/or maintained [11]. So although they had ostensibly acted to improve security, the likelihood was that the organizations were still vulnerable to attack.

The upshot of this is that the process of managing security remains an ongoing task after implementation, and there are a number of things that need to be done in this respect [12]:

- Monitor effectiveness of the technical measures;
- Monitor compliance by staff, via manual and computer records;
- Maintain protection by acting on monitored information where necessary;
- Provide general and specialist training for all staff;
- Periodically reassess whether countermeasures are still relevant to current threats;
- Ensure that new systems are developed/procured in accordance with policy.

The key realizations here are that the risks to which an organization and its systems are exposed will not remain static, and existing security cannot simply be relied on to run itself.

2.4 Conclusions

This chapter has shown that security is a multifaceted problem, with many definitions, many perspectives, and many threats, vulnerabilities, and potential impacts that need to be considered. Unsurprisingly, this leads to a complex issue in terms of the resulting security management. As already indicated earlier; the question of "what is information security" does not have a single correct answer, and this chapter has demonstrated that security needs to be approached from a variety of perspectives using a range of approaches. The concepts introduced will be built on throughout the rest of this book.

It is important to remember that security is not just about the implementation of a technical solution; consideration also needs to be given to the human element. The correct combination of technical, physical, personnel, procedural, and legal measures must be taken in combination to ensure that all aspects of security are considered and that appropriate policies, guidelines, rules, regulations, and procedures are followed by everyone who interacts with the IT systems.

References

[1] British Standards Institution, Information technology—Security techniques—Code of practice for information security management, BS ISO/IEC 17799, June 16, 2005.

[2] Microsoft, *Microsoft Security Glossary*, October 29, 2002, http://www.microsoft.com/security/glossary.mspx#computer_security (accessed June 14, 2007).

[3] Software Engineering Institute, Carnegie Mellon University, *Glossary*, http://www.sei.cmu.edu/str/indexes/glossary/index_body.html (accessed June 14, 2007).

[4] Department of Trade and Industry, *Information Security Breaches Survey 2006*, URN 06/803, April 2006.

[5] Gordon, L. A., M. P. Loeb, W. Lucyshyn, and R. Richardson, *2006 CSI/FBI Computer Crime and Security Survey*, Computer Security Institute, 2006.

[6] CERT/CC Vulnerability Statistics, http://www.cert.org/stats/ (accessed June 14, 2007).

[7] Anderson, J. P., *Computer Security Threat Monitoring and Surveillance*, James P. Anderson Co., Fort Washington, PA, 1980.

[8] Furnell, S. M., "Categorising Cybercrime and Cybercriminals: The Problems and Potential Approaches," *Journal of Information Warfare*, Vol. 1, No. 2, 2002, pp. 35–44.

[9] Furnell, S. M., *Computer Insecurity: Risking the System*, London: Springer, 2005.

[10] Ward, J., *BS 7799 Training Course*, Symantec UK Ltd., 2004.

[11] 3i Group plc., *E-security—2002 and Beyond*, London, England, January 2002.

[12] Davey, J., *Managing Security*, Deliverable 11, HC1028 Implementing Secure Healthcare Telematics Applications in Europe (ISHTAR) project, March 10, 1997.

CHAPTER 3

Business-Integrated Information Security Management

Juhani Anttila, Jorma Kajava, Rauno Varonen, and Gerald Quirchmayr

This chapter deals with two well-known business management frameworks, the plan-do-check-act (PDCA) model and the process management model, which are also basic elements in the international standardization of information security management. Both models were originally developed to support the quality of overall business management in any type of organization. They entered international standardization at first in the ISO 9000 standards for quality management and later standards for information security management. This chapter emphasizes business management viewpoints and highlights the multifarious possibilities the recognized managerial models offer for information security management that is seamlessly integrated (embedded) with general business management activities. Additionally, some significant aspects of modern business environments are considered in this context. Building on this approach, we then discuss the role of information security standardization and business continuity management from the business management's point of view.

3.1 Business-Integrated Information Security Management

Apart from being an interesting issue from the management viewpoint, information security is in many cases crucial for actual business. In this sense, information security management is fully analogous to other highly specialized key areas ensuring competitive business performance and success. These areas include the management of finances, human resources, product quality, business risks, and innovation. In all these areas, it is useful for organizations to use established and recognized management approaches and practices.

Information security management can be defined as coordinated activities used to direct and control an organization with regard to information security. To ensure success, products and business activities should be considered with a view to the dimensions of information security:

1. Integrity, meaning that the information used is accurate;
2. Availability, meaning access to relevant information when it is needed, for as long as necessary;

3. Confidentiality, meaning that the information used is not manipulated by anyone.

Due to its importance for sustainable business performance and success as well as for business credibility, information security management belongs squarely under the responsibility of business management, and all information security-related activities should be integrated seamlessly with normal business management activities. General managerial principles and methodology may be used in all specialized management areas, and information security management is no exception [1–3].

3.2 Applying The PDCA Model to Manage Information Security

A well-known general model for all areas of management, including information security, is the so-called PDCA model, also referred to as the Deming/Shewhart cycle (see Figure 3.1). This model became popular especially through W. Edwards Deming's lectures on managerial quality given over several decades (1950s–1990s). However, the model was originally created by Walter Shewhart (also an American) in the 1920s. Later on, Shoji Shiba from Japan did a remarkable job by combining the original PDCA model with ideas taken from knowledge management and Buddhist philosophy. Also Joseph Juran's so-called trilogy model contains the same elements as the PDCA model. In addition, the PDCA model has consistent links with traditional systems theory and systems dynamics. To sum up, one may note that the PDCA model offers a great variety of different applications, possibilities, and uses [4].

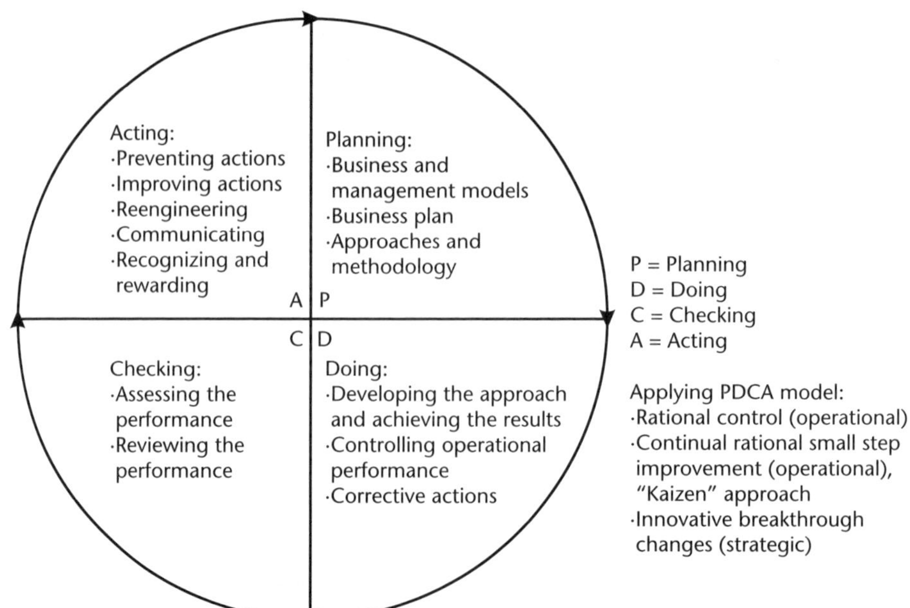

Figure 3.1 PDCA model for management and its application to strategic and operational business management.

The PDCA model is strongly related to the international standards of information security management, especially ISO/IEC 27001:2005, ISO/IEC 17799:2005, and ISO/IEC 17799–1:1996 [5–7].

These standards, being the most widely recognized reference documents indicating a professional approach, emphasize the integrative nature of information security management. On top of that, the OECD Guidelines for the Security of Information Systems and Networks—Towards a Culture of Security [8] states that information security management should be part of all organizations' and societies' business cultures.

Describing consistent management, the PDCA model consists of four consecutive activities, as shown in Figure 3.1:

1. **P:** Planning business activities—what should be done and what results should be achieved;
2. **D:** Doing business activities in accordance with the plans;
3. **C:** Checking what has been done and what results have been achieved;
4. **A:** Acting rationally by taking into account the observations and results of the checking phase.

In organizational environments, the PDCA model is to be applied in three different scopes:

1. *Control*: Managing daily operations in business processes to achieve the specified results. Normally, rectifying nonconformities is carried out in connection with control.
2. *Prevention and operational improvements*: Solving acute problems, preventing nonconformities, and finding/implementing operational step-by-step improvements in business processes.
3. *Breakthrough improvements*: Innovating and implementing strategically significant changes in the way of doing business.

Top business leaders (senior executives) are responsible for breakthrough improvements, while activities related to control, prevention, and small-step improvements are responsibilities of operational managers. All people within organizations should be aware of the importance of information security, allowing them to take it into account in their normal everyday business tasks in a simple and natural way without any additional measures. Information security experts should have a supporting role.

To ensure information security, organizations should carry out a variety of information security–specific measures in planning, doing and checking business activities/results, and reacting to new situations. In addition, they should perform actions aimed at correcting, preventing, and improving their activities, and—should the need arise—even reengineer their business processes. The international standards mentioned earlier contain information detailing methodological approaches to information security tasks. However, although the ISO/IEC 27001 standard explicitly refers to the PDCA model, its application is rather unsystematic and inexplicit for the purposes of information security management (e.g., the standard does not cover all three different application scopes—see Figure 3.1).

There is no reason why control, prevention, small-step improvements, and breakthrough improvements should not be relevant also for information security management. As a matter of fact, normal quality management practices can even be said to obligate the use of such methods also in the field of information security. Although general managerial practices and methods can be applied, they should be incorporated with professional information security expertise. This, in turn, implies a close and effective cooperation between business executives and information security experts.

3.3 Information Security Management Through Business Process Management

Used already in ancient plant and construction activities, the process approach is often referred to when describing natural development. Through industrialization, processes became an everyday concept in the so-called process industry. From the 1980s on, the structured analysis and design technique (SADT) has used the process approach to describe the internal interlinked activities of computers. In large-scale business, however, this approach has only been used comprehensively for fewer than 20 years, although a number of practical methods have been developed for business management during that time. These approaches have drawn especially from systems theory and system dynamics. In the 1990s, the process concept was introduced into the ISO 9000 quality management standard, and in recent years it has entered the international standards of information security management [3, 4]. Very recently, business process automation using service oriented architecture (SOA) technologies has become an important part of business process management.

Processes are connected to all kinds of daily tasks or activities in organizations. In fact, the process concept originally denoted any type of activity or operation. However, the drive to increase the effectiveness and efficiency of business operations has made structural aspects of business processes an interesting management issue. In some cases, this has led to the danger that structural aspects, such as formal process diagrams, have been overemphasized in process management, which, due to its significance, is by nature a comprehensive business management issue. Today, however, many organizations find that truly effective and efficient process management implies a radical change from established management thinking and structures.

All business results—including information security—are achieved through managing business processes and projects. Basic (or core or key—different terms are used in different organizations) business processes imply continuously running interlinked business activities, and projects are singular processes for unique business tasks. Both the strategic and operational management level is involved in this process approach, with the strategic level focusing on managing the network of interlinked business processes (i.e., the whole business system) and the operational level on managing single processes and projects.

In integrating information security practices, it is extremely important to understand information security issues in the context of business processes. This is because, in practice (operationally), information security originates from processes. That is based on the fact that all process activities are nowadays very strongly information intensive, and information flows between these activities and between different actors and even distant operational locations (see example in Figure 3.2). Thus, information security is influenced directly in real time through process arrangements, tools, and people in practical work. Further, it is also affected by what appropriate and systematic practices are applied to manage these.

Process management (see Figure 3.3) indicates how strategic and operational business objectives are realized through business processes in accordance with the PDCA principle. Measurements are used to provide feedback for process management, which in fact comprises three PDCA loops:

1. The loop of control and corrective actions;
2. The loop of prevention;
3. The loop of real improvements through innovative redesigning and reengineering of process(es).

Both the entire process network (the business system) and the individual business processes should be managed according to this systematic model. Management of the comprehensive process network includes the normal responsibilities of business management, such as using business plans, action plans, business performance assessments, and regular business reviews. It is essential that the business system is understood specifically as a network of business processes and not only as functional units (organizational *silos*). The management of individual processes consists

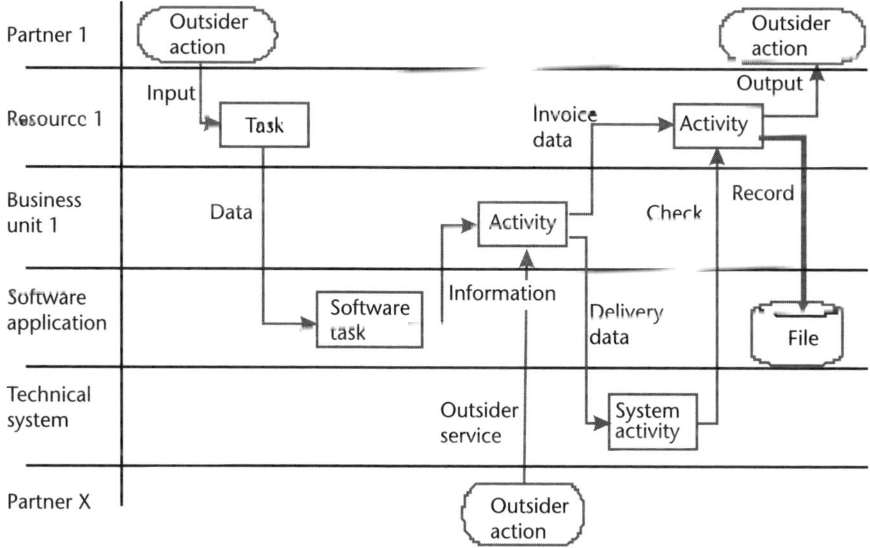

Figure 3.2 Information security is realized in the activities and information flows of business processes (e.g., order/delivery process).

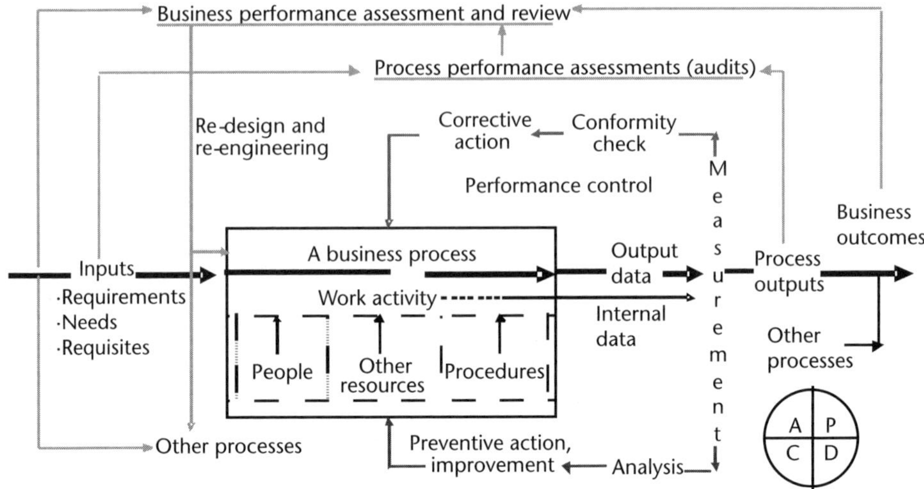

Figure 3.3 PDCA loops in business process management.

of process planning, operation control, performance improvement, and quality assurance, while process management is based on the process plan, on process performance assessment, and on monitoring process performance indicators. Of all process management tools, the most important is the process plan.

To take account of information security issues, one should understand what phenomena within single business processes and between different processes are critical from the information security point of view. After that one may be able to define suitable performance indicators and set quantitative target values for information security according to the relevant needs and expectations. A key management issue is to monitor these indicators in real time and to initiate—as needed—necessary measures for correction, prevention, or performance improvement in accordance with the PDCA model. General guidance may be found in the standards mentioned earlier for defining information security control methods for business processes.

Also information security performance should be considered from both the strategic and operational viewpoints. The strategic performance management of processes consists of the organization's vision-based and strategy-based measures and evaluations of its overall process performance, whereas operational process performance measures for daily management are focused on diagnostics and analysis required for corrective and preventive actions.

Process performance is a fuzzy concept in general and the information security performance of processes is no less hazy. An evaluation of process performance comprises the strategic assessment of business performance in its entirety (process network) and the operational assessment of individual processes. Assessment results are useful for company-internal process improvement and quality assurance. Quality assurance includes all measures through which an organization demonstrates to its stakeholders that it is capable of effectively fulfilling all agreed requirements. Business integration means that these evaluations and quality assurance measures should also include information security issues.

3.4 Factors Affecting the Use of Systematic Managerial Tools in Business-Integrated Information Security Management

Information security management—or business management as a whole—using the PDCA model and process management requires responding to the realities of the organization's business environment, especially the need to manage variety and operational agility set new requirements for all management practices. Being fundamentally uncertain and ambiguous, today's business environment includes the following [4]:

1. Emerging and self-organizing networks of actors affecting business;
2. Numerous heterogeneous global actors in virtual networks on the marketplace;
3. That everything is linked with everything else, and all linkages are not known;
4. Paradoxical freedom of the actors ("both-and" instead of "either-or");
5. Significance of immaterial issues (information, knowledge, services);
6. Increased speed of activities and change;
7. Significance of transaction phenomena;
8. Immense pressure and stress of business leaders.

All these aspects pose great difficulties and challenges for information security management.

Three different kinds of activities—mechanistic, organic, and dynamic—are inherent in all business processes and activities. Mechanistic aspects are highly disciplined tasks (e.g., access control to certain information in the order-delivery process), organic aspects relate, for example, to necessary business interactions with internal and external operational partners, and dynamic aspects reflect spontaneous human activities in on-time situations. At the moment, business performance and competitiveness are mainly based on organic and dynamic actions. However, business management involves addressing all three dimensions appropriately in process networks. This includes both the network as a whole and the individual processes that make it up. The various aspects of process activities should be considered in process documentation in an appropriate way.

Most business processes of today are complex responsive processes of relating, implying that every process involves dealing with complexity. This cannot be accomplished using simplistic tools. Thus, only the mechanistic parts of process activities allow information security to be controlled adequately by strict rules and instructions or by automatic technological solutions. Organic and dynamic business situations involve people skills, competences, awareness, initiatives, commitments, and responsibilities, as well as a general security culture [9–11].

The success of information security management depends on quality of management, which determines how business management is really carried out and how systematic tools are used across the whole organization. At several levels, management actions are performed regarding the whole organization, its business units or functions, business processes, individuals, and teams. On this view, leadership

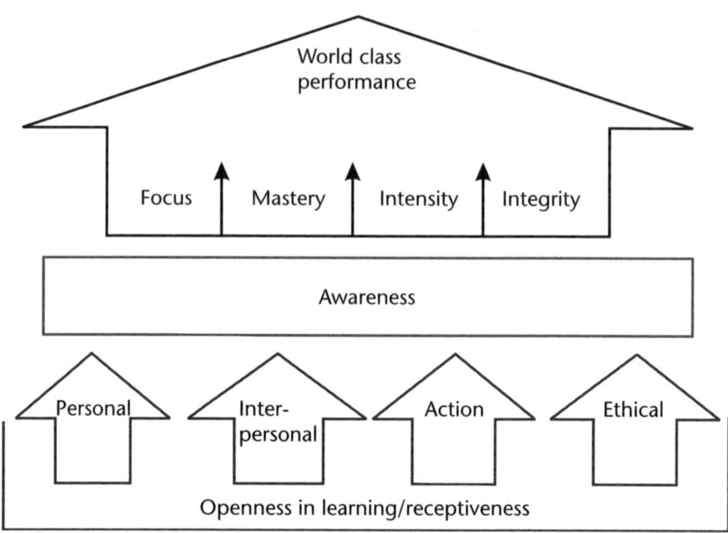

Figure 3.4 Quality of leadership performance. A successful business executive's primary quality feature is awareness.

emphasizes managers' or superiors' personal and human aspects in managing the business actions (see Figure 3.4).

A critical issue for process performance is the operation of individual actors within processes and how they understand their information security–related roles and responsibilities. No conflict should exist between a person's activities within a business process and his/her internal mental processes, as a conflict may cause a significant threat to information security. Chances of preventing and resolving such conflicts in an effective manner depend largely on the organization's social networking culture and its human resource management practices, including procedures of compensation and rewarding as well as incentives and recognition. Some problems may be avoided by replacing human activities within business processes by automatic IT solutions.

3.5 Information Security Management Standardization and International Business Management

From the viewpoint of information security management, standards have been important references. For a fairly long time, we have relied heavily on the British Code of Practice for Information Security Management, which has become the de facto standard within the IT community for many years. Recent years have seen the emergence of international standards, such as ISO/IEC 17799 and ISO/IEC 27001, for information security management [10]. Standards serve to define the main concepts, principles, and components of information security management.

Many organizations still regard information security as something that can be taken care of by investing in systems and technology. Nonetheless, oiling the wheels of the organization is a critical part of success and often involves improvements in information security, too. However, this approach often leaves central factors out:

individual employees and business managers. People working in organizations are definitely the critical factor in the implementation of information security. All managers should manage their organizations' business activities with regard to information security among their other business duties. On the other hand, enhanced security may also result in reduced privacy and a deterioration of the working atmosphere.

The latest development in the international standardization of information security management emphasizes the seamless integration of information security management with general business management. ISO/IEC 27001, especially, recommends applying to information security management the recognized managerial principles of the general management models according to the ISO 9000 standards [12]. The PDCA model and the process management model are of utmost importance in this context. However, the way in which these models have been applied to the standardization of information security is inadequate, and therefore their application requires additionally fundamental and multidisciplinary background knowledge. It is assumed that these models and all the possibilities they offer are familiar to information security experts. This is not the case at all.

To ensure information security, organizations should carry out a number of different measures aimed specifically at enhancing information security when planning, carrying out, and checking business activities/results and reacting to different situations. To that end, organizations should perform corrective, preventive, and improvement actions, and should the need arise—which may be quite normal in today's rapidly changing business environments—be prepared to comprehensively reengineer their business processes. The international standards mentioned earlier contain information detailing methodological approaches to information security tasks.

In order to incorporate information security issues into business activities, one should understand what phenomena within single business processes and between different processes are critical from the information security point of view. Next, one should consider relevant needs and expectations, define suitable performance indicators, and set quantitative information security targets in accordance with normal, proven process management methodologies. A key management issue is to monitor these process performance indicators in real time and to initiate—if required—such necessary measures to prevent, correct, or improve. For business processes, the previously mentioned standards provide only very general guidance for defining information security control methods.

Particular challenges are created when the standards are being applied in organizations that are gearing their activities toward the global market. It is an interesting question whether the introduction of the ISO/IEC 17799 and ISO/IEC 27001 standards does really provide help for busting international cooperation barriers and for establishing improved information security. Knowing the principles underlying the best practices that the standards are based on, one is rather doubtful whether standards of that type alone can solve the problems. Admittedly, "A Code of Practice for Information Security Management" has been the linchpin of information security management for a number of years. But a lot of criticism has been leveled at it, and a number of shortcomings have been pointed out. As a result, something more innovative is needed. However, for the time being, the approach adopted in ISO/IEC 17799 and ISO/IEC 27001, in spite of its weaknesses, is one of the few widely recognized international standard bases for information security in IT business.

Organizations confront many serious information security and compliance requirements relating to corporate governance and IT governance today. Significant standards for them particularly include:

1. Sarbanes-Oxley Act (SOX), 2002, for public company accounting reform and investor protection;
2. Basel II for the financial industry to manage risks and measure capital requirements;
3. FISMA, 2002, to bolster computer security within the government networks;
4. HIPAA, 1996, for health insurance portability and accountability;
5. Gramm-Leach-Bliley Act (GLBA) to protect consumers' personal financial information held by financial institutions.

All these requirements documents emphasize the role and behavior of business management and may be realized in practice only through a consistent management of business processes.

Apart from the lack of a failsafe standard, the complex nature of information security poses problems for international business cooperation. To solve the problems, some researchers and consultants have proposed ready-made technological or procedural solutions. The difficulty in their widespread application lies in the fact that information security is a convoluted, multifaceted phenomenon. Each information security–related event is in a sense unique, partially because organizations in different countries differ widely in terms of technology, corporate culture, hardware base, security awareness, and so on. Factors such as these have to be taken into account in information security management at the organizational level.

In information security, we are used to small advances occurring as a response to experienced threats. We should, however, also be prepared to make a giant leap when needed. Therefore, one should also understand the standards of information security management to support this kind of development.

In general, standards are needed and aimed to enhance quality of products (both goods and services) and business processes, to facilitate the interoperability of information systems and of collaborating organizations and people. With the advent of ubiquitous computing, international standards and their innovative application will play an even greater role in our everyday lives. For these purposes we have two different categories of standards that complement each other:

1. *Technical standards*: The interoperability of different systems can only be achieved by means of standards. Major communications solutions, for example, can only be based on a common standard.
2. *Managerial standards*: Standards serve to improve business performance. If the standards only fix the best practices (i.e., approaches that have already been tried in practice) and require their application, then they hinder the development on the never-ending road toward excellent information security performance.

However, the final decisions in actual business cases depend on the behavior of individual business managers and operators during their everyday work within business processes.

3.6 Business Continuity Management

The management of business continuity is gaining a more important role within IT security management. That is why leading IT and consulting companies all over the world have recently developed new methods and technologies and have included business continuity planning and management in their customer service portfolios. As we have seen with many of the early e-commerce sites, once an intrusion is successful, the only alternative to stop the attacker from continuing to cause serious damage usually is the complete shutdown of the attacked system. Looking at several forms of attacks that are aimed at achieving exactly this result (DoS, DDoS), the survivability of systems [www.cert.org] and business continuity management [13] are becoming crucial. With a 24×7 service expectation from the customer, the complete shutdown of an attacked system is not an acceptable option anymore. It is therefore necessary to develop integrated solutions based on the technology available today. A business-integrated managerial approach to IT security is still rather the exception than the rule; this leaves systems vulnerable to attackers. Several layers of defenses are often considered too expensive or too difficult to administer, which leads to denial of service attacks still being successful.

As discussed in [14], a more comprehensive approach to IT security than the one mostly followed in today's practice is needed if systems are to survive successful or partially successful intrusions. Similar to fault tolerance, intrusion tolerance will become a necessity for the reliable operation and management of information systems and IT [15]. It is therefore very astonishing that business continuity management was largely ignored until the Y2K problem demonstrated the possible impact of failing IT infrastructures. It did, however, take the series of terrorist attacks in the United States, Europe, and Asia and the recent failed attempt to cause disruption to IT services in the London financial district to make the management of larger corporations realize the potential size and gravity of exposure to unmanaged risk. Business continuity planning and the survivability of IT infrastructures have now become a major issue. Warnings related to cyber terrorism and other forms of attacks are now taken far more seriously than only a decade earlier. Changing the corporate culture, however, is still a major effort, and management needs to be convinced that it pays off to invest in business continuity, layered system defenses, and survivable systems.

The different stages of the business continuity management (BCM) life cycle described by David Smith [15] show the necessity for integration with business processes and clearly demonstrate the link between technological and business management issues related to business continuity management. The six stages of the business continuity management life cycle according to [15] are the development of an understanding of the business; followed by the definition of the BCM strategies, development, and implementation of a BCM response; the building and embedding of a BCM culture; the exercising, maintenance, and audit phase; rounded up by the BCM program.

The next logical step forward after a business impact analysis will be to have a closer look at business processes and see how security aspects can be integrated directly into management of business processes. One way toward achieving this goal is discussed in [16].

Following a guide to BCM by the Bank of Japan [17], the necessary activities leading to successful BCM can be grouped as follows:

1. Formulating a framework for robust project management:
   - Basic policy;
   - Firm-wide control section;
   - Project management procedures;
2. Identifying assumptions and conditions for business continuity planning:
   - Disaster scenarios;
   - Critical operations;
   - Recovery time objectives;
3. Introducing action plans:
   - Business continuity measures;
   - Robust backup data;
   - Procurement of managerial resources;
   - Decision-making procedures and communication arrangements;
   - Practical manuals;
4. Testing and reviewing;
5. Other issues:
   - Precluding and mitigating disaster damage;
   - Location of backup facilities;
   - Use of outside service providers.

In practice, testing and reviewing are the steps that are still not carried out on a regular basis by the majority of organizations, simply because this costs time, money, and effort. As long as nothing happens, it is always difficult to argue why such tests and reviews are needed. The problem is that once disaster strikes, it usually is too late. Typical trouble experienced when not testing are failing backups of data and software and backup sites taking several times longer to become operational than planned for.

Other very similar guides are made available by leading consulting groups, such as Gartner, Deloitte, Touche, Thomatsu [18], and PWC. Guides issued by financial institution do however have the advantage of being backed by a traditionally security-aware source. As banks have always had a highly developed security culture that could, with some transformations, also be migrated into the e-business world, it was to be expected that e-commerce sites operated by banks would be the most secure. To this point banks have kept their promises and today are seen as examples of how to best approach IT security, business continuity planning, and management. In addition, many of the leading IT companies (e.g., IBM, HP, Fujitsu [19], Siemens, and Hitachi) have over the past years put advanced backup and recovery technology on the market. This technology is covering the field from distributed advanced storage systems to overall system availability through hot swaps. Some governmental institutions (e.g., in the United Kingdom) have made their continuity guides available online [20].

3.7 Conclusions

To be effective, information security management should be carried out in integration with normal business management practices. Distinct solutions for information security are abnormal, ineffective, and ultimately frustrating. Integration is realized through practical managerial tools. Among these, the PDCA model is a proven, recognized methodology for all types of management. It is also suitable for information security management, but its use has been rather vague, and the advantages if offers have not been fully tapped in practice.

Numerous business cases demonstrate that, although process management is a very simple thing in principle, its practical implementation seems incredibly difficult, because it always puts a strain on the organization's leadership issues. The development and management of business processes is a long-term effort and should take into account the realities of the business environment. This also explains why the application of process management to information security management is such a struggle. Another point is that security professionals are unfamiliar with the foundations and practical approaches of process management.

As we have seen in the case of information security standards and business continuity, the necessary methods, guidelines, and the technology supporting them are available. None of them does, however, come to bear unless we achieve a fully business-integrated information security management, create the required awareness inside companies and organizations, and provide the necessary training.

Business continuity management life cycles show the necessity for integration with the management of business processes and do clearly demonstrate the link between technological and managerial issues related to business continuity management. When information security is integrated in a natural way with a crucial aspect of business management, it is also being genuinely interested by business leaders.

References

[1] Anttila, J., "Managing and Assuring Information Security in Integration with the Business Management of a Company," in *Proceedings, Information Security, Small Systems Security & Information Security Management*, Vol. 2, IFIP TC 11 wG 11.1 & 2, Vienna and Budapest, 1998, http://www.qualityintegration.biz/Tonava.html.

[2] Anttila, J., "Business-Integrated Information Security Management," European Intensive Programme on Information and Communication Technologies Security IPICS'2003, Oulu, Finland, 2003, http://www.qualityintegration.biz/InformationSecurityIntegration.html.

[3] Anttila, J., "A Comprehensive Process Approach for Overall Performance Excellence," Quality Conference in Ostrava, and Workshops in Mumbai and Tallinn, 2002, http://www.qualityintegration.biz/BusinessProcessManagement.html.

[4] Anttila, J., General Managerial Tools for Business-Integrated Information Security Management, 2006, http://www.qualityintegration.biz/InformSecPDCA.html.

[5] ISO/IEC 27001:2005: Information Technology—Security Techniques—Information Security Management Systems—Requirements, ISO, Geneva, 2005.

[6] ISO/IEC 17799:2005: Information Technology—Security Techniques—Code of Practice for Information Security Management, ISO, Geneva, 2005.

[7] ISO/IEC 11770-1:1996: Information Technology—Security Techniques—Key Management—Part 1: Framework, ISO, Geneva, 1996.

[8] OECD Guidelines for the Security of Information Systems and Networks—Towards a Culture of Security, OECD, Paris, 2002.

[9] Kajava, J., et al., "Exploring the Use of an E-Learning Environment to Enhance Information Security Awareness in a Small Company," in *Proceedings of the Computational Intelligence and Security CIS2006 Conference*, Guangzhou, PRC, 2006, http://www.qualityintegration.biz/GuangzhouW.html.

[10] Kajava, J., et al., "Senior Executives Commitment to Information Security—From Motivation to Responsibility," in *Proceedings of the Computational Intelligence and Security CIS2006 Conference*, Guangzhou, PRC, 2006, http://www.qualityintegration.biz/GuangzhouC.html.

[11] Anttila, J., et al., "Fulfilling the Needs for Information Security Awareness and Learning in Information Society," in *Proceedings of The 6th Annual Security Conference*, Las Vegas, NV, 2007, http://www.isy.vcu.edu/%7Egdhillon/secconf/secconf07/PDFs/21.pdf.

[12] ISO 9000/9001/9004: Quality Management Systems, ISO, Geneva, 2000.

[13] Business Continuity Management—Exploiting Agility in an Uncertain World, BT Guide, 2003.

[14] Quirchmayr, G., "Survivability and Business Continuity Management," in *Proceedings of Australian Information Security Workshop*, ACSWE Frontiers 2004, *Australian Computer Science Communications*, Vol. 26, Nr. 7, 2004.

[15] Smith, D., "Business Continuity and Crisis Management," *Management Quarterly*, January 2003.

[16] Quirchmayr, G., and J. Slay, "A BPR-Based Architecture for Changing Corporate Approaches to Security," in *Proceedings of the 5th Australian Security Research Symposium*, Perth, Australia, July 11, 2001.

[17] Business Continuity Planning at Financial Institutions, Bank of Japan, 2003.

[18] Gartner, Deloitte, Touche, Thomatsu, A New Paradigm for Business Continuity Management, 2002.

[19] Fujitsu Siemens, FlexFrame™ for Oracle, Technical Overview, White Paper, September, 2006.

[20] City of London, 2006, www.londonprepared.gov.uk/business/london_first.pdf.

CHAPTER 4

User Authentication Technologies

Nathan L. Clarke, Paul S. Dowland, and Steven M. Furnell

This chapter introduces the principles of user authentication technology. It begins by examining the need for reliable identification and authentication of users, and then introducing the three primary foundations of authentication techniques (i.e., something the user *knows*, *has*, or *is*). Subsequent sections expand on each of these approaches, with attention to passwords and alternative secret-knowledge approaches, passive and active token methods, and physiological and behavioral biometric techniques. Each section considers underlying principles, specific methods, and potential attacks. The chapter also examines operational considerations, with discussion of the factors that affect the selection of one authentication technique over another.

The majority of IT systems require a means to discriminate between legitimate users and would-be impostors. As a result, valid users need to be given accounts on the systems that they need to use. Although users with the same or similar access needs could conceivably be given access via a group-based account, proper regard for security dictates that they must be identified *individually* in order to enable access controls to be defined on a user-specific basis, as well as to enable individual accountability for activities. This requires each user to be assigned an identity (user ID) within the system. However, recognizing the potential for impostor attacks, any claimed identities must then be *authenticated* to prevent masqueraders. As such, a user authentication mechanism serves to provide a first line of system protection and can safeguard against abuse by external parties or unauthorized insiders.

A variety of methods and techniques can be employed in order to achieve authentication. At the highest level, these can be categorized into three main approaches, based on Wood [1]:

- Something the user *knows* (e.g., password or PIN);
- Something the user *has* (e.g., a card or other token);
- Something the user *is* (i.e., a biometric characteristic).

These are not mutually exclusive, and multiple methods may be combined in some contexts. For example, withdrawing money from a bank ATM machine requires possession of a card and knowledge of the accompanying PIN. Meanwhile, paying for goods in a shop or restaurant may require the card and the reproduction of an appropriate signature (the latter being an example of a biometric characteristic).

There is no single "best" method of authentication, and the selection of an appropriate approach will depend on the context in which it must be used and a level of tradeoff between the security required and the convenience for the end user. This chapter gives consideration to each of the approaches, beginning with the use of secret knowledge.

4.1 Authentication Based On Secret Knowledge

The use of an item of secret knowledge represents the most commonly encountered means of authentication, thanks to the predominance of password and PIN-based approaches.

4.1.1 Principles of Secret Knowledge Approaches

Characteristics of a good method are that it should involve secrets, which have the following traits:

- Resilient to guesswork by impostors;
- Easily recalled by the legitimate user;
- Simple and convenient to provide when required.

The last characteristic refers to the fact that legitimate users ought not to find that the process of entering the secret knowledge is overly time consuming or prone to errors that lead them to be rejected.

4.1.2 Passwords

Passwords are by far the most commonly used means of authentication in IT systems and have the advantage of being conceptually simple for both system designers and end users. In fact, passwords are capable of providing a fairly effective level of protection if they are used correctly. However, there are various means by which users can compromise the protection that passwords seek to provide [2]:

- Making poor choices, which leave them vulnerable to password cracking tools and social engineering (see Section 4.2.4);
- Sharing them with friends and colleagues, such that the supposedly secret knowledge becomes (at best) a shared secret and no longer remains under the control of the legitimate owner (i.e., colleagues may share it with other people, without the original owner's knowledge);
- Writing the information down (e.g., on notes stuck onto monitors or underneath keyboards), with the consequence that others may discover it;
- Maintaining the same password for long periods, thus increasing the window of opportunity for an impostor in the event of the password being discovered (or having previously been shared);

- Using the same password on multiple systems, with the consequence that a breach on one system potentially renders the others vulnerable.

The problem of poor password selection typically reflects the fact that users make choices that are easier for them to remember, without necessarily considering the potential for them to be known, guessed, or cracked by other people. Common failings here include:

- Reusing login name as the password;
- Basing passwords on personal information, such as the name of a spouse, relatives, or pets, or deriving them from other details such as address or car registration—all of which may be known to close friends and family or could be discovered by a determined attacker;
- Use of strings such as "qwerty" and "fred," on the basis that the close proximity of the keys makes them easy to type;
- Use of words such as "secret," "password," "sex," "love," and "money"—all of which are popular choices and therefore predictable to a would-be impostor;
- Using other words that may be found in a dictionary;
- Selecting passwords that are too short, thus reducing the number of tries that would be needed if someone were to try to discover them by testing all the possible character combinations (an attack known as *brute force*, which is mentioned again later in Section 4.1.4).

The last two points relate to a particular threat faced by password-based systems; namely, the availability of automated tools that take a password in its encrypted form (which may have been acquired by copying it from the target system or eavesdropping as it was sent over the network) and then attempt to find the plaintext characters that will encrypt to the same thing. Because users are prone to basing their passwords on recognizable words that would be found in a dictionary, such tools maintain preencrypted copies of these words, enabling them to be cracked via a lookup rather than having to use brute force methods (making repeated attempts iterating through all combinations of potential passwords—this often involves starting with "a" and working through to "z" followed by "aa" through to "zz," and so forth). A well-known example of such a tool is L0phtCrack, a screenshot from which is shown in Figure 4.1. In this particular case, the password for one of the user accounts has already been determined, and the Administrator password is in the process of being attacked by brute force. Another point that can be noted from the screenshot is that the program is using a default dictionary of 29,156 words, and so any passwords using words held within this could be cracked without resorting to brute force. A final observation is that the "summary" at the bottom right corner refers to users' accounts having been "audited" rather than cracked. This reflects the fact that the tool is intended for use by authorized persons (e.g., system administrators, who may have a legitimate need to assess whether their users are choosing appropriate passwords), rather than specifically serving the attacker community.

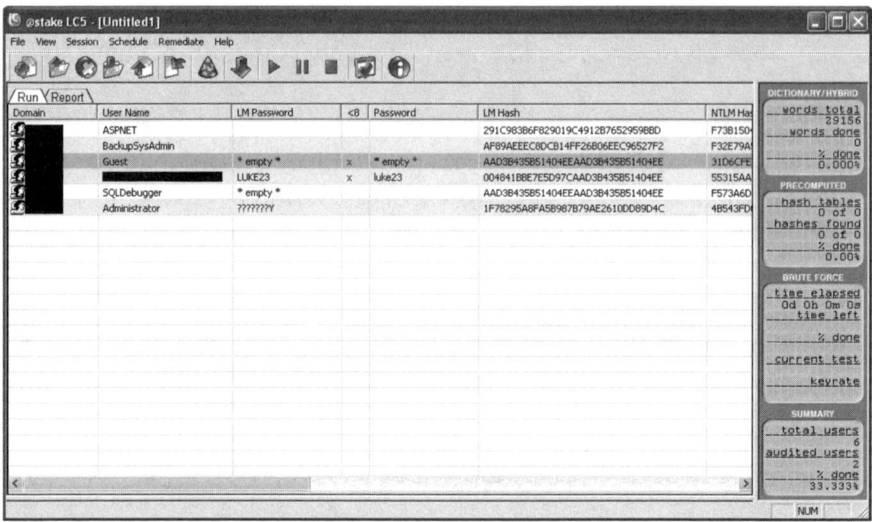

Figure 4.1 The L0phtCrack tool.

Having recognized the problem, password selection can be improved by choosing longer passwords that include a combination of character types. As Table 4.1 illustrates, using numbers and symbols as well as letters increases the number of possible character permutations that an automated password-cracking program must try in order to crack a password by brute force. Doing this does, however, run the risk of making the password harder to remember, and so it is useful to have a scheme to simplify the process for the user. One example would be to choose the first letters of words within a phrase that they find easy to remember. For example, the phrase "the quick brown fox jumped over the lazy dog" yields a password of "tqbfjotld," to which numbers and/or symbols can then be added in order to give a password that appears meaningless, but should be remembered easily.

Good practice would suggest that passwords ought to be changed every 30 days, and that on highly sensitive systems a lifetime of 7 to 14 days might be more appropriate. However, having chosen a password, users are often unwilling to

Table 4.1 Password Permutations Based On Length and Composition

Length (characters)	*Number of Permutations*		
	Alphabetic	*Alphanumeric*	*Alphanumeric + 10 Other Symbols*
1	26	36	46
2	676	1,296	2,116
3	17,576	46,656	97,336
4	456,976	1,679,616	4,477,456
5	11,881,376	60,466,176	205,962,976
6	308,915,776	2,176,782,336	9,474,296,896
7	8,031,810,176	78,364,164,096	435,817,657,216
8	208,827,064,576	2,821,109,907,456	20,047,612,231,936
9	5,429,503,678,976	101,559,956,668,416	922,190,162,669,056
10	141,167,095,653,376	3,656,158,440,062,980	42,420,747,482,776,600

change it, with the consequence that many passwords remain unchanged for months or even years at a time. The risk here is that if the password has been compromised, then an impostor has an ever-increasing opportunity to make use of it.

In a properly configured system, there are a number of things that a system administrator can do to encourage and enforce appropriate password usage. For example, in Windows XP, the administrator can set the password policy to enforce each of the following:

- *Password history*: Remembers previously used passwords, and prevents them from being reused until X passwords have also been used (where the administrator can specify X to be between 1 and 24). An alternative means of enforcing a history, which is used in some other systems, is to specify the minimum reuse period as a time value (e.g., the number of months before a password can be reused).
- *Maximum password age*: Specifies how long a password can be used before it has to be changed (with the OS allowing expiration periods of between 1 and 999 days).
- *Minimum password age*: Specifies how long a password has to be used before it can be changed again. This prevents users from immediately changing their password back after having been prompted to change it.
- *Minimum password length*: This represents one step toward preventing users from selecting weak and easily crackable passwords, by forcing them to use at least the specified length.
- *Password complexity requirements*: This again relates to prevention of weak choices, and if enforced, users are required to select passwords with the following characteristics:
  - Not contain all or part of the user's account name;
  - Be at least six characters in length;
  - Contain characters from three of the following four categories:
    - English uppercase characters (A through Z);
    - English lowercase characters (a through z);
    - Base 10 digits (0 through 9);
    - Non-alphanumeric characters (e.g., !, $, #, %).

Complying with such policies might not be a problem if we were dealing with just one or two systems that required passwords. In reality, however, users face a fundamental problem as a result of the volume of password-based systems that they are required to use. Consequently, if users choose (or are obliged) to follow the good practice regarding password selection and update, and also avoid reusing the same password in different places, then a frequent result is that they feel the need to compromise protection in another way—namely by writing the password down. If the user is security-aware, then more appropriate solutions here can include writing down cryptic reminders rather than the passwords themselves, or making use of password management utilities (whereby passwords for other systems are recorded for reference, but under the protection of a suitably strong master password or some other form of strong authentication).

Recognizing that users are prone to forgetting their passwords, it is also relevant to consider how this circumstance can be managed. In a workplace context, the solution will typically involve a reset initiated by the system administrator or a helpdesk (depending on the size of the organization and the number of users being supported). However, in web contexts it is necessary to consider more automated methods. In this respect, many systems offer the option to store a prompt to aid recollection. However, the selection of such prompts clearly needs to be done with care, to prevent them from giving a sufficient clue to someone else, as well as to ensure that they will actually remain meaningful if the legitimate user has occasion to use them. In some cases, however, users will still be unable to remember their password and will therefore require some means to recover or reset it. In this respect, many sites provide mechanisms that involve emailing a web link to the registered user's email account (the basis being that only the legitimate user should have access to the email account, thus ensuring that only the correct person can then follow the link to reset or recover the password). From a security perspective, resetting the password is preferable to divulging the existing one; if an impostor does happen to be at work (i.e., they have already compromised the user's email account) then resetting the password forces them to chose a new one—and thus alert the legitimate user once they realize they can no longer login. By contrast, revealing the existing password could enable an impostor to compromise other accounts as well (based on the aforementioned problem that people often use the same password for multiple systems).

Unfortunately, the disadvantage of following all the good advice is that some people will perceive it to remove the advantage of simplicity that made passwords attractive in the first place. As such, another approach may be to consider alternative types of secrets, which legitimate users may recall more easily.

4.1.3 Alternative Secret-Knowledge Approaches

Given the problems with passwords, a variety of other secret-knowledge methods have emerged as potential alternatives. The basic objective is to utilize secrets that legitimate users will find easier to remember and/or harder to compromise, while still providing a sufficient safeguard against impostor guesswork, brute force, and other forms of attack.

Potential approaches include moving the basis of the techniques away from purely recall-based approaches used by standard PINs and passwords toward methods that rely on less demanding concepts, such as recognition and provision of personal information. Two broad categories are examined here—the use of question and answer challenges, and the use of graphical methods in place of textual information.

4.1.3.1 Question and Answer Approaches

The user is asked to answer a series of questions, with correct answers leading to successful authentication. Such questions may be based on cognitive or associative information [3].

Cognitive passwords are based on questions to which the answers depend on the user's personal opinions, interests, and life history. Examples of potential questions here would include the following:

1. What is your mother's maiden name?
2. Where were you born?
3. What was the name of your best friend at school?
4. What is your favorite music?
5. What is your favorite food?
6. What was the name of your first pet?

The list includes a mixture of factual and opinion-based questions. From a reliability perspective, the former style (i.e., questions 1, 2, 3, and 6) would potentially be more effective than the latter (i.e., questions 4 and 5), as user's opinions could change over time and cause them to provide incorrect responses if they do not remember (or update) their original answers in the system. In addition, care must be taken to use questions that all users should be able to answer. For example, question 6 in the list would be unsuitable for a user who has never had a pet. One solution here is to allow users to select their own questions, but this clearly runs the risk of them taking a similarly cavalier approach to that which is used to select their passwords (and as a result they could choose questions to which others may also know the answers). A preferable approach would typically be to have a large number of well-chosen, preset questions, and allow the user to choose and answer a subset of these when they enroll on the system. In either scenario, the risk with cognitive questions is that they are vulnerable to an impostor who has intimate knowledge of the legitimate user and his or her background.

In contrast to the cognitive questions, the associative password approach relies more on the inner workings of the user's mind. The basis is that, rather than being asked for "factual" information, the user is asked to respond with the words that they associate with a series of keyword prompts. With appropriately chosen keywords, the idea is that different users will make different associations, and an impostor would have to think like the legitimate user in order to gain access. The principle is illustrated in Table 4.2, which presents five keywords and five examples of possible associations for each one.

There are two main risks with associative prompts. First, with a poorly chosen keyword, there will often be a small range of most probable answers, which therefore, enable the most likely answers to be determined by guesswork (e.g., for the

Table 4.2 Advantages and Disadvantages of User Authentication Methods

Keyword	*Examples of Possible Associative Responses*
Blue	Sky, Sea, Color, Whale, Thunder
Dog	Cat, Bone, Animal, Bark, Food
Anger	Management, Argument, Fight, Rage, Pain
Monster	Frankenstein, Dalek, Hitler, Scary, Mash
Bag	Carry, Hand, Sack, Shopping, Purse

keyword "Blue" in the table, the first couple of responses are likely to be the ones chosen by most people). Second, the words that legitimate users themselves associate with the keywords might change over time or according to their mood, thus increasing the chances of them failing to authenticate correctly. Indeed, it is for the latter reason that trial activities have found associative passwords to be less effective, as well as less popular with users [4].

In operation, both the cognitive and associative password approaches would typically maintain a large bank of questions (to which legitimate users provided answers when they originally registered on the system), and then randomly select a small number of these (e.g., 3 to 5 questions) as challenges at each login. In this way, even if an impostor knew or discovered the answers to some questions, it would be unlikely that he could provide the complete set required at a particular login.

The questions used in these contexts must require answers that are suitably distinctive to the legitimate user, in order to prevent everyone having similar answers, or their responses being too easy to discover or guess. The use of such questions has the potential advantage of using easily memorable (but nonetheless secret) information, but can involve a rather lengthy exchange between the user and the system in order to gain acceptance. As such, the approaches may have more applicability as a fallback or secondary level of authentication, rather than as a direct replacement for traditional passwords in all contexts.

4.1.3.2 Visual and Graphical Methods

The approaches discussed up to this point have all relied on the user's ability to recall textual information. However, with most devices now able to display pictures, a range of alternative methods can be considered, in which the secrets are based on visual information.

Three main approaches fall into this category:

- Those that rely on the user to remember a sequence of images (e.g., the Déjà Vu system requires users to remember a series of photos or abstract images [5]; the PassImages approach, depicted in Figure 4.2, uses pictures of everyday items [6]; and the commercially available Passfaces system relies on recognition of a series of faces [7]);
- Those that rely on them to remember something about an image (e.g., Blonder patented a graphical password in which the user can select a number of areas in a picture as a password, with the basis of the authentication "secret" being that the user had to correctly recall the location and the order of the regions [8]);
- Those that require the user to draw an image (e.g., Jermyn et al. presented a method in which the "password" was realized as a simple picture drawn on a grid [9]).

The theoretical basis in many of the graphical methods is that images are easier to remember than strings of characters. In particular, the methods that present images to the user (rather than requiring users to draw them) enable users to rely on *recognition* of things they have previously seen rather than precise *recall* of poten-

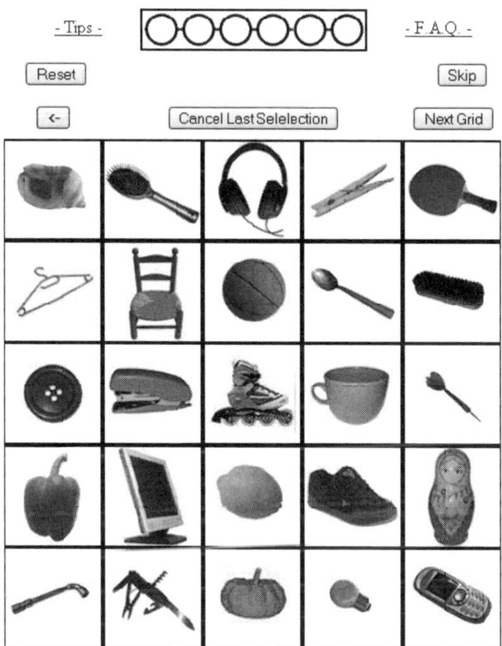

Figure 4.2 The PassImages approach.

tially arbitrary textual passwords. Previous studies, outside the area of security, have demonstrated that people have a developed memory for pictures and are highly skilled in recognizing previously seen images. For example, in a study by Standing et al., a sequence of 2,560 photographs was presented to an audience, and subsequent testing revealed recognition rate of around 90 percent (i.e., enabling participants to discriminate between previously seen and unseen images) [10]. As a possible consequence of such abilities, several of the aforementioned graphical authentication systems have reported positive findings in terms of both users' ability to remember the required information and their acceptance of the methods in practice.

In spite of the available alternatives, the continued dominance of passwords can be attributed to one or more of the following reasons:

- They are readily understood by the user community (even though they may not be used correctly). As such, users do not need to be given much in terms of guidance or training before being able to use them (once they have experienced the use of a password-based approach on one system, the experience is basically transferable to others).
- They are quicker to use than many of the alternative approaches and hence are considered convenient from this perspective.
- They are generally applicable to a range of systems and services and are therefore a straightforward option from the perspective of system designers and developers.

By contrast, other approaches would have constraints in at least one of these respects and may not be viable options at all in some contexts (e.g., some of the graphical approaches may not be applicable to devices with small or restricted display capabilities).

4.1.4 Attacks Against Secret-Knowledge Approaches

Secret-based approaches may be vulnerable to a number of types of attack.

- *Guesswork*: Depending on the target users, the attacker's knowledge of them, and the type of secret involved, it may be possible for the required information to be identified by informed guessing.
- *Brute force*: Trying potentially all of the different permutations of a secret until the correct answer is found. As previously mentioned in Section 4.1.2, a typical approach is to use automated software to find a match to an encrypted version of a password.
- *Social engineering*: Attempting to acquire the necessary login details via a direct interaction with the target user (which could take place in person, by phone, or electronic methods).
- *Phishing*: Use of bogus email messages (possibly in conjunction with fake websites) to trick users into divulging their details. A typical approach would be to set up a site that masquerades as the genuine service, and then get users to login in order to collect their details.
- *Shoulder surfing*: Attempting to observe keyboard and screen activity from close proximity while a legitimate user logs in.
- *Manipulated input devices*: Techniques such as the use of fake keypads at ATM machines and software/hardware-based keyloggers installed on PCs can provide a means for would-be attackers to capture the secret details during entry by legitimate users.

It should be noted that even if they do not yield the authentication secret itself, techniques such as social engineering, phishing, and shoulder surfing may also be used to assist the guesswork and brute force approaches. For example, using social engineering to find out a bit more about the target user could provide some direction for subsequent guesswork. Alternatively, if someone manages to partially shoulder surf a user's login and determine that his password begins with a "p" and involved eight key presses, it will provide a good starting point for a brute force attack and dramatically reduces the number of possibilities that would need to be tried.

The susceptibility of these techniques will vary depending on the type of secret knowledge involved and the way in which the authentication system has been implemented. For example, the ability to apply social engineering techniques to certain types of graphical passwords may be far more limited than the ability to discover the answers to cognitive questions by engaging the user in conversation. By contrast, shoulder surfing techniques could be more effective against graphical methods, as the impostor would be able to observe the screen rather than needing to look at what the user was typing on the keyboard.

4.2 Authentication Based On Tokens

The inability for people to remember long and random passwords, and follow good password practice, has been a key driver in the modern use of token-based authentication. Although not as common as password-based systems for logical access control, few can go without utilizing tokens in one shape or another: credit cards, phone cards, mobile phone SIM cards, workplace identification, and passports.

4.2.1 Principles of Token-Based Approaches

The concept of token authentication is that the token must be present in the legitimate user's possession in order for the user to be successfully authenticated to the system. The principal assumption relied on for token-based authentication to operate securely is that the legitimate (and only the legitimate) person has possession of the token.

Characteristics of a good token include:

- Every token must be unique;
- Tokens must be difficult to duplicate;
- Tokens must have a portable and convenient form factor;
- Cheap and readily replaceable for lost, stolen, or mislaid tokens.

4.2.2 Token Technologies

There are a wide variety of different tokens, designed either to serve a different purpose (i.e., electronic payment card versus logical access control) or as an improvement over the existing technology (e.g., smartcards versus magnetic cards). However, broadly, tokens can be classified by the fundamental nature of their operation:

- *Passive tokens* simply store a secret that is presented to an external system for processing and validation.
- *Active tokens* also store a secret; however, this is never released to an external system. Instead, the token is capable of processing the secret with additional information and presents the result of the verification to an external system.

Passive tokens are the original form of token, the most familiar of which are the magnetic stripe plastic cards used for credit/debit cards, driving licenses, physical door access, and library cards. The other main category of passive token is proximity based. Proximity tokens provide authentication and subsequent access (traditionally to physical but increasingly to logical systems) by close physical proximity between a token and reader—with the reader subsequently attached to the system that you wish to be authenticated on. The distance required between the token and reader can vary from centimeters to a few meters; the former operating in a *passive* mode being powered by self-induction and without an internal power source such as a battery, and the latter referred to as *active* mode, which includes a power source (hence the greater range). Obviously the choice of which mode to utilize would depend on the security requirements of the application, with shorter distance proximity cards providing a stronger verification of who has accessed a resource; quantity and cost, with active cards tending to be more expensive; and longevity, with active cards having a finite operational time due to the reliance on the battery.

Active tokens began to appear largely due to the weaknesses of the passive-based approach. The token does not present the secret knowledge to an external system but typically uses it to generate another piece of information: usually a one-time password, which is subsequently presented for authentication. Implemented in this manner, active tokens are immune from network sniffing and replay attacks, as the password used to get access is different every time. The key assumption in this approach is both the token and system *must* be synchronized so that the one-time password generated on the token is identical to the one-time password generated by the system. In order to manage this synchronization issue, one-time password tokens primarily come in two forms: counter-based and clock-based. Counter-based tokens were first to become commercially available and were simple in principal. Every time you needed a one-time password, you pressed a button on the token, you would generate a new password, and the internal counter would increment. Every time a new password was entered into the system, the system counter was also incremented. So as long as you only pressed the button when a one-time password was required, the systems remained synchronized. Unfortunately, due to the nature of the token, often carried in pockets and bags, the button would often be pressed by mistake causing the two systems to be out of synchronization and subsequently not work—requiring resynchronization by the system administrator. To get over this user issue, clock-based tokens were created that replied on the clocks on both systems being the same. These tokens were generally far more effective; however, time synchronization was more difficult to achieve over the long term due to fractional differences in clocks. Despite these limitations, one-time password tokens are increasing in popularity, with many banks and financial organizations providing them to their customers for increased security (particularly with regard to online banking) [11].

The most popular active-based token is the smartcard. Regarded as an upgrade to the traditional magnetic stripe–based passive tokens, smartcards have been extensively deployed by banks throughout Europe and increasingly in the United States. As the secret information remains within the card, cryptographically protected, the ability for an attacker to effectively sniff the network and utilize that information for financial benefit has been dramatically reduced. In addition, given the small form factor of the smartcard chip, performing side-channel attacks (e.g., analysis of power signals within the token) has become far more difficult and requires a significant level of technical expertise and equipment. Many modern smartcards now integrate tamper-resistant and tamper-evidence controls within them.

The final type of active token is pluggable tokens based on USB and PCMCIA computer connections. USB-based tokens have become the dominant choice and simply need to be plugged into the computer for access to be granted. The main advantage of pluggable tokens over one-time passwords is that they support more complicated authentication protocols.

4.2.3 Two-Factor Authentication

The use and applicability of a token-based approach depends on what it is being used to give access to and the level of security required to protect access. Unfortunately, tokens by themselves are not inherently good at ensuring the system gives access to the correct person rather an impostor. The system merely authenticates the presence of the token, not the person with whom it belongs. Therefore, through theft or duplication of the token, it would be relatively straightforward for an attacker to compromise a system.

It is therefore not surprising that tokens are most commonly utilized in a two-factor authentication mode. Two-factor authentication refers to the pairing of two of the basic authentication approaches: password and biometric, password and token, and token and biometric. The use of the token-password approach is very well established, with many commercial companies deploying large-scale systems. Credit cards and debit cards are excellent examples, where the card is paired with a PIN. The token-biometric approach is newer and increasingly been used in logical access control (e.g., as a pluggable USB key with integrated fingerprint reader).

Although the use of two-factor authentication increases the overall security by providing an additional layer of complexity, it is important to realize that these systems are not infallible.

4.2.4 Attacks Against Tokens

A major threat to tokens is theft of the token itself. In systems that rely only on the token, security is effectively breached the instance the theft takes place. Although strong procedural policies can be put in place by organizations to ensure lost or stolen cards are reported and access prohibited, it would be ill conceived to rely on the users to provide effective access control.

The main problem with passive tokens is the increasingly simple ability to duplicate the contents of the token. For instance, the equipment required to copy or *skim* magnetic stripe cards has dramatically reduced in cost (less than $10). The addition of two-factor authentication on bank cards for purchases (as well as for use in an ATM) has assisted in reducing fraud; however, the lack of requiring to enter a PIN for Internet-based purchases has simply moved the fraud from the store to online. Credit card companies are addressing this issue through schemes such as *Verified by Visa* and *MasterCard SecureCode*, which require additional secret knowledge–based verification.

Although active tokens have provided an additional layer of protection, a number of attacks exist against them, such as Internet protocol (IP) hijacking of sessions by sniffing network traffic while the one-time passwords are transmitted, man in the middle attacks, and side channel analysis of smartcards by power consumption. However, the nature of the attacks is far more sophisticated than was previously encountered, effectively raising the technological bar required by attackers to compromise the system.

4.3 Authentication Based On Biometrics

The use of biometrics, or specifically unique human characteristics, has existed for hundreds of years in one form or another, whether it is a physical description of a person or perhaps more recently a photograph. Consider for a moment what it is that actually allows you to recognize a friend in the street or allows you to recognize a family member over the phone. Typically this would be their face and voice, respectively, both of which are biometric characteristics. However, the definition of biometrics within the IT community is somewhat broader than just requiring a unique human characteristic(s)—it describes the process as an automated method of determining or verifying the identity of a person.

4.3.1 Principles of Biometric Technology

Many commercial biometric techniques exist, with new approaches continually being developed in research laboratories. The underlying process, however, remains identical. Biometric systems can be used in two distinct modes, dependent on whether the system wishes to *determine* or *verify* the identity of a person.

- *Verification*: determining whether a person is who he claims to be;
- *Identification*: determining who the person is.

The particular choice of biometric will greatly depend on which of these two methods is required, as performance, usability, privacy, and cost will vary accordingly. Verification, from a classification perspective, is the simpler of the two methods, as it requires a one-to-one comparison between a recently captured sample and a reference sample, known as a template, of the claimed person. Identification requires a sample to be compared against every reference sample, a one-to-many comparison, contained within a database, in order to find if a match exists. Therefore the unique characteristics used in discriminating people need to be more distinct or unique for identification than for verification. The majority of biometrics is not based on completely unique characteristics. Instead a compromise exists between the level of security required and thus more discriminating characteristics and the complexity, intrusiveness, and cost of the system to deploy. It is unlikely, however, in the majority of situations that a choice would exist between which method(s) to implement. Instead, different applications or scenarios tend to lend themselves to a particular method. For instance, PC login access is typically a verification task, as the user will select his username. However, when it comes to a scenario such as claiming benefits, an identification system is necessary to ensure that the person has not previously claimed benefits under a pseudonym. Typically the characteristics of a good biometric comprise of the following factors:

- *Unique*: the ability to successfully discriminate people. More unique features will enable more successful discrimination of a user from a larger population than techniques with less distinctiveness.
- *Universal*: the ability for a technique to be applied to a whole population of users. Do all users have the characteristics required (e.g., what about users without fingerprints)?

- *Permanent*: the ability for the characteristics to not change with time. An approach in which the characteristics change with time will require more frequent updating of the biometric template and result in increased cost to maintain.
- *Collectable*: the ease with which a sensor is able to collect the sample. Does the technique require physical contact with the sensor or can the sample be taken with little or no explicit interaction from a person? What happens when a user has broken his arm and is subsequently unable to present his finger to the system?
- *Acceptable*: the degree to which the technique is found to be acceptable by a person. Is the technique too invasive? Techniques not acceptable will experience poor adoption and high levels of circumvention and abuse.
- *Unforgeablity*: the ability not to duplicate or copy a sample. Approaches that utilize additional characteristics, such as liveness testing, can improve the protection against forging the sample.

A generic biometric system is illustrated in Figure 4.3, showing both of the key processes involved in biometric systems: enrollment and authentication. Enrollment describes the process by which a user's biometric sample is initially taken and used to create a reference template for use in subsequent authentication. As such, it is imperative that the sample taken during enrollment is from the authorized user and not an impostor, and that the quality of the sample is good. The actual number of samples required to generate an enrollment template will vary according to the technique and the user. Typically, the enrollment stage will include a quality check to ensure the template is of sufficient quality to be used. In cases where it is not, the user is requested to re-enroll onto the system.

Authentication is the process that describes the comparison of an input sample against one or more reference samples—one in the context of a verification system, many with an identification system. The process begins with the capture of a biometric sample, often from a specialized sensor. The biometric or discriminatory information is extracted from the sample, removing the erroneous data. The sample is then compared against the reference template. This comparison performs a correlation between the two samples and generates a measure or probability of similarity. The threshold level controls the decision as to whether or not the sample is valid, by determining the required level of correlation between the samples. This is

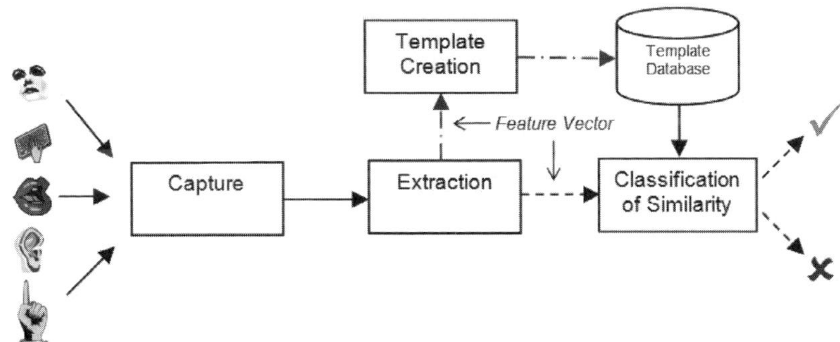

Figure 4.3 A generic biometric process.

an important consideration in the design of a biometric, as even with strong biometric techniques, a poorly selected threshold level can compromise the security provided. Finally, the decision is typically passed to a policy management system, which has control over a user's access privileges.

Given that all biometrics work on the basis of comparing a biometric sample against a known template, which is securely acquired from the user when he is are initially enrolled on the system, the template-matching process gives rise to misclassifications, of both the authorized user and impostors. These misclassifications result in a characteristic performance plot between the two main errors governing biometrics: the *false acceptance rate (FAR)*, or rate at which an impostor is accepted by the system, and the *false rejection rate (FRR)*, or rate at which the authorized user is rejected from the system. The error rates share a mutually exclusive relationship—as one error rate decreases, the other tends to increase, giving rise to a situation in which neither of the error rates is typically both at zero percent [12]. Figure 4.4 illustrates an example of this relationship. This mutually exclusive relationship translates into a tradeoff for the system designer between high security and low user convenience (tight threshold setting) or low security and high user convenience (slack threshold setting). Typically, a threshold setting that meets the joint requirements of security and user convenience is usually set. A third error rate, the *equal error rate (EER)*, equates to the point at which the FAR and FRR meet and is typically used as a means of comparing the performance of different biometric techniques. These performance rates when presented are the averaged results across a test population, therefore presenting the typical performance a user might expect to achieve. Individual performances will tend to fluctuate dependent on the uniqueness of the particular sample.

The actual performance of different biometric techniques varies considerably with the uniqueness and sophistication of the pattern classification engine. In addition,

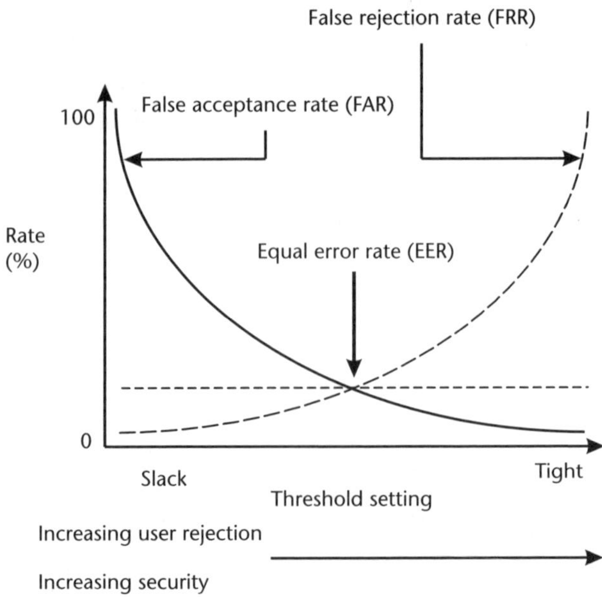

Figure 4.4 Mutually exclusive relationship between the FAR and FRR.

published performances from companies often portray a better performance than typically achieved given the tightly controlled conditions in which they perform their test.

In addition to the FAR and FRR, two further error rates are often used to indicate the collectability and acceptability of a biometric technique. The *failure to enroll* rate is the rate at which the system fails to enroll a person, and the *failure to acquire* rate is the rate at which the system fails to acquire a sample. Both error rates are typically associated with problems capturing the sample due to poor human-computer interaction (or lack of education of how to use the system) and the sensitivity of the capture device. Reasons for failure are numerous but can include dirty or scarred fingerprints; swiping the finger too quickly across a sensor; poor illumination for facial recognition; poor facial orientation (looking away from the camera during capture); changes in signature composition over time; and ensuring the correct distance and orientation of the face for iris recognition.

4.3.2 Biometric Technologies

Biometric approaches are typically subdivided into two categories: physiological and behavioral. Physiological biometrics are based on classifying a person according to some physical attribute, such as his fingerprints, face, or hand. Conversely, behavioral biometrics utilize some unique behavior of the person, such as his voice or the way in which he writes his signature. It is often argued that many biometrics could fit into both categories (e.g., although fingerprints are physiological biometrics, the way in which a user presents a finger to the sensor, and the subsequent image that is captured, is dependent on behavior). However, it is common to select the category based on the principal underlying discriminative characteristic—in this particular example, the fingerprint itself.

Most core biometric techniques commercially available are physiologically based and tend to have a more mature and proven technology. In addition, physiological biometrics typically have more discriminative and characteristic invariant features and as such are often utilized in both verification and identification systems. Given the more discriminative characteristics, the performance of physiological systems tends to be significantly better than their behavioral counterparts. However, the cost of implementation is often higher, as most physiological approaches require specialist hardware or sensors to capture the biometric sample.

4.3.2.1 Facial Recognition

Utilizing the distinctive features of a face, facial recognition has found increasing popularity in both computer/access security and crowd surveillance applications, due in part to the increasing performance of the more recent algorithms and its transparent nature (i.e., authentication of the user can happen without his explicit interaction with a device or sensor). The actual features utilized tend to change between proprietary algorithms but include measurements that tend not to change over time, such as the distance between the eyes and nose, areas around cheekbones, and the sides of the mouth. A number of commercial products are currently on the market such as Imagis Technologies ID-2000 [13] and Identix Face IT [14], with newer products based on three-dimensional facial recognition [15].

4.3.2.2 Facial Thermogram

A noncommercialized biometric, facial thermogram, utilizes an infrared camera to capture the heat pattern of a face caused by the blood flow under the skin. The uniqueness is present through the vein and tissue structure of a user's face. However studies to date have not quantified its reliability within an identification system. Recent studies have shown the external factors such as surrounding temperature play an important role in the performance of the recognition [16]. This technique has significant potential as recognition can be provided transparently, night or day, and if implemented within a facial recognition system (as a multimodal biometric—the combination of two or more biometric samples or techniques to constructively improve system performance [17]) would improve overall authentication performance.

4.3.2.3 Fingerprint Recognition

The most popular biometric to date, fingerprint recognition, can utilize a number of approaches to classification, based on minutiae (irregularities within fingerprint ridges) and on correlation (to authenticate a person) [17]. The image capture process does require specialized hardware, based on one of four core techniques: capacitive, optical, thermal, and ultrasound, with each device producing an image of the fingerprint. Fingerprint recognition is a mature and proven technology with very solid and time-invariant discriminative features suitable for identification systems. Although the uniqueness of fingerprints is not in question, fingerprint systems do suffer from problems such as fingerprint placement, dirt, and small cuts on the finger, and they are inherently an intrusive authentication approach, as the user is required to physically interact with the sensor. To date, fingerprint recognition has been deployed in a wide variety of scenarios from access control to logical security on laptops, mobile phones, and PDAs.

4.3.2.4 Hand Geometry

The second most widely deployed biometric is hand geometry. The technique involves the use of a specialist scanner that takes a number of measurements, such as length, width, thickness, and surface area of the fingers and hand. Different proprietary systems take differing numbers of measurements, but all the systems are loosely based on the same set of characteristics. Unfortunately, these characteristics do not tend to be unique enough for large-scale identification systems, but they are often used for time and attendance systems [18]. The sensor and hardware required to capture the image tend to be relatively large and arguably unsuitable for many applications such as computer-based login [19].

4.3.2.5 Iris Recognition

The iris is the colored tissue surrounding the pupil of the eye and is composed of intricate patterns with many furrows and ridges. The iris is an ideal biometric in terms of both its uniqueness and stability (variation with time), with extremely fast and accurate results [20]. Traditionally, systems required a very short focal length for capturing the image (e.g., physical access systems), increasing the intrusiveness of the approach. However, newer desktop-based systems for logical access are acquiring images at distances up to 18 inches. Cameras, however, are still sensitive to eye alignment, causing inconvenience to users.

4.3.2.6 Retina Scanning

Retina scanning utilizes the distinctive characteristics of the retina and can be deployed in both identification and verification modes. An infrared camera is used to take a picture of the retina, highlighting the unique pattern of veins at the back of the eye. Similarly to iris recognition, this technique suffers from the problems of user inconvenience, intrusiveness, and limited application, as the person is required to carefully present his eyes to the camera at very close proximity. As such, the technique tends to be most often deployed within physical access solutions with very high security requirements.

Additional physiological biometrics have been proposed such as odor, vein, and fingernail bed recognition, with research continuing to identify body parts and other areas with possible biometric applications [21].

Behavioral biometrics classify a person on some unique behavior. However, as behaviors tend to change over time (e.g., due to environmental, societal, and health variations), the discriminating characteristics used in recognition also change. This is not necessarily a major issue if the behavioral biometric has built-in countermeasures that constantly monitor the reference template and new samples to ensure its continued validity over time without compromising the security of the technique. In general, behavioral biometrics tend to be more transparent and user convenient than their physiological counterparts, however, at the expense of a lower authentication performance.

4.3.2.7 Keystroke Analysis

The way in which a person types on a keyboard has been shown to demonstrate some unique properties [22]. The process of authenticating people from their typing characteristic is known as keystroke analysis (or dynamics). The particular characteristics used to differentiate between people can vary, but often includes the time between successive keystrokes, also known as the inter-key latency, and the hold time of a key press. The unique factors of keystroke analysis are not discriminative enough for use within an identification system, but can be used within a verification system. Authentication itself can be performed in both *static* (text-dependent) and *dynamic* (text-independent) modes. However, although much research has been undertaken in this field due to its potential use for computer security, only one commercial product to date has been deployed and is based on the former (and simpler) method of static verification. BioPassword [23] performs authentication based on a person's username and password. A major downside to keystroke analysis is the time and effort required to generate the reference template. As a person's typing characteristics are more variable than say a fingerprint, the number of samples required to create the template is greater, requiring the user to repetitively enter a username and password until a satisfactory reliability level is obtained.

4.3.2.8 Service Utilization Profiling

Service utilization describes the process of authenticating people based on their specific interactions with applications and or services [24]. For instance, within a PC, service utilization would determine the authenticity of the people dependent on which applications they used, when and for how long, in addition to also utilizing other factors. Although the variance experienced within a user's reference template could be far larger than with other biometrics, it is suggested that sufficient discriminative traits exist within our day-to-day interactions to authenticate a person. Although not unique and distinct enough to be used within an identification system, this technique is non-intrusive and can be used to continuously monitor the identity of users while they work on their computer system. However, this very advantage also has a disadvantage with regard to users' privacy, as their actions will be continually monitored, and such information has the potential to be misused (e.g., the capturing of passwords). Although no authentication mechanisms exist utilizing this technique, a number of companies are using service utilization as a means of fraud protection [25, 26].

4.3.2.9 Signature Recognition

As the name implies, signature recognition systems attempt to authenticate people based on their signature. Although signatures have been used for decades as a means of verifying the identity of a person on paper, their use as a biometric is more recent. Authentication of the signature can be performed statically or dynamically. Static authentication involves utilizing the actual features of a signature, whereas dynamic authentication also uses information regarding how the signature was produced, such as the speed and pressure of the stylus. Numerous commercial applications exist, including for use within computer access and point-of-sale verification [27, 28]. However, due to the behavioral aspect of the technique and the variability between signatures, it is not recommended for use within an identification system.

4.3.2.10 Voice Verification (or Speaker Recognition)

A natural biometric and arguably a strong behavioral option, voice verification utilizes many physical aspects of the mouth, nose, and throat but is considered a behavioral biometric, as the pronunciation and manner of speech is inherently behavioral [21]. Although similar, it is important not to confuse voice verification with voice recognition, as both systems perform a distinctly different task. Voice recognition is the process of recognizing *what* a person says, whereas voice verification is recognizing *who* is saying it. Voice verification, similarly to keystroke analysis, can be performed in static (text-dependent) and dynamic (text-independent) modes, again with the former being a simpler task than the latter. Numerous companies provide various applications and systems that utilize voice verification (e.g., Nuance [29] provides static authentication for call centers and Anovea [30] provides static authentication for logical access solutions). To the author's best knowledge, no true commercial dynamic-based approaches based on anything a user might say exist. Pseudo-dynamic approaches do exist, which request the user to say a random two numbers that have not been particularly trained for during enrollment [31]. But, as the user is told what two numbers to repeat, the technique cannot be truly dynamic.

The list of biometrics provided should not be considered exhaustive, as new techniques and measurable characteristics are constantly being identified. It does, however, outline the main biometric approaches to date with an insight into the newer techniques. For further updates on new biometric techniques refer to the International Biometrics Group [32] and Biometrics Catalogue [33].

4.3.3 Attacks Against Biometrics

Attacks against biometrics involve compromising at least one stage of the biometric process. As Figure 4.5 illustrates, an attacker could attack the biometric sample (i.e., the person), the sensor, the biometric algorithm, and any and all communications links between elements.

The acquisition of a biometric sample via forceful means is an inevitable consequence, largely due to its simplicity versus the remaining more technical attacks. Fortunately, to date, few such attacks have occurred; however, one notable incident in Malaysia did result in a person losing his thumb [34].

Far more common attacks are those against the sensor itself. Spoofing is a technique whereby the attacker is able to provide a fake biometric sample to the sensor and convince it that it is indeed a legitimate sample. The nature of spoofing will depend on the individual biometric technique, but common examples are presenting a photograph to a facial recognition system, creating a latex fingerprint from a latent fingerprint found on an object the legitimate person has used (although this sounds very sophisticated, it is in fact a relatively straightforward procedure), and playing a voice sample previously recorded from the user. There have also been documented cases of attackers simply breathing on a fingerprint sensor and obtaining access—this works due to the latent oils from the finger remaining on the sensor from the previous person [35]. As a consequence, biometric vendors are adding a liveness component to their systems to ensure the sample being provided is acquired

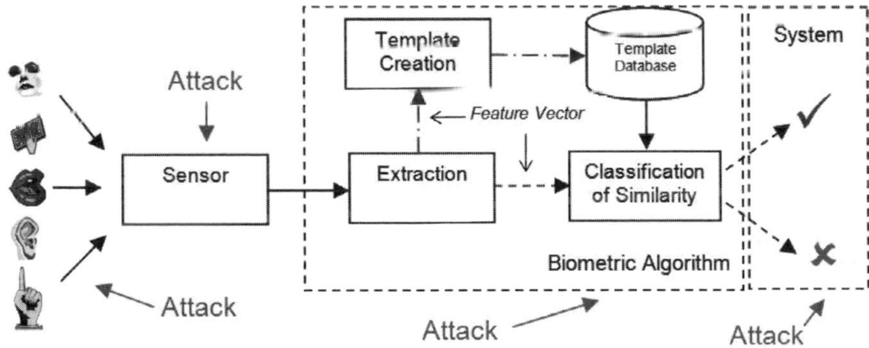

Figure 4.5 Attacks against the biometric process.

from a live host. Approaches taken for fingerprint systems include temperature, pulse, blood pressure, perspiration, and electric resistance. To date, many of the liveness tests have fallen short in providing effective protection, with ingenious attackers circumventing even these measures. This will, however, improve in time and become an inherent component in all biometric products.

Attacks against the biometric algorithm can be among the most technically difficult to achieve. A thorough analysis and understanding of the data extraction and classification elements can highlight possible weaknesses in the algorithm that can be exploited. All users using a biometric technique will have other people that present samples that are more similar to their own than others. An understanding of what types of characteristics are more similar and what features provide more discriminative information can provide insights into developing feature vectors that can be injected into the system, potentially bypassing the sensor. Should the attacker be permitted an indefinite number of attempts, they could simply attempt a brute-force attack against the feature vector.

The ability and difficulty for an attacker to compromise an element will greatly depend on the implementation of the individual biometric and the degree to which it relies on a network. All-in-one standalone solutions that house the sensor, extraction, classification, and storage of the template within a single device, with strong tamper-resistant protection, have far fewer opportunities for compromise than network-based solutions, in which the communication between elements is open to capture, modification, and replay attacks. No particular attack specific to biometrics is required here, and traditional methods for attacking network communications and the computer systems requiring authentication can be utilized (see later in this book for more information on network-based attacks).

4.4 Operational Considerations

This chapter has identified a range of factors that affect both the viability and success inherent in deploying authentication solutions. In selecting a suitable authentication mechanism, consideration needs to be given to the performance of the chosen techniques; their cost in terms of deployment, training (where applicable), and operation; reliability (considering issues including FAR and FRR, as discussed earlier), as well as problems enrolling and authenticating individuals; security (choosing a technique that provides an appropriate level of security), as well as considering the end-user acceptance of new techniques. For example, looking at eight years worth of results from the CSI/FBI Computer Crime and Security Survey [36–43], it is apparent that biometrics have made relatively little impact when compared to other methods and the proportion of respondents using them has remained fairly static (note: the surveys prior to 2004 did not include the category for smart cards or other tokens) with only a slight increase in the last few years.

The results from Figure 4.6 show that there has been relatively little change in the usage of alternative forms of authentication. There are of course many potential reasons for these, and Table 4.3 summarizes the relative advantages and disadvantages of the three main approaches to authentication.

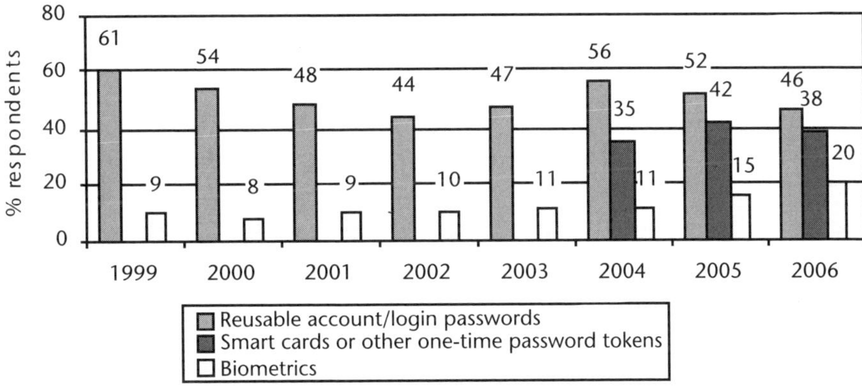

Figure 4.6 Use of authentication technologies. (*Source*: CSI/FBI surveys.)

Table 4.3 Advantages and Disadvantages of User Authentication Methods

Method	*Advantages*	*Disadvantages*
Secret knowledge	Familiar to users Does not require additional hardware No upfront deployment costs	Easily compromised and undermined by users Causes significant administrative overhead when users forget passwords Hidden costs (e.g., queries to IT helpdesk or system administrator) can be significant
Tokens	Cannot be shared as easily as a password (users will deny themselves access while someone else has their token) Users likely to be aware if their account is at risk (e.g., if token cannot be found)	Requires additional hardware investment, for both tokens and readers Potential delays in restoring access if users lose their tokens
Biometrics	Stronger authentication that cannot be shared with others Some methods have the potential for implementation in a transparent, nonintrusive manner (e.g. face recognition, keystroke analysis), allowing continuous authentication potential	Potential for error, including false acceptance of impostors and rejection of legitimate users Many methods demand additional hardware User performance in behavioral methods may be influenced by factors such as stress and fatigue Users may object to some methods (e.g., fingerprint recognition has criminological overtones; retinal scanning causes concern over use of lasers; keystroke analysis may lead to suspicion that productivity is being monitored)

4.5 Conclusions

This chapter has introduced the principles of authentication technologies and has presented a range of approaches that can be taken to provide a secure, reliable authentication process. The classic concept of authentication based on secret knowledge is well established but has been shown to be frequently compromised by user actions or by the manner in which it is implemented. This chapter has provided a comprehensive review of classic secret-based methods, identifying the weaknesses and proposing best practice. It has also introduced variations on this approach using knowledge or visual recognition methods that can be used as either replacements or in combination with other methods of authentication. Tokens and biometrics have been discussed as potential alternative methods, along with an analysis of their strengths and weaknesses together with an overview of a range of potential attack methodologies. It can be seen that approaches to authentication vary considerably, the level of acceptance and utilization of techniques vary considerably, and there is no single one-fit solution that can be applied in all environments.

References

[1] Wood, H. M., "The Use of Passwords for Controlled Access to Computer Resources," *NBS Special Publication 500-9*, U.S. Department of Commerce, May 1977.

[2] Furnell, S., "Authenticating Ourselves: Will We Ever Escape the Password?" *Network Security*, March 2005, pp. 8–13.

[3] Haga, W. J., and M. Zviran, "Question-and-Answer Passwords: An Empirical Evaluation," *Information Systems*, Vol. 16, No. 3, 1991, pp. 335–343.

[4] Irakleous, I., et al., "An Experimental Comparison of Secret-Based User Authentication Technologies," *Information Management & Computer Security*, Vol. 10, No. 3, 2002, pp. 100–108.

[5] Dhamija, R., and A. Perrig, "Déjà Vu: A User Study Using Images for Authentication," *Proceedings of the 9th USENIX Security Symposium*, 2002.

[6] Charrau, D., S. M. Furnell, and P. S. Dowland, "PassImages: An Alternative Method of User Authentication," *Proceedings of 4th Annual ISOneWorld Conference and Convention*, Las Vegas, NV, March 30–April 1, 2005.

[7] Passfaces Corporation, "The Science Behind Passfaces," White Paper, http://www.realuser.com/published/The%20Science%20Behind%20Passfaces.pdf (accessed May 4, 2007).

[8] Blonder, G. E., "Graphical Password," United States Patent 5559961, September 24, 1996.

[9] Jermyn, I., et al., "The Design and Analysis of Graphical Passwords," *Proceedings of the 8th USENIX Security Symposium*, August 1999.

[10] Standing, L., J. Conezio, and R. Haber, "Perception and Memory for Pictures: Single-Trial Learning of 2500 Visual Stimuli," *Psychonomic Science*, Vol. 19, No. 2, 1970, pp. 73–74.

[11] Savvas, A., "HSBC Rolls Out Security Tokens for Online Business Customers," Computer Weekly.com, http://www.computerweekly.com/Articles/2006/04/11/215324/hsbc-rolls-out-security-tokens-for-online-business-customers.htm, 2006.

[12] Cope, B. J. B., "Biometric Systems of Access Control," *Electrotechnology*, April/May 1990, pp. 71–74.

[13] Imagis Technologies, http://www.imagistechnologies.com/.

[14] Identix, "Identix. Your Trusted Biometric Provider," http://www.identix.com/.

[15] Biovisec, "3D Facial Recognition," http://www.biovisec.com/.

[16] Socolinsky, D., and A. Selinger, "Thermal Face Recognition in an Operational Scenario," *Proceedings of CVPR 2004*, Washington DC, 2004.

[17] Maltoni, D., et al., *Handbook of Fingerprint Recognition*, New York: Springer, 2003.

[18] Ashbourne, J., *Biometric: Advanced Identity Verification, The Complete Guide*, London: Springer, 2000.

[19] IR Recognition Systems, "Hand Geometry," Ingersoll-Rand, http://www.handreader.com/.

[20] Daugman, J., "How Iris Recognition Works," University of Cambridge, http://www.cl.cam.ac.uk/users/jgd1000/irisrecog.pdf, 1998.

[21] Woodward, J., N. Orlans, and P. Higgins, *Biometrics: Identity Assurance in the Information Age*, New York: McGraw-Hill, 2003.

[22] Spillane, R., "Keyboard Apparatus for Personal Identification," IBM Technical Disclosure Bulletin, 17(3346), 1975.

[23] BioPassword, "Security at your Fingertips." Bio Net Systems, http://www.biopassword.com/bp2/welcome.asp.

[24] Rodwell, P. M., S. M. Furnell, and P. L. Reynolds, "A Conceptual Security Framework to Support Continuous Subscriber Authentication in Third Generation Networks," *Proceedings of Euromedia 2001*.

[25] Graham-Rowe, D., "Something in the Way She Phone," New Scientist.com, http://www.newscientist.com/hottopics/ai/somethingintheway.jsp, 2001.

[26] Rogers, J., "Data Mining Fights Fraud—Company Operations," *Computer Weekly*, http://articles.findarticles.com/p/articles/mi_m0COW/is_2001_Feb_8/ai_70650704, 2001.

[27] PDALok. "PDALok—The Only Way to Lock Your Handheld," Romsey Associates, http://www.pdalok.com/default.htm.

[28] Communication Intelligence Corp., "Electronic Signatures," http://www.cic.com.

[29] Nuance, "The Voice Automation Expert," http://www.nuance.com/, 2004.

[30] Anovea, "Anovea—Authentication Technology," http://www.anovea.com/.

[31] VeriVoice, "The Sound Solution for Integrating a Robust Voice Verification Technology," http://www.verivoice.com/.

[32] International Biometrics Group, "Independent Biometrics Expertise," IMG, http://www.biometricgroup.com/.

[33] Biometrics Catalogue, Biometrics Catalogue, http://www.biometricscatalog.org.

[34] New Scientist, "Finger Chopped Off to Beat Car Security," *New Scientist*, Issue 2494, April 2005.

[35] Harrison, A., "Hackers Claim New Fingerprint Biometric Attack," Security Focus, http://www.securityfocus.com/news/6717, 2003.

[36] Power, R., *1999 CSI/FBI Computer Crime and Security Survey*, Computer Security Institute, 1999.

[37] Power, R., *2000 CSI/FBI Computer Crime and Security Survey*, Computer Security Institute, 2000.

[38] Power, R., *2001 CSI/FBI Computer Crime and Security Survey*, Computer Security Institute, 2001.

[39] Power, R., *2002 CSI/FBI Computer Crime and Security Survey*, Computer Security Institute, 2002.

[40] Richardson, R., *2003 CSI/FBI Computer Crime and Security Survey*, Computer Security Institute, 2003.

[41] Gordon, L. A., et al., *2004 CSI/FBI Computer Crime and Security Survey*, Computer Security Institute, 2004.

[42] Gordon, L.A., et al., *2005 CSI/FBI Computer Crime and Security Survey*, Computer Security Institute, 2005.

[43] Gordon, L.A., et al., *2006 CSI/FBI Computer Crime and Security Survey*, Computer Security Institute, 2006.

CHAPTER 5

Authorization and Access Control

Torsten Priebe

This chapter examines issues of authorization and access control, from first principles through advanced implementations. In general, *authorization* deals with defining which access rights (e.g., read, write, delete, execute) regarding a certain *object* a certain *subject* possesses. *Access control* examines whether a subject possesses the authorization to access a desired object when an access request occurs; as a result, the action is granted or denied. Authorization and access control address *confidentiality* (read access) and *integrity* (write access) security requirements.

Different access control models and techniques have been proposed to counter the various threats against the security. The models are independent of a specific domain. They can be used by a database management system (DBMS) to provide database security or an operating system for file system security. The same models can also be applied on application level (i.e., an application program can use an API to check a user's permissions and react correspondingly).

In this chapter we will discuss the most prominent access control models. First, classic access control models are discussed, based on their coverage by Pernul [1, 2]. In a nutshell, discretionary access control (DAC) specifies the rules under which subjects can, at their discretion, create and delete objects, and grant and revoke authorizations for accessing objects to others. In addition to controlling the access, mandatory access control (MAC) regulates the flow of information between objects and subjects. Other more exotic models are also presented in short. The personal knowledge approach concentrates on enforcing the basic law of many countries for the informational self-determination of humans and the Clark and Wilson model tries to represent common commercial business practice in a computerized security model.

Afterward, the role-based access control model (RBAC) is investigated in more detail, as it has evolved to a de facto and proposed NIST standard [3]. Finally, more recent developments toward attribute-based access controls (ABAC) and the extensible access control markup language (XACML) [4] are covered.

5.1 Discretionary Access Control (DAC)

Discretionary access control (DAC) represents the simplest form of an access control model. Access privileges on security objects are explicitly assigned to security subjects. DAC is implemented in most operating systems for file system security and database management systems (DBMSs).

DAC is based on two principles: the first is the principle of *ownership of information*, and the second the principle of *delegation of rights*. Ownership of information means that the creator of an information item becomes its owner. The owner may grant access to others and define the type of access (read, write, execute, and so on) given to them (i.e., granting and revoking of access privileges is under the *discretion* of the users themselves). Delegation of rights means that a user who has received a certain access right may be allowed to pass this right on to other users.

DAC has been studied for a long time. From 1970 through 1975, there was a good deal of interest in the theoretical aspects of the model. Technically speaking, DAC is based on the concepts of a set of *security objects* O, a set of *security subjects* S, a set of *access privileges* T defining what kind of access a subject has to a certain object, and in order to represent content- or time-based access rules a set of *predicates* P. For example, applied to relational databases, O is a finite set of values $\{o_1, \ldots, o_n\}$ representing relation schemas (i.e., tables); applied to operating systems, O is a set of files on the file system. S is a finite set of potential subjects $\{s_1, \ldots, s_m\}$ representing users, groups of them, or transactions operating on behalf of users. Access types (privileges) are the set of operations such as select, insert, delete, or update in case of databases, and read, write, execute in case of operating systems. Predicate $p \in P$ defines the access window of subject $s \in S$ on object $o \in O$ (e.g., time restrictions). The tuple $<o,s,t,p>$ is called access rule and a function f is defined to determine if an authorization $f(o,s,t,p)$ is valid or not:

$$f: O \times S \times T \times P \rightarrow \{\text{true},\text{false}\}$$

For any $<o,s,t,p>$, if $f(o,s,t,p)$ evaluates into *true*, subject s has authorization t to access object o within the range defined by predicate p.

As mentioned earlier, an important property of discretionary security models (found in relational database systems; see Chapter 6) is the support of the principle of delegation of rights in which a right is the (o,t,p)-portion of the access rule. A subject s_i who holds the right (o,t,p) may be allowed to delegate that right to another subject s_j $(i \neq j)$.

5.1.1 Implementation Alternatives

Most systems supporting DAC store access rules in an *access control matrix*. In its simplest form, the rows of the matrix represent subjects, the columns represent the objects, and the cells (i.e., the intersection of a row and a column) contain the access types that the subject has authorization for with respect to the object. An empty cell in the matrix denotes the absence of assigned authorizations.

Table 5.1 shows a sample access control matrix. In the example, subject s_1 may read and write o_1 (e.g., a certain file on the file system) but has no access to o_2. Subject s_2 is permitted to read o_2, but has no access to o_1. Since an IT system usually consists of a rather large number of subjects and objects, the resulting matrix quickly becomes very large. For performance reasons, fast access to the matrix is

Table 5.1 Sample Access Control Matrix

	o_1	o_2	...
s_1	rw	—	
s_2	—	R	
...			

necessary. Often however "unfavorable" sparse matrices (many empty fields) develop using up an unnecessarily large amount of memory. Matrix compression techniques may help coping with that problem.

Another possibility is the use of an *access control list (ACL)*. Two implementations are possible: either a list is assigned to each object, containing all users that have access to the object (and their permitted actions), or a list is assigned to each subject, storing all access rights granted to the subject for accessing certain objects. Access control lists can be understood as a matrix, which is divided into their individual columns or rows (without considering the empty cells).

5.1.2 Discussion of DAC

Discretionary access control (DAC) is a well-known technique with only few open research issues, supported by many commercial software systems (e.g., DBMSs). Although very common, discretionary models suffer from major drawbacks when applied to systems with security critical content. In particular we see the following limitations:

- *Enforcement of the security policy*: DAC is based on the concept of ownership of information. In contrast to enterprise models, in which the whole enterprise is the "owner" of information and responsible for granting access to stored data, DAC systems assign the ownership of information to the creator of the data items and allow the creator subject to grant access to other users. This has the disadvantage that the burden of enforcing the security requirements of the enterprise is the responsibility of the users themselves and cannot be controlled by the enterprise without involving high costs.
- *Cascading authorization*: If two or more subjects have the privilege of granting or revoking certain access rules to other subjects, this may lead to cascading revocation chains. As an example consider subjects s_1, s_2, s_3, and access rule (s_1,o,t,p). Subject s_2 receives the privilege (o,t,p) from s_1 and grants this access rule to s_3. Later, s_1 grants (o,t,p) again to s_3 but s_2 revokes (o,t,p) from s_3 because of some reason. The effect of these operations is that s_3 still has the authorization (from s_1) to access object o even if subject s_2 has revoked it. This has the consequence that subject s_2 is not aware of the fact that authorization (s_3,o,t,p) is still in effect. For more details of cascading authorizations in SQL database systems see Chapter 6.

- *Trojan horse attacks*: In systems supporting DAC, the identity of the subjects is crucial, and if actions can be performed using another subject's identity, then DAC can be subverted. A *Trojan horse* can be used to grant a certain right (o,t,p) of subject s_i on to s_j $(i \neq j)$ without the knowledge of subject s_i. Any program that runs on behalf of a subject acts with the identity of the subject and therefore has all of the DAC access rights of the subject's processes. If a program contains a Trojan horse with the functionality of granting access rules on to other users, this cannot be restricted by discretionary access control methods.
- *No information flow control*: DAC addresses direct access to objects by subjects; however, it does not control the flow of information to other subjects. If a user has access to a certain object (e.g., database table) he can easily create a copy of it. As he is then the owner of the new (copied) object, he can grant other users access to it.

For a more detailed discussion on discretionary controls in databases see Fernandez et al. [5] and for open research questions Lunt [6].

5.2 Mandatory Access Control

While DAC is concerned with defining, modeling, and enforcing access to information, MAC is in addition concerned with the flow of information within a system. The term MAC is based on the fact that access control is based on a set of predefined rules, which are *mandatory*, rather than user definable.

In the following we will present the need-to-know model and the military security model as the most prominent examples for MAC.

5.2.1 Need-to-Know Model

The *need-to-know model* is based on the principle that a subject can access only those objects that are necessary for fulfilling his work duties (i.e., he is required to have a "need to know" the content of the object). As the access permissions of subjects are assigned according to their competence and work area, the model is sometimes also called *policy of competence*.

In order to apply the need-to-know model, both security subjects and objects need to be assigned one or more "topic areas," called compartments. For a subject s to be able to access an object o, the set of compartments of the object $Comp(o)$ must be a subset of need-to-know of the subject $NTK(s)$.

Additionally, the model can be extended to prevent an undesired information flow by defining that write access is only permitted if the need-to-know of the subject is a subset of the compartments assigned to the object. Altogether, access is controlled by the following MAC rules:

1. Subject s may read object o, if $Comp(o) \subseteq NTK(s)$;
2. Subject s may access o for writing, if $NTK(s) \subseteq Comp(o)$.

Consider the following example:

- Compartments: {medical data (M), financial data (F), private data (P)};
- *Comp*(o) = {M,F};
- *NTK*(s_1) = {F}, *NTK*(s_2) = {P}, *NTK*(s_3) = {P,M}, *NTK*(s_4) = {F,M,P}.

The example leads to the following results:

- Subject s_1 may only write object o;
- Subject s_4 may only read object o;
- Subjects s_2 and s_3 may not access object o.

5.2.2 Military Security Model

The *military security model* requires that security objects and subjects are assigned to certain security classes represented by a label. The label for an object is called its *classification*, while a label for a subject is its *clearance*. The classification represents the sensitivity of the labeled data, while the clearance of a subject its trustworthiness to not disclose sensitive information to others. A security label consists of two components: a hierarchical set of sensitivity levels (e.g., top_secret > secret > classified > unclassified) and a nonhierarchical set of categories, representing classes of object types of the universe of discourse (like the compartments of the need-to-know model).

Clearance and classification levels are totally ordered, and resulting security labels are only partially ordered—thus, the set of classifications forms a lattice in a mathematical sense. In this lattice security class, c_1 is comparable with and dominates (≥) c_2 if the sensitivity level of c_1 is greater than or equal to that of c_2 and the categories in c_1 contain those in c_2 (like in the need-to-know model). The model grew out of the military environment where it is practice to label information. However, this custom is also common in many companies and organizations in which similar labels, like "confidential" or "company confidential," are used.

The military security model is usually based on the *Bell and LaPadula* security paradigm [7] and formalized by two rules. The first protects the information from unauthorized disclosure, and the second protects from contamination or unauthorized modification by restricting the information flow from high to lower trusted subjects.

1. *Simple security property*: Subject s is allowed to read object o if *Clear*(s) ≥ *Class*(o).
2. **-property:* Subject s is allowed to write object o if *Class*(o) ≥ *Clear*(s).

Figure 5.1 shows read (r) and write (w) access rights for two subjects on different security levels according to the Bell and LaPadula model.

The Bell LaPadula model is particularly concerned about keeping secrets. It controls the information flow as the *-property protects information from being "written down" along the hierarchy of sensitivity levels (i.e., lower level data may be read but not written).

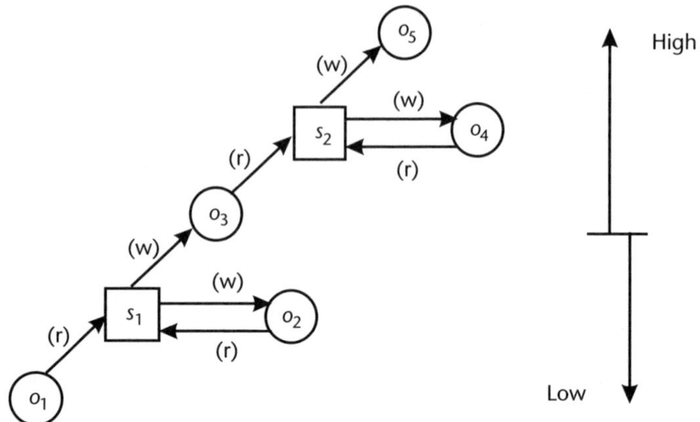

Figure 5.1 Bell and LaPadula model.

In contrast to the Bell LaPadula model, which avoids information from flowing downward from higher to lower classified subjects, the Biba model is concerned about the integrity of the system (i.e., trying to avoid higher classified objects from being corrupted by lower classified subjects). In order to achieve this, Biba defines *integrity levels* analogous to the sensitivity levels of Bell and LaPadula. The following two rules are set up in order to prevent information from flowing up the hierarchy of integrity levels:

1. *Simple integrity property*: Subject s may write object o, if $I(s) \geq I(o)$.
2. *Integrity *-property*: If s has read access to object o with $I(o)$, then s may only have writing access to object p, if $I(o) \geq I(p)$.

In Figure 5.2, the Biba model is sketched compared to the Bell and LaPadula model. Note that if the Biba model is combined with the Bell and LaPadula model, the integrity levels $I(s)$ and $I(o)$ of the Biba model are not necessarily the same as the security levels of the Bell and LaPadula model.

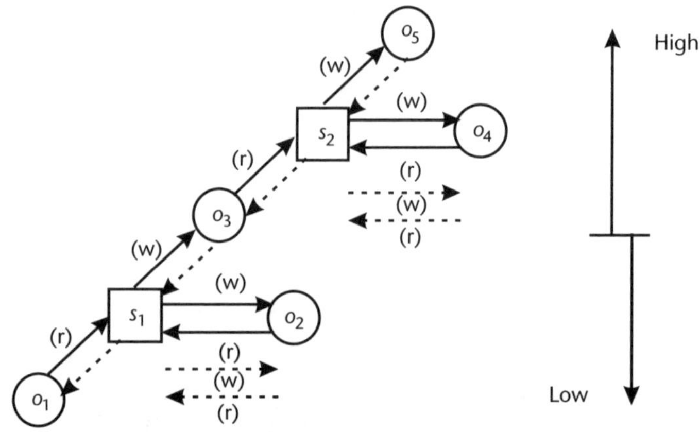

Figure 5.2 Comparison of Biba and Bell and LaPadula models.

5.2.3 Discussion of MAC

MAC leads to *multilevel secure (MLS) databases* because its content may appear different to users with different clearances. This is because of two reasons: first, not all clearances may authorize all subjects to all data, and, second, the support of MAC may lead to *polyinstantiation* of attributes or tuples. Polyinstantiation (i.e., multiple instances of a data item referring to a single fact of reality but differing by the classification label) is supported in multilevel databases and is necessary to support users with cover stories. For a more detailed discussion of the multilevel secure database model, see Chapter 6.

5.3 Other Classic Approaches

In this section we have included three other access control approaches we believe are relevant to information systems. The personal knowledge approach is an attempt to enforce the privacy aspect of information, the Chinese wall policy, a confidentiality policy unique to the commercial sector, and the Clark and Wilson model, which emphasizes the preservation of the integrity of the information system with respect to security.

5.3.1 Personal Knowledge Approach

The *personal knowledge approach* [8] is focused on protecting the privacy of individuals stored in an information system. The main goal of this model is to meet the right of humans for informational self-determination as requested in constitutional laws of many countries. In this context, privacy can be summarized as the basic right for an individual to choose which elements of his private life may be disclosed. The personal knowledge approach is built around the concept of a person and its knowledge. Each person represented in the database knows everything about himself, and if he wants to know something about someone else represented in the database that person must be asked. To achieve this high goal, the personal knowledge approach combines techniques of relational databases, object-oriented programming, and capability-based operating systems. More technically, it is based on the following constructs:

- *Persons*: Persons either represent information about individuals stored in the database or are the users of the information system. Each person is characterized by its personal knowledge. Technically, a person is seen as an encapsulated object that "knows" everything about himself (his application domain) and about his relationships to other persons known to the system. These are the only two components of "personal knowledge," and for any person it is not permitted to remember permanently anything else. Within the system each person is uniquely identified by a surrogate.
- *Acquaintances*: Persons are acquainted with other persons. The set of acquaintances of a person describes the environment of that person and is the set of objects with which the person is allowed to communicate. Communication is

performed by means of messages that may be sent from a person to its acquaintances for querying about their personal knowledge or for asking to perform an operation (e.g., to update their knowledge). Acquaintances of a person are represented in the knowledge body of the corresponding person object by their surrogate.

- *Roles and authorities*: Depending on the authority of the sender, the receiver of a message may react in different ways. The authority of a person with respect to an acquaintance is based on the role the person is currently acting in. While the set of acquaintances of a person may change dynamically, authorities and roles are statically declared in the system.
- *Remembrance*: Each person remembers the messages he is sending or receiving. This is established by adding all information about recent queries and updates together with the authorities available at that time to the "knowledge" of the sender and receiver person. Based on this information, auditing can be performed and all transactions can be traced by just asking affected persons.

Security (privacy) enforcement based on the personal knowledge approach is based on two independent features. First, after login, each user is assigned as an instance of a person object type and thus holds individually received acquaintances and statically assigned authorities on roles. Second, if the user executes a query or an update operation, the query is automatically modified such that resulting messages are sent only to the acquaintances of the person. Summarizing, the personal knowledge approach is fine-tuned to meet the requirements of informational self-determination. Thus, it is best suitable as the underlying security paradigm for information systems in which information about humans not available to the public is kept (e.g., hospital information systems or databases containing census data).

5.3.2 Clark and Wilson Model

Similar to the personal knowledge approach, the *Clark and Wilson model* is also based on roles. Clark and Wilson [9] argue that their model is based on concepts already well established in the pencil-and-paper office world. The approach consists of three basic principles: there are security subjects, (constraint) security objects, and a set of well-formed transactions. Users are restricted to execute only a certain set of transactions permitted to them, and each transaction operates on an assigned set of data objects only. This approach is interpreted as follows:

- *Roles*: Security subjects are assigned to roles. Based on their roles in an organization, users have to perform certain functions. Each business role is mapped into a function of the information system, and ideally at a given time a particular user is playing only one role. A function corresponds to a set of (well-formed) transactions that are necessary for the users acting in the role. In this model it is essential to state which user is acting in what role at what time, and for each role what transactions are necessary to be carried out.
- *Well-formed transactions*: A well-formed transaction operates on an assigned set of data and needs to be formally verified that all relevant security and

integrity properties are satisfied. In addition it should provide logging and atomacy and serializability of resulting subtransactions in a way that concurrency and recovery mechanisms can be established. It is important to note that in this model, the data items referenced by the transactions are not specified by the user operating the transaction. Instead, data items are assigned depending on the role the user is acting in.

- *Separation of duty*: This principle requires that each set of users being assigned a specific set of responsibilities based on the role of the user in the organization. The only way to access the data in the database is through an assigned set of well-formed transactions specific to the role each of the users play. The roles need to be defined in a way that makes it impossible for a single user to violate the integrity of the system (i.e., certain transactions must not be performed by the same user). For example, the design, implementation, and maintenance of a well-formed transaction must be assigned to a different role than the execution of the transaction.

A first attempt to implement the concept of a well-formed transaction was made by Thomsen and Haigh [10]. The authors have compared the effectiveness of two mechanisms for implementing well-formed transactions, lock-type enforcement (for lock data views, see Stachour and Thuraisingham [11]), and the Unix setuid mechanisms. With lock-type enforcement, the accesses of the user processes to data can be restricted based on the domain of the process and the type of data. `Setuid` and `setgid` features allow a user who is not the owner of a file to execute the commands in the file with the permission of the owner. Although the authors conclude that both mechanisms are suitable for implementing the Clark and Wilson concept of a well-formed transaction, no further studies and implementation projects are known.

5.3.3 Chinese Wall Policy

The *Chinese wall policy* is a further well-known security model that combines discretionary and mandatory aspects of access control. In 1989 it was presented by Brewer and Nash [12], who derived it from the British law for stock brokers consulting different firms.

Consultants are supposed to be the users of the information system and deal with public and confidential company information of their clients. As an example we consider a set of companies partly competing with each other and a group of consultants. It is forbidden that a consultant work for a company if he already has insider knowledge of a competitor. Thus, the main goal of this policy is to prevent information flows from competing companies to the same consultant because this may result in a conflict of interest in the consultant's analysis, and moreover insider knowledge about two similar types of companies presents the potential for consultants to use such knowledge for personal profit.

The solution to this problem provided by Brewer and Nash is the following: At the beginning, new consultants start with no mandatory restrictions on their access rights. Let us consider the case that a new consultant accesses information about a

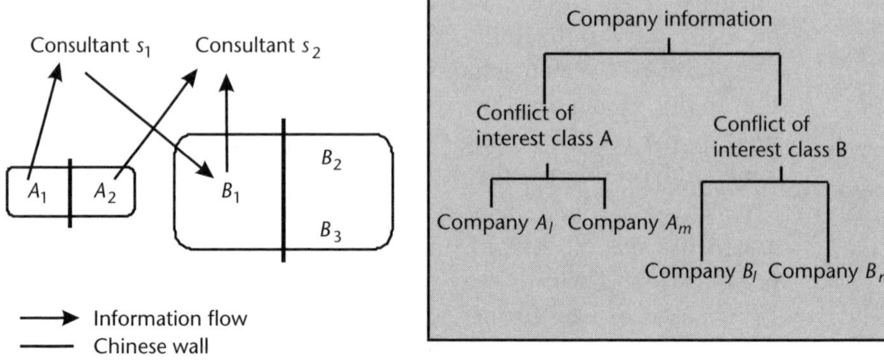

Figure 5.3 The Chinese wall policy.

bank A_1. Thereafter, this consultant is mandatorily denied access to information about any other bank (i.e., about any other organization with which a conflict of interest with respect to bank A_1 exists). However, there are still no mandatory restrictions regarding the consultant's access to an oil company, an insurance company, and so forth. Controlling the access is performed by building "Chinese walls" between the information concerning companies already accessed and all other companies within the same conflict-of-interest class.

Following this policy, it is useful to distinguish between public information and company information. Public information involves desirable features, such as public bulletin boards, electronic mail, and public databases. Company information must be categorized into mutually disjoint conflict-of-interest classes. Each company must belong to one conflict-of-interest class. The Chinese wall policy requires that a consultant not be able to read information for more than one company in any given conflict-of-interest class.

The concept of the Chinese wall policy is depicted in Figure 5.3. The example shows two conflict-of-interest classes. Assume A_1 and A_2 are banks and B_1 through B_3 are oil companies. Consultant s_1 who has consulted (and received information from) bank A_1 may not consult bank A_2. However, he may work with oil company B_1 as a transfer of information from A_1 to B_1 would not cause a conflict of interest.

5.4 Role-Based Access Control

The concept of *role-based access control (RBAC)* has evolved to the de facto standard in enterprise multiuser systems involving a large (but structured) number of users with different rights and obligations as well as a large amount of sensitive data. In contrast to earlier access control approaches, in particular DAC, RBAC simplifies the administration of access permissions by removing the direct links between subjects (users) and objects. It introduces an indirection by means of roles, which can be derived from the organizational structure encapsulating organizational functions or duties (e.g., secretary, operator) and then assigned to the users.

Experiences in organizational practice show that the definition of roles as part of the organizational structure is relatively stable, while the assignment of users to roles in comparison changes more frequently. This assignment can be modified without affecting the definition of a role itself, while on the other hand the role's definition can be modified without interfering with the assignment to users. Furthermore, by assigning different users to the same role and different roles to the same user, redundancy is reduced.

RBAC realizes the security principle of "least privilege" by assigning only those authorizations that are absolutely necessary for the fulfillment of the subject's duties. A subject cannot obtain more authorizations from a second subject (i.e., in contrast with DAC, there is no delegation of rights).

In Ferraiolo et al. [3] a NIST standard for RBAC has been proposed as a reference model. The RBAC reference model is divided into four submodels, which embody different subsets of the functionality:

- *Core RBAC*: Core RBAC covers the essential RBAC features such that permissions are assigned to roles and roles are assigned to users.
- *Hierarchical RBAC*: Hierarchical RBAC adds the notion of role hierarchies and inheritance.
- *Constraint RBAC*: Constraint RBAC adds constraints to implement static separation of duty (restriction of the user-role membership) and dynamic separation of duty (restriction of the role activation by users).
- *Consolidated RBAC model*: Hierarchical and constraint RBAC are combined in the so-called Consolidated model.

5.4.1 Core RBAC

Core RBAC comprises five basic element sets: *Users* are active elements that can be human beings as well as software artifacts. *Roles* correspond to fields of activities of users. These activities are bound to *permissions* needed to carry them out. Permissions should be only assigned to a role if absolutely necessary to fulfill the corresponding activities. A permission is a certain *operation* (e.g., read) that can be executed on a certain *object*. Since a user can have several functions in an organization, several different roles can be assigned to her. A *session* represents the context in which a sequence of activities is executed; a user can be active in more than one session at a time (e.g., in different windows) and activate as many roles as desired, at minimum one, for the period of a session.

The elements of the core RBAC model and their relationships are shown in Figure 5.4. Formally, the core RBAC model can be described by the following tuple:

$$RBAC = (U,O,R,P,S,UA,PA,user\_session,session\_role)$$

The individual items of the tuple are as follows:

- A set of users (subjects) U;
- A set of objects O;
- A set of roles R;
- A set of access permissions P;

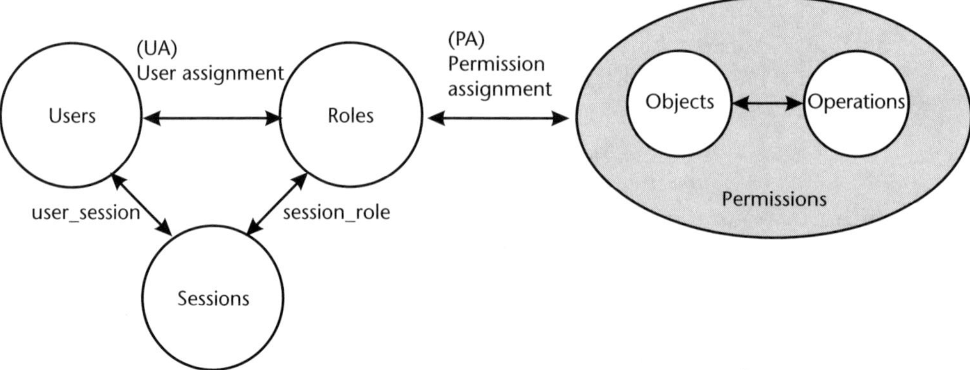

Figure 5.4 Core RBAC.

- A set of sessions *S*, updated dynamically when a session is started or terminated by a user;
- A relation *UA* (user assignment) that describes which roles from the set of *R* are assigned to a certain user;
- A relation *PA* (permission assignment) that describes which authorizations from the set *P* are assigned to a role;
- A relation *user_session* describing pairs (*u*,*s*)—that is, a user can be active in multiple sessions at a time;
- A relation *session_*role describing pairs (*s*,*r*)—that is, a user can activate a subset of her possible roles (from *UA*) in a session.

5.4.2 Hierarchical RBAC

In addition to core RBAC, *hierarchical RBAC* introduces a partial order between the roles (see Figure 5.5), an inheritance relation in which senior roles acquire the permissions of their juniors and junior roles acquire the user membership of their seniors. There may be general permissions that are needed by a large number of users and consequently may be assigned to a more general role.

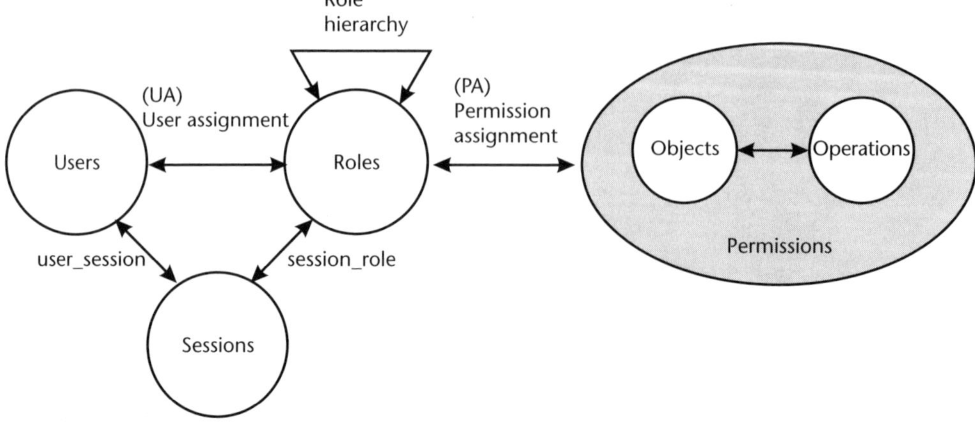

Figure 5.5 Hierarchical RBAC.

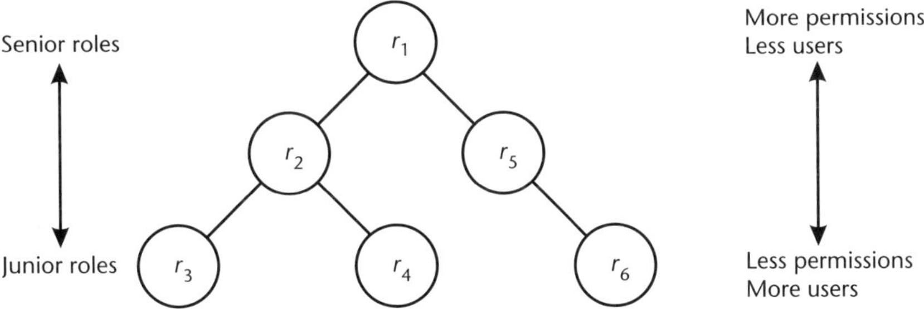

Figure 5.6 Sample role hierarchy.

The role hierarchy corresponds to a bidirectional inheritance relation as depicted in Figure 5.6. From the bottom up, each subject that is member in a senior role by direct assignment indirectly receives the permissions assigned to the derived roles (i.e., the permissions are inherited upward). From the top down, every user who is assigned to a senior role is indirectly a member of a more junior role as well (i.e., the user assignment is inherited downward).

5.4.3 Constraint RBAC

Constraint RBAC assumes that there are relations or exclusions between some fields of activity and allows defining separation of duty constraints to enforce conflict-of-interest policies.

As shown in Figure 5.7, constraints are possible on the user assignment and role hierarchy relations as well as the *session_role* relation for implementing static and dynamic separation of duty:

- *Static separation of duty (SSD)*: A static separation of duty is useful in which fields of activities are in a conflict of interest that could be broken by the simultaneous membership of a user in multiple roles. To address this,

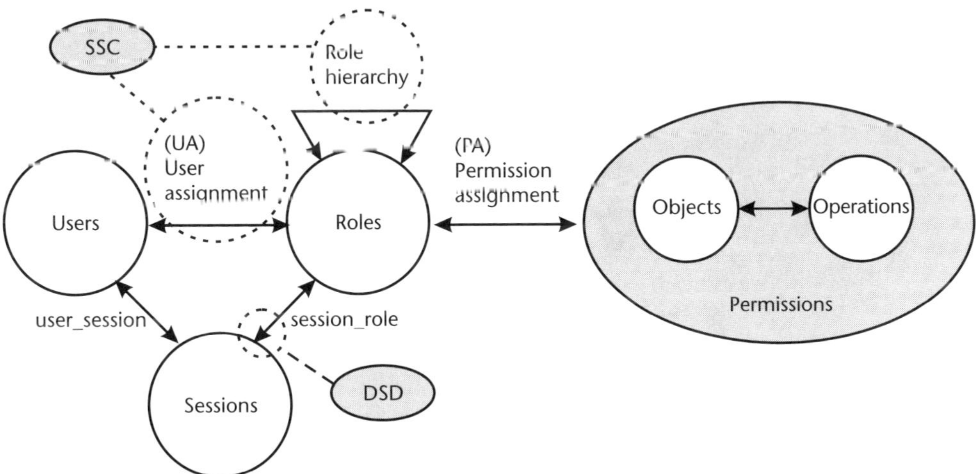

Figure 5.7 Constraint RBAC.

constraint RBAC allows constraints to be defined on the *UA* relation. In the consolidated model (hierarchical and constraint RBACs), constraints are also possible on the role hierarchy to avoid conflicts of interest between roles in the same inheritance hierarchy.

- *Dynamic separation of duty (DSD)*: Constraints can be applied as well to the *session_role* relation in order to limit the dynamic allocation of roles to sessions (i.e., in contrast to core RBAC, in which a user can activate as many of the assigned roles as desired, in Constraint RBAC only those roles that do not conflict with any dynamic limitation constraint can be activated). For example, it might be important that only one superuser can be active in a system at a time. Another user who is also assigned to the superuser role cannot activate that role in his session, unless the other user deactivates it.

5.4.4 Discussion of RBAC

RBAC simplifies the administration of authorizations. However, for very large open systems such as digital libraries, e-commerce, e-government systems, or hospital systems, the role hierarchies can become very complex. When the number of protected objects also increases, a manual assignment of authorizations becomes very expensive and error-prone. Furthermore, in many situations access depends on contents of an object and the environment the subject is acting in. In these applications we need to deal with users not previously registered. Access is granted based on a user's assignment to a (possibly short-term) project or even based on its current location. The attribute-based access control model presented in the next section provides a more convenient and efficient way to manage access rights for these situations.

5.5 Attribute-Based Access Control

The basic idea of *attribute-based access control* is utilize (possibly dynamic) properties of subjects and objects as the basis for authorization, rather than directly (and statically) defining access rights between users (or roles) and objects. On the user side, an attribute (in this context also called credential) could be his position within an organization, quite similar to a role. Especially for external users, however, acquired credentials (e.g., subscriptions, customer status) or attributes such as age or shipping address may need to be used instead. Attributes can be very dynamic (contrary to rather statically defined roles). For example, in a mobile computing environment, the current location of a user could be an attribute. For coding, the user attributes (e.g., X.500 [13]) can be used. For the security objects, the content (e.g., of documents) can be described by means of metadata (e.g., based on the Dublin core metadata standard). Such metadata elements should be used for authorization purposes.

Figure 5.8 shows the basic concept of the attribute-based access control model. Subjects and objects are represented by a number of attributes. The permissions are no longer assigned between subjects and objects but between subject and object

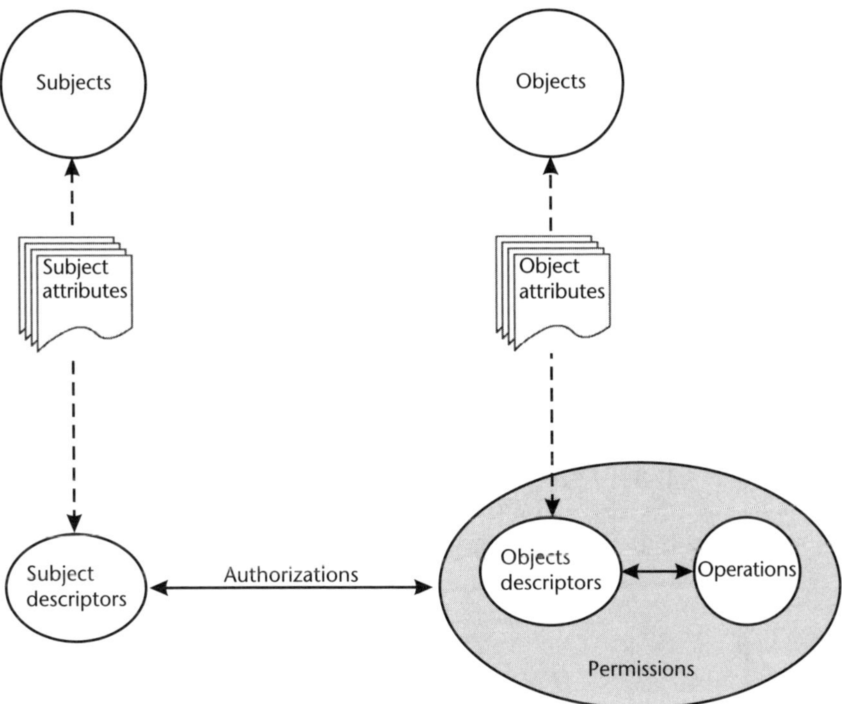

Figure 5.8 Elements of attribute-based access control.

descriptors, something like "virtual" subjects and objects that can possibly relate to multiple real identities.

A number of attribute-based access control approaches have been proposed in the literature. Adam et al. [14] present the digital library access control model (DLAM) for security in digital libraries, which defines access rights on the basis of user attributes and concepts associated with objects. Other work has its origins within the area of the public key and privilege management infrastructures and is based on X.509 attribute certificates [15]. First ideas of using attribute certificates for access control can be found in [16]. Meanwhile, several projects and systems in this area have developed (e.g., PERMIS and Shibboleth). The XML dialect for defining access control policies, XACML [4], is also based on user and object attributes. A more recent example of an attribute-based model in the area of digital rights management (DRM) is $UCON_{ABC}$ [17].

5.5.1 ABAC—A Unified Model for Attribute-Based Access Control

A first standardized reference model for attribute-based access control was presented in Priebe et al. [18] as a security pattern using the name metadata-based access control (MBAC). This model has been developed further under the name ABAC [19]. Figure 5.9 shows the ABAC model using a slightly updated terminology as a UML class diagram.

Note that in the model the actual subjects and objects are not listed, as ABAC abstracts from individual identities. Authorizations are defined between *subject* and

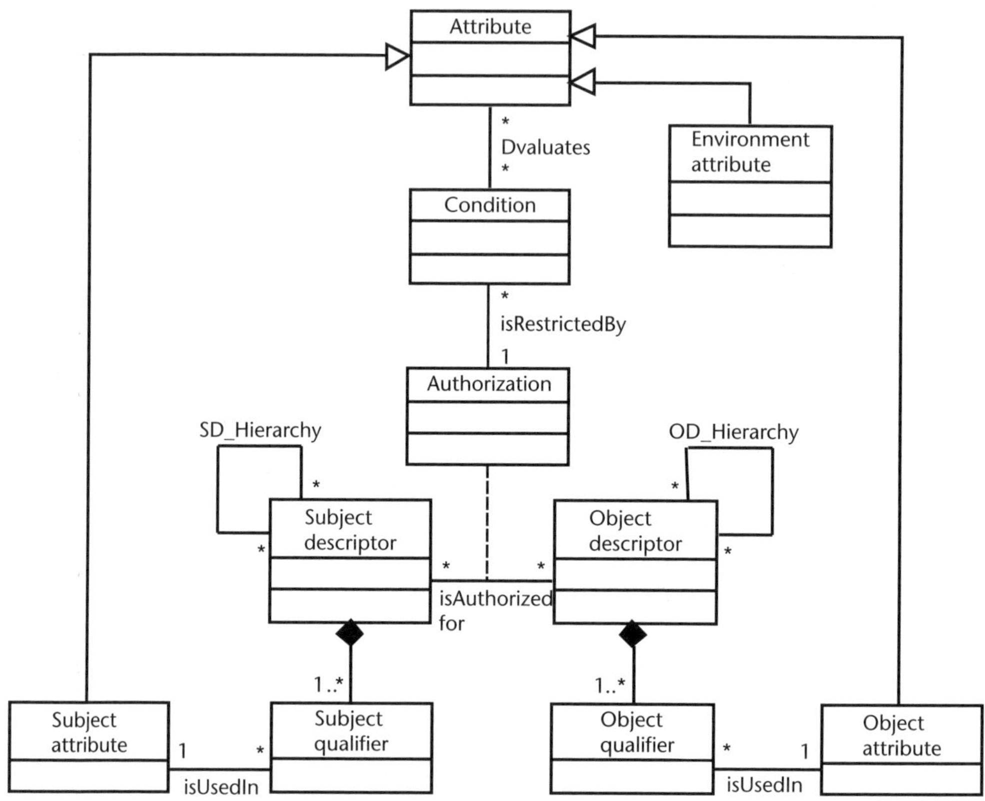

Figure 5.9 ABAC model in UML.

object descriptors. A subject descriptor consists of a number of *subject qualifiers* (e.g., age > 18, ZIP code begins with "93"). The same is true for the object descriptors. Here *object qualifiers* are defined on the basis of object attributes (e.g., concerns a certain project, published by a certain publisher). Subject and object descriptors are similar to groups; however, the assignment of individuals to these groups is done not explicitly but implicitly based on the attribute values.

In order to simplify the policy specification, subject and object descriptor hierarchies have been introduced. Similar to role hierarchies in hierarchical RBAC (see Section 5.4.2) specialized descriptors can be derived from more general ones (inheriting their qualifiers).

Finally, it is possible to specify additional *conditions* on authorizations, similar to predicates in DAC (see Section 5.1) or constraints in constraint RBAC (see Section 5.4.3). By using conditions, subject attributes can be compared with object attributes, which cannot be achieved by the descriptors and qualifiers alone in which attributes can only be compared with constant values. Environment attributes (e.g., current time) can be included into the access control decision (e.g., in order to restrict access to regular business hours). Examples for ABAC access control policies are shown in Figure 5.10 and covered in detail in Section 5.5.2.

The unified ABAC model is based on concepts found in DLAM [14], XACML [4], and $UCON_{ABC}$ [17]. As the terminology of these models differs to some degree, Table 5.2 summarizes the different terms used.

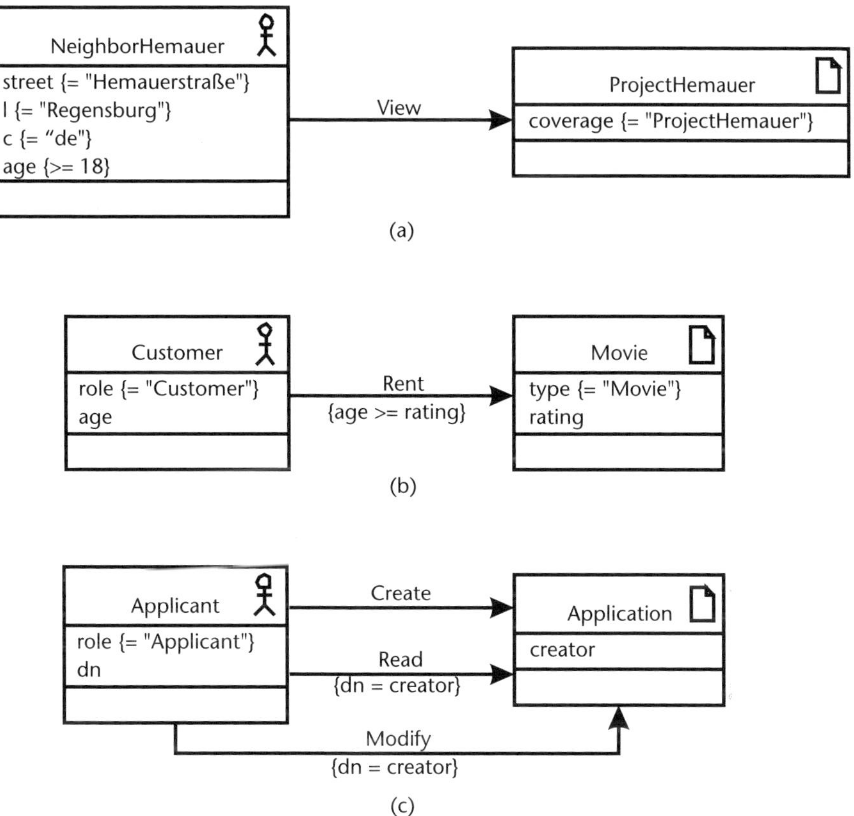

Figure 5.10 Sample ABAC policies in UML.

5.5.2 Designing ABAC Policies with UML

Due to the higher flexibility (and complexity) of attribute-based access control models, the text form is not very suitable for designing and documenting security policies. Also the XML-based notation of XACML [4] (see Section 5.5.4) is difficult to read for humans. Therefore, we suggest a graphic notation based on UML.

As specified in Table 5.3, subject and object descriptors are represented as classes with a (textual or graphical) stereotype in a UML class diagram. The subject

Table 5.2 ABAC Terminology

ABAC	*DLAM*	*UCON$_{ABC}$*	*XACML*
Subject	User	Subject	Subject
Subject attribute	Attribute	Subject attribute	Subject attribute
Subject descriptor	Credential specification	—	Target subject
Subject qualifier	Credential expression	—	Subject match
Object	Object	Object	Resource
Object attribute	Concept	Object attribute	Resource attribute
Object descriptor	Entity specification	—	Target resource
Object qualifier	Conceptual expression	—	Resource match
Authorization	Authorization	Authorization	Rule
Environment attribute	—	Environment attribute	Environment attribute
Condition	—	Condition	Condition

Table 5.3 UML Elements and Stereotypes for Designing ABAC Policies

ABAC Element	*UML Element*	*Stereotype*	*Symbol*
SubjectDescriptor	Class	<<SubjectDescriptor>>	
SubjectAttribute	Attribute	—	—
SubjectQualifier	Constraint	—	—
ObjectDescriptor	Class	<<ObjectDescriptor>>	
ObjectAttribute	Attribute	—	—
ObjectQualifier	Constraint	—	—
Authorization	Association	—	—
Condition	Constraint	—	—

and object attributes of the subject and object qualifiers are represented as associated class attributes, if available with operator and value as a UML constraint. An authorization is drawn as a directed association between the subject and object descriptor classes named with the permitted access type (e.g., read). A condition can be attached to an authorization in form of a UML constraint.

Figure 5.10 shows three sample ABAC policies in the presented UML notation. The attribute names are based on X.500 and Dublin core. The first example (a) assumes an e-government context. Only adult neighbors are allowed to access documents regarding a certain building project on Hemauerstrasse in Regensburg. For this purpose, a subject descriptor "NeighborHemauer" is defined, which demands the Hemauerstrasse as street and Regensburg in Germany as place, represented by the X.500 attributes "locality" (l) and "country" (c). A further subject qualifier specifies that the age must be greater or equal to 18. On the object side an object descriptor "ProjectHemauer" is defined, which represents all objects, which are linked to the project "ProjectHemauer" via the Dublin core element "coverage." For example, a 30-year-old subject "John Doe," which lives on Hemauerstrasse in Regensburg, is classified automatically as "NeighborHemauer" based on its attribute values and thus receives access to the regarded documents. A manual role assignment is not necessary.

Example (b) assumes an e-commerce system for renting movies. A customer has the role "Customer" and an age attribute; he may rent movies only if his age is greater or equal to the rating of the film. This restriction is defined as a condition. Note that in role-based access control, a role for each rating class would have to be defined.

Finally, example (c) shows how another basic problem of RBAC can be solved by means of ABAC conditions, namely the requirement that the creator of an object enjoys special rights. In RBAC a distinct role would have to be defined for each creator. The example represents an excerpt of a policy for a job service company. Applications may be submitted by each applicant; however, an applicant may read and change only his own applications. For this purpose, the X.500 attribute "distinguished name" (dn) of the applicant is compared with the Dublin core attribute "creator" of the application.

5.5.3 Representing Classic Access Control Models

The goal of this section is to show that ABAC can also be seen as a generalization of traditional access control models, and that these can be represented using ABAC. In the following sections, we will present how ABAC policies that implement DAC, MAC, and RBAC can be defined.

5.5.3.1 Discretionary Access Control

As discussed in detail in Section 5.1, in DAC for each pair (subject, object), the access rights the subject has on the object are explicitly specified.

In ABAC, subject and object descriptors need to be defined for this purpose (i.e., DAC policies can be mapped to ABAC by defining an object descriptor for each object and a subject descriptor for each subject), using a qualifier on a fully identifying attribute. In case of X.500 that would be the "distinguished name" (dn); in case of Dublin core, the "identifier." Figure 5.11 shows a simple DAC example as an ABAC policy.

Extended DAC concepts are delegation of the authorization assignment and restrictions by means of predicates (see Section 5.1). Delegation cannot directly be expressed by means of the ABAC model; if necessary, the principle can, however, be implemented on the application level by appropriate access types and authorizations. Predicates in DAC mostly correspond to conditions in ABAC. The restriction of authorizations based on the current time has already been named as an example.

5.5.3.2 Mandatory Access Control

The MAC model is based on predefined systemwide rules with the Bell and LaPadula model as the most prominent representative (see Section 5.2.2). Subjects and objects are assigned to security classes (e.g., top_secret > secret > classified > unclassified) and access and information flow are controlled by means of two rules: A subject may read an object, if its clearance is greater than or equal to the classification of the object. A subject may write an object, if its clearance is smaller than or equal to the classification of the object.

For the implementation in ABAC, a single subject descriptor with a "clearance" attribute is defined for all subjects. Similarly, an object descriptor with a "classification" attribute is defined for the objects. It is presumed that subjects and objects are classified accordingly via these attributes. The two Bell and LaPadula rules can easily be defined by means of ABAC conditions

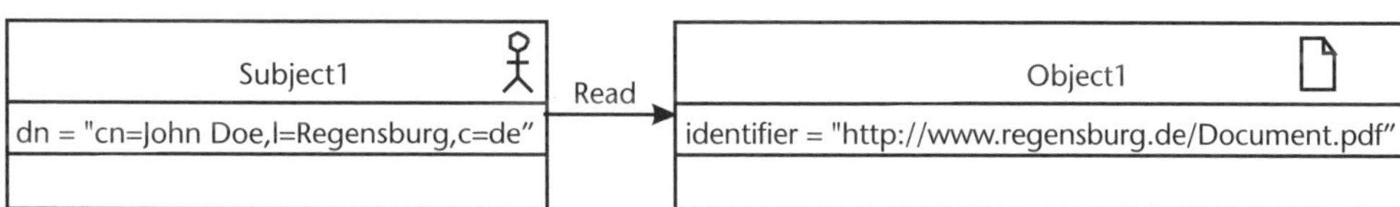

Figure 5.11 DAC sample as ABAC policy.

5.5.3.3 Role-Based Access Control

The RBAC model (see Section 5.4), is based on the concept of roles as intermediaries between subjects and objects.

The implementation of the RBAC model in ABAC is done by introducing a "role" attribute; its values on the subject side represent the role assignment. Besides, for each role an appropriate subject descriptor is defined. Object descriptors are defined for each individual object (e.g., based on the Dublin core element "identifier"), like for DAC. Examples of using a role attribute in ABAC have already been given in Figures 5.10(b) and 5.10(c).

5.5.4 Extensible Access Control Markup Language

The *extensible access control markup language (XACML)* [4] is the result of an OASIS standardization effort proposing an XML-based access control language. XACML defines a generic authorization architecture as well as an XML dialect for specifying attribute-based access control policies. All relevant aspects of the introduced ABAC model can be represented in XACML.

A XACML authorization architecture consists of several logical components. First, a reference monitor concept is used to intercept access requests. This component is called a *policy enforcement point (PEP)*. A PEP transmits access requests to a *policy decision point (PDP)* for retrieval and evaluation of applicable policies. Policies are specified and stored in *policy administration points (PAPs)*. In case a PDP needs attributes of subjects, objects, or the environment, which are missing in the original request, *policy information points (PIPs)* deliver the data needed for the evaluation. As in general not all of these components will use the same message format for communication, a context handler is employed as a mediator. After evaluation, the result is sent back to the requester and is enforced at the PEP. One or more obligations can be executed then (e.g., creating a log entry or sending an email).

As mentioned, the XACML specification furthermore provides a policy language and a request/response language. The request/response language defines the declaration of access control requests for a specific object and responses to these requests. The policy language is based on three main elements: `PolicySet`, `Policy`, and `Rule`. A `PolicySet` can contain a set of single policies or another `PolicySet`. `Policies` consist of single `Rules` that have a `Condition`, an `Effect`, and a `Target`. `Conditions` can be used beyond the `Target` to further specify the applicability of a `Rule` using predicates, while `Effects` denote the result of a `Rule`, usually "Permit" or "Deny." To find the relevant policy for an access control request, every `PolicySet`, `Policy`, and `Rule` has a `Target`, which is evaluated at the time of access request. A `Target` consists of a specification of sets of subjects, objects, operations, and environment using their respective attributes, which can be evaluated with match functions. When the relevant `Policies` and `Rules` are found, they are evaluated independently of each other; contradicting evaluation results can be resolved using rule- as well as policy-combining algorithms.

The XACML architecture is depicted in Figure 5.12. An access control decision and enforcement is performed according to the following steps:

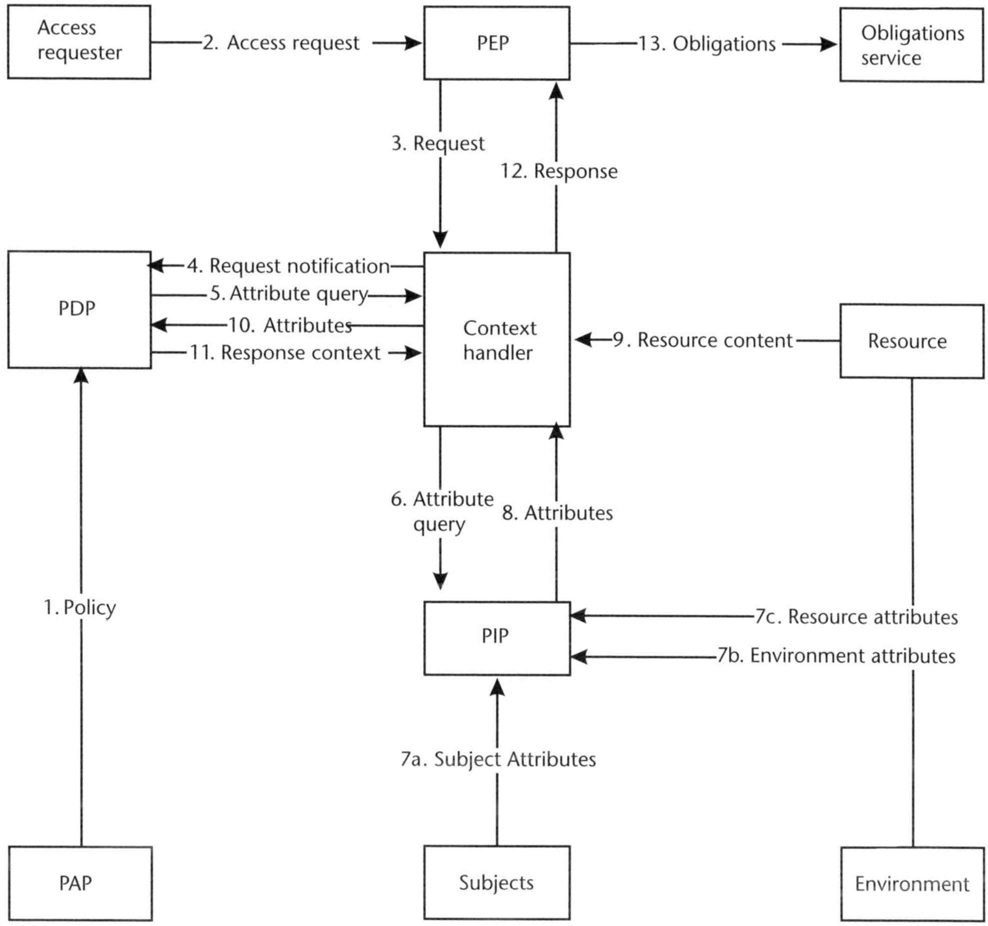

Figure 5.12 XACML architecture.

1. The PAP provides an XACML policy, which has been created beforehand by a security administrator, to the PDP. Figure 5.13 shows example (a) from Figure 5.10 as an XACML policy. In the example, we assume users have to be at least 18 years old and live on Hemauerstrasse in Regensburg, Germany, in order to access resources covering "ProjectHemauer."
2. The user (access requester) sends a resource request to the policy enforcement point. In our example, we assume that the requester is 30-year-old "John Doe" living on Hemauerstrasse in Regensburg and that he requests access to a resource identified by the URL www.regensburg.de/Document.pdf.
3. The PEP forwards this request (which may already contain user, resource, and environment attributes) to the context handler. It is assumed that the user is identified by his email address john.doe@regensburg.de (XACML usually uses email addresses rather than X.500 distinguished names) and that he has included "street," "locality" (l), "country" (c), and "age" attributes.

4. The context handler creates an XACML request and sends it to the PDP.
5. In case the PDP needs additional subject, resource, and environment attributes, they are requested from the context handler.
6. The context handler requests those attributes from a policy information point.

```
<?xml version="1.0"?>
<!DOCTYPE Policy [
  <!ENTITY xsd "http://www.w3.org/2001/XMLSchema#">
  <!ENTITY xacml "urn:oasis:names:tc:xacml:1.0:">]>

<Policy xmlns="&acml;policy"
  PolicyId="SamplePolicy" RuleCombiningAlgId=
  "&xacml;rule-combining-algorithm:permit-overrides">

<Target>
  <Subjects><AnySubject/></Subjects>
  <Resources><AnyResource/></Resources>
  <Actions><AnyAction/></Actions>
</Target>

<Rule RuleId="SampleRule" Effect="Permit">
  <Target>
    <Subjects><Subject>
      <SubjectMatch MatchId="&xacml;function:string-equal">
        <AttributeValue DataType="&xsd;string"
          >Hemauerstrasse</AttributeValue>
        <SubjectAttributeDesignator DataType="&xsd;string"
          AttributeId="urn:example:street"/>
      </SubjectMatch>
      <SubjectMatch MatchId="&xacml;function:string-equal">
        <AttributeValue DataType="&xsd;string"
          >Regensburg</AttributeValue>
        <SubjectAttributeDesignator DataType="&xsd;string"
          AttributeId="urn:example:l"/>
      </SubjectMatch>
      <SubjectMatch MatchId="&xacml;function:string-equal">
        <AttributeValue DataType="&xsd;string">de</AttributeValue>
        <SubjectAttributeDesignator DataType="&xsd;string"
          AttributeId="urn:example:c"/>
      </SubjectMatch>
      <SubjectMatch MatchId=
        "&xacml;function:integer-less-than-or-equal">
        <AttributeValue DataType="&xsd;integer">18</AttributeValue>
        <SubjectAttributeDesignator DataType="&xsd;integer"
          AttributeId="urn:example:age"/>
      </SubjectMatch>
    </Subject></Subjects>

    <Resources><Resource>
      <ResourceMatch MatchId="&xacml;function:string-equal">
        <AttributeValue DataType="&xsd;string"
          >ProjectHemauer</AttributeValue>
        <ResourceAttributeDesignator DataType="&xsd;string"
          AttributeId="urn:example:coverage"/>
      </ResourceMatch>
    </Resource></Resources>

    <Actions><Action>
      <ActionMatch MatchId="&acml;function:string-equal">
        <AttributeValue DataType="&xsd;string">view</AttributeValue>
        <ActionAttributeDesignator DataType="&xsd;string"
          AttributeId="&acml;action:action-id"/>
      </ActionMatch>
    </Action></Actions>
  </Target>
</Rule>

<Rule RuleId="DenyOtherActions" Effect="Deny"/>

</Policy>
```

Figure 5.13 Sample XACML policy.

7. The PIP collects the requested attributes, if possible, from the subject, resource, and environment. In this case, the "coverage" attribute of the resource is acquired.
8. The PIP delivers the attributes back to the context handler.
9. Optionally, the context handler attaches the resource itself to the request.
10. The extended request is sent to the PDP. Figure 5.14 shows the complete sample request in XACML syntax (including subject and resource attributes).
11. The PDP evaluates the policy and sends the response context (including the access control decision) back to the context handler. In our example, the decision is "Permit."

```
<?xml version="1.0"?>
<!DOCTYPE Policy [
  <!ENTITY xsd "http://www.w3.org/2001/XMLSchema#">
  <!ENTITY xacml "urn:oasis:names:tc:xacml:1.0:">]>

<Request xmlns="&xacml;context">
  <Subject>
    <Attribute AttributeId="&xacml;subject:subject-id"
       DataType="&xacml;data-type:rfc822Name">
      <AttributeValue>john.doe@regensburg.de</AttributeValue>
    </Attribute>
    <Attribute AttributeId="urn:example:street"
      DataType="&xsd;string">
      <AttributeValue>Hemauerstrasse</AttributeValue>
    </Attribute>
    <Attribute AttributeId="urn:example:l" DataType="&xsd;string">
      <AttributeValue>Regensburg</AttributeValue>
    </Attribute>
    <Attribute AttributeId="urn:example:c" DataType="&xsd;string">
      <AttributeValue>de</AttributeValue>
    </Attribute>
    <Attribute AttributeId="urn:example:age"
      DataType="&xsd;integer">
      <AttributeValue>30</AttributeValue>
    </Attribute>
  </Subject>
  <Resource>
    <Attribute AttributeId="&xacml;resource:resource-id"
       DataType="&xsd;string"><AttributeValue
       >http://www.regensburg.de/Document.pdf</AttributeValue>
    </Attribute>
    <Attribute AttributeId="urn:example:coverage"
      DataType="&xsd;string">
      <AttributeValue>ProjectHemauer</AttributeValue>
    </Attribute>
  </Resource>
  <Action>
    <Attribute AttributeId="&xacml;action:action-id"
      DataType="&xsd;string">
      <AttributeValue>view</AttributeValue>
    </Attribute>
  </Action>
</Request>
```

Figure 5.14 Sample XACML request.

12. The context handler translates the response context back to the native format of the PEP and forwards it.
13. The PEP satisfies possible obligations.
14. If access is granted, the PEP allows access to the resource (not shown). Otherwise, access is refused. In our example, access is granted.

5.5.5 Discussion of ABAC

ABAC and the XACML standard have been developed in order to cope with the security requirements of large, open, and potentially distributed systems. However, the higher flexibility of attribute-based approaches comes along with higher complexity in the specification and maintenance of the policies.

In order to reduce the complexity, XACML 2.0, which became a standard in 2005, introduces a number of so-called profiles that predefine certain XACML elements for certain situations. For example, there is a profile for defining RBAC policies. Further profiles exist for digital signatures, multiple resources, hierarchical resources, SAML, and privacy.

Furthermore, recent research works try to combine ABAC with semantic web technologies in order to simplify the specification and maintenance of policies [20, 21]. The idea behind these approaches is that the attributes a user possesses do not necessarily match those used by the developers of a web-based information system or service. Ontologies can be used to provide a mapping between different attribute schemes.

5.6 Conclusions

In this chapter different access control models and techniques have been discussed. First, classic access control models, DAC and MAC in particular, have been presented. Other more exotic models were also presented in short. Afterward, the RBAC was investigated in more detail, as it has evolved to a de facto standard. Finally, more recent developments toward ABAC and the XACML standard were covered.

The chapter has focused on approaches controlling the access of users (security subjects) to system resources (security objects), also called user-based access security (UAS). Throughout the chapter, the access to information stored in database systems was mainly used as an example (see Chapter 6 for more details on database security). The presented approaches can, however, also be applied on an application level (e.g., in J2EE or .NET environments). These software environments support—besides UAS—also code-based access security (CAS), dealing with permissions assigned to code fragments based on their trust level (e.g., application code from an untrustworthy source may be denied access to the file system or network sockets). For a discussion of CAS, the reader is pointed to Java and .NET literature (e.g., [22]).

References

[1] Pernul, G., "Database Security," in *Advances in Computers*, Vol. 38, M. C. Yovits (ed.), Academic Press, 1994.

[2] Pernul, G., "Information Systems Security: Scope, State-of-the-art, and Evaluation of Techniques," in *Int'l. Journal of Information Management*, Vol. 15, No. 3, Butterworth-Heinemann, 1995.

[3] Ferraiolo, D. F., et al., "Proposed NIST Standard for Role-Based Access Control," in *ACM Transactions on Information and Systems Security*, Vol. 4, No. 3, 2001.

[4] *eXtensible Access Control Markup Language (XACML)*, OASIS eXtensible Access Control Markup Language Technical Committee, http://www.oasis-open.org/committees/tc_home.php?wg_abbrev=xacml.

[5] Fernandez, E. B., R. C. Summers, and C. Wood, *Database Security and Integrity*, Reading, MA: Addison-Wesley, 1981.

[6] Lunt, T. F., "Security in Database Systems: A Researcher's View," in *Computers & Security*, Vol. 11, North Holland: Elsevier, 1992.

[7] Bell, D. E., and L. J. LaPadula, *Secure Computer System: Unified Exposition and Multics Interpretation*, Technical Report MTR-2997, Bedford, MA: MITRE Corp., 1976.

[8] Biskup, J., and H. H. Brüggemann, "The Personal Model of Data: Towards a Privacy-Oriented Information System," in *Computers & Security*, Vol. 7, North Holland: Elsevier, 1988.

[9] Clark, D. D., and D. R. Wilson, "A Comparison of Commercial and Military Computer Security Policies," in *Proc. 1987 IEEE Symposium on Research in Security and Privacy.*

[10] Thomsen, D. J., and J. T. Haigh, "A Comparison of Type Enforcement and Unix Setuid Implementation of Well-Formed Transactions," in *Proc. 6th Ann. Comp. Security Applications Conf. (ACSAC'90)*, IEEE Computer Society Press, 1990.

[11] Stachour, P. D., and B. Thuraisingham, "Design of LDV: A Multilevel Secure Relational Database Management System," in *IEEE Trans. KDE*, Vol. 2, No. 2, 1990.

[12] Brewer, D. F. C., and M. J. Nash, "The Chinese Wall Security Policy," in *Proc. 1989 IEEE Symposium on Research in Security and Privacy.*

[13] X.520: The Directory—Selected Attribute Types, ITU-T Recommendation, 1996.

[14] Adam, N. R., et al., "A Content-Based Authorization Model for Digital Libraries," in *IEEE Transactions on Knowledge and Data Engineering*, Vol. 14, No. 2, 2002.

[15] X.509: The Directory—Public Key and Attribute Certificate Frameworks, ITU-T Recommendation, 2000.

[16] Oppliger, R., G. Pernul, and C. Strauss, "Using Attribute Certificates to Implement Role-Based Authorization and Access Control," in *Proceedings of the 4th Conference on "Sicherheit in Informationssystemen" (SIS 2000)*, Zürich, Schweiz, 2000.

[17] Park, J., and R. Sandhu, "The $UCON_{ABC}$ Usage Control Model," in *ACM Transactions on Information Systems Security*, Vol. 7, No. 1, 2004.

[18] Priebe, T., et al., "A Pattern System for Access Control," in *Proc. 18th Annual IFIP WG 11.3 Working Conference on Data and Application Security*, Sitges, Spain, 2004.

[19] Priebe, T., et al., "ABAC—Ein Referenzmodell für attributbasierte Zugriffskontrolle," in *Proc. 2. Jahrestagung Fachbereich Sicherheit der Gesellschaft für Informatik (Sicherheit 2005)*, Regensburg, Germany, 2005.

[20] Damiani, E., et al., "Extending Policy Languages to the Semantic Web," in *Proc. 4th International Conference on Web Engineering (ICWE 2004)*, Munich, Germany, 2004.

[21] Priebe, T., W. Dobmeier, and N. Kamprath, "Supporting Attributed-Based Access Control with Ontologies," in *Proc. of the First International Conference on Availability, Reliability and Security (ARES 2006)*, Vienna, Austria, 2006.

[22] Piliptchouk, D., *Java Versus .NET Security*, Cambridge, MA: O'Reilly Media, 2004.

CHAPTER 6

Data-Centric Applications

Günther Pernul

This chapter presents three cases on the application of the classical authorization and access control models in data-centric applications. It starts with protection mechanisms for relational databases, in particular view-based authorization and discretionary access controls. Although implemented in today's commercially available database management systems (DMBSs), the mechanisms comprise some structural limitations and risks. Next is a section on the multilevel secure relational database model, combining relational database technology with mandatory access controls. Although this protection technique addresses a higher level of threats, other types of deficiencies and limitations occur, limiting its practical applicability to only certain types of applications. Section 6.3 contains the description of a case on role-based access controls for a federation of different databases. The major design choices are discussed, and a solution is presented. While cases 1 and 2 describe state-of-the-art technology, case 3 is more research oriented and describes achievements attained in a former research project.

6.1 Security in Relational Databases

The general concept of database security is very broad and entails such things as moral and ethical issues imposed by the public and society, legal issues in which control is legislated over the collection and disclosure of stored information, or more technical issues such as how to protect the stored information from loss or unauthorized access, destruction, misuse, modification, or disclosure. In this case study, such a broad perspective of security in relational databases is not taken. Instead, main focus is directed toward aspects related to authorization and access controls. This is legitimate because the basic security services, such as user identification, authentication, and auditing, normally fall within the scope of the underlying operating system (OS) (if supported by the DBMS, most products only call corresponding functions on the OS level). In addition, enforcing database integrity and consistency constraints are subject of the *semantic data model* supported by the DBMS or are dependent on the physical design of the DBMS software (namely, the transaction and database recovery manager).

Structured query language (SQL) has developed as the major standard for relational databases, and, as most of the DBMS products in use today are based on the relational data model, the following discussion mostly concerns the framework of SQL databases. For expressing and enforcing security constraints in SQL databases,

Project			
Title	*Subject*	*Client*	*Manager*
Alpha	Development	A	124
Beta	Research	B	125
Celsius	Production	C	123

Employee			
SSN	*Name*	*Department*	*Salary*
123	Bob	Sales	80.000
124	Susan	Research	60.000
125	Josef	IT	50.000
126	Calvin	Assembling	22.000
127	Joseph	Research	38.000

Assignment	
SSN	*Title*
123	Celsius
124	Alpha
124	Beta
125	Alpha
125	Beta
125	Celsius
126	Celsius

Figure 6.1 Database snapshot *project management.*

there are two main concepts: structuring the granularity of the security objects by means of database views and the concept of GRANT and REVOKE in SQL. We will first introduce a simple example database used throughout the case and later both of the concepts will be described in following sections.

Figure 6.1 contains a snapshot of the project management database. We consider tables `Employee`, `Project`, and `Assignment`. Between `Employee` and `Project` a <n:m> and a <1:n> relationship exists, meaning that an employee may be assigned to more than one project, a project may be carried out by several employees, and each project has a responsible manager (from `Employee`).

Before going into the details of the SQL security mechanisms, a few words are also necessary about a special kind of attack that by its name appears to be related to SQL security. The attack, called *SQL injection*, refers to a vulnerability in web-based systems in which a user input is not filtered by the application for special characters and directly passes into a SQL query. The purpose of the attack is to change the SQL statement to one of the attacker's choice, thereby compromising a website, the database, or even the whole server. SQL injections are a special case of a more general software vulnerability that can occur whenever one programming or scripting language is embedded in another and will not be further treated in this chapter.

6.1.1 View-Based Protection

A database view is a virtual table based on the result of a SQL SELECT query. In contrast to base tables, a view is not materialized, meaning that it is not physically stored in the database but may be only instantiated with data on demand. The content of a view may reference any number of other tables and views.

```
CREATE VIEW table_name [(col_name1, col_name2, ...)]
  AS SELECT ...
```

Views work very well for authorization and access control. Instead of authorizing a user with access restrictions on the basis of the physically stored base tables, the user may be authorized on the view tables only. Views may implement some sort of filter function, automatically filtering out data the user is not authorized to see. In the following, some examples referring to the project management database are given outlining the power of the view concept.

```
(1) CREATE VIEW Earning_little
    AS
       SELECT *
       FROM Employee
       WHERE Salary < 55.000
```

Earning_little			
SSN	*Name*	*Department*	*Salary*
125	Josef	IT	50,000
126	Calvin	Assembling	22,000
127	Joseph	Research	38,000

```
(2) CREATE VIEW Emp AS
       SELECT
         SSN, Name, Department
       FROM Employee
```

Emp		
SSN	*Name*	*Department*
123	Bob	Sales
124	Susan	Research
125	Josef	IT
126	Calvin	Assembling
127	Joseph	Research

```
(3) CREATE VIEW Emp_research
    AS SELECT
       SSN, Name, Department
       FROM Employee WHERE
       Department = "Research"
```

Emp_research		
SSN	*Name*	*Department*
124	Susan	Research
127	Joseph	Research

```
(4) CREATE VIEW Dep_involved
    AS SELECT Title, Subject,
       Department
    FROM Project, Assignment,
       Employee
    WHERE
       Employee.SSN =
       Assignment.SSN AND
       Assignment.Title =
       Project.Title
```

Dep_involved		
Title	*Subject*	*Department*
Alpha	Development	Research
Alpha	Development	IT
Beta	Research	Research
Beta	Research	IT
Celsius	Production	Sales
Celsius	Production	IT
Celsius	Production	Assembling

Example (1) is a horizontal view on the base table `Employee`. The view `Earning_little` could be the authorization object for a user whose access should be restricted to employees earning less than $55,000 only. Similar is the situation described in example (2). The view `Emp` is a vertical subset of `Employee` restricting users from accessing the column Salary. Example (3) is a mixed view, combining vertical and horizontal view building. The view `Dep_involved` is a mixed view, too; however, it additionally shows a means to change the abstraction level of the security object. Recall that stored in the database are the individual assignments of employees to projects while the view only shows the departments involved in the projects. Similar is the situation in example (5). The view `Dep_avg_salary` shows the average salary paid in the different departments, while the base table stored in the database contains the individual salary for each employee. It must be noted that such statistical controls only work meaningfully if data about enough individuals are available and that statistical controls are subject to database tracking attacks. Securing a database by only using aggregate queries is a difficult problem, since intelligent users can use a combination of aggregate queries to derive information about a single individual. The view `My_employees` is an example that also parameters may be passed to the view creation process. In this example the user-ID is used to define a context-based control (i.e., depending on the manager, the view contains only data referring to employees working in projects led by the user).

```
(5) CREATE VIEW Dep_avg_salary
    AS SELECT
       Department, AVG(Salary)
       FROM Employee
       GROUP BY Department
```

Dep_avg_salary	
Department	*Salary*
Sales	80,000
Research	49,000
IT	50,000
Assembling	22,000

```
(6) CREATE VIEW My_Empl AS
     SELECT SSN, Name,
         Department, Salary
    FROM Project, Assignment,
       Employee
    WHERE Employee.SSN =
       Assignment.SSN AND
       Assignment.Title =
       Project.Title AND
    Manager = USER
```

My_Empl			
SSN	*Name*	*Department*	*Salary*
125	Josef	IT	50,000
126	Calvin	Assembling	22,000
123	Bob	Sales	80,000

Although very simple, the SQL view statement is a powerful mechanism to structure the security object. By means of examples, we have shown value-independent controls (2); value-dependent controls (1), (3), and (4); statistical controls (5); and context-dependent controls (6). At a first glance, one might get the impression that views are almost ideal for enforcing access restrictions. However, there are also some constraints and structural limitations, which will be discussed in Section 6.1.3.

6.1.2 SQL Grant/Revoke

Discretionary access controls (see Section 5.1) are based on two basic principles: the first is the principle of ownership of information, and the second is that the owner possesses all privileges to use the object he owns. This also includes the privilege of delegation of rights, meaning that privileges may propagate among users and the user may grant access privileges he owns to other users. Moreover it is also possible to delegate this function to other subjects. This is done by using the SQL GRANT statement (Figure 6.2).

To give an example, suppose the creator of the base table `Employee` wants the user Sam to insert or delete employees:

```
GRANT INSERT, DELETE ON Employee TO Sam WITH GRANT OPTION
```

Sam is now allowed to use the `Employee` table and to pass that privilege to others. Granting to PUBLIC would mean that any user in the system will be allowed to perform the specified operations.

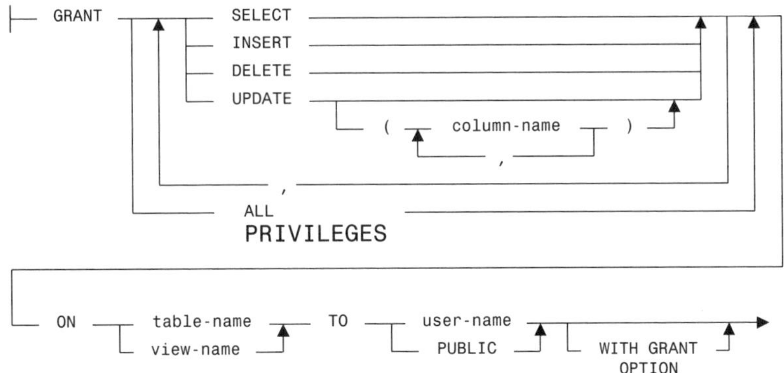

Figure 6.2 Syntax of the SQL GRANT statement.

To record the authorizations, the ANSI/ISO SQL standard introduces a so-called privilege dependency graph. The privilege dependency graph is a directed graph such that each node represents a privilege and each arc represents the fact that an authorization directly depends on another (i.e., was granted with the GRANT OPTION set). In each node a privilege dependency graph stores the following information: the subject who granted the authorization (grantor), the subject to whom the authorization is granted (grantee), the type of privilege being granted, the object (usually table or view, or a specific column in the case of a column privilege) to which the authorization refers, a flag indicating whether the subject has the GRANT OPTION for the privilege, and a timestamp representing the time at which the authorization was granted.

Consider the privilege dependency graph given in Figure 6.3. Authorization a7 depends on two other authorizations (a4 and a6). We say a7 is supported by a4 and a6. Even if user B would not have granted the privilege to user D at timestamp

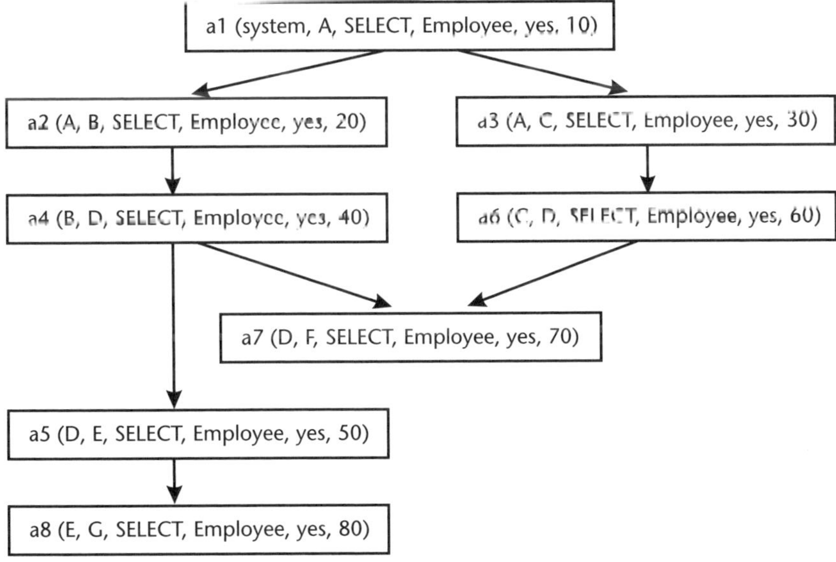

Figure 6.3 Example of a privilege dependency graph.

40, user D would still be able to pass the privilege on to user F at timestamp 70, because by then he had received the same privilege by user C.

The principles of delegation of rights and decentralized administration of authorization lead to some problems when revoking privileges. Coming back to the previous example, what is supposed to happen if authorization a4 is revoked? User D would lose the privilege granted by user B. Should he still have access to table `Employee`, because he had received the same privilege by user C? What is supposed to happen with authorization a7 that was granted by user B? Should it also be revoked? There are some semantic inconsistencies with the revoking of previously granted privileges, and as a result the REVOKE statement was only later included in the ANSI/ISO SQL standard.

A simple example of the REVOKE statement would be as follows. After this statement is issued, Sam will no longer be able to insert or delete employees:

```
REVOKE INSERT, DELETE ON Employee FROM Sam
```

What is supposed to happen if Sam had granted his privilege to others using the GRANT OPTION he had received? If the system would just revoke the privilege from Sam but not from the others, their authorizations would be abandoned. Looking at the privilege dependency graph, they would no longer have a supporting authorization.

The syntax of the REVOKE statement is shown in Figure 6.4. According to the diagram, either RESTRICT or CASCADE should be specified. These options deal with privileges that have been passed on using the GRANT statement WITH GRANT OPTION. If such privileges do not exist, it should be able to omit RESTRICT and CASCADE (assumed in the example).

If the RESTRICT option with a REVOKE command is used, the system will check if after performing the revocation any abandoned authorizations will arise. If that is the case, the REVOKE statement will fail and an error message will be displayed. In the example, revoking a4 with a RESTRICT option will fail, but not

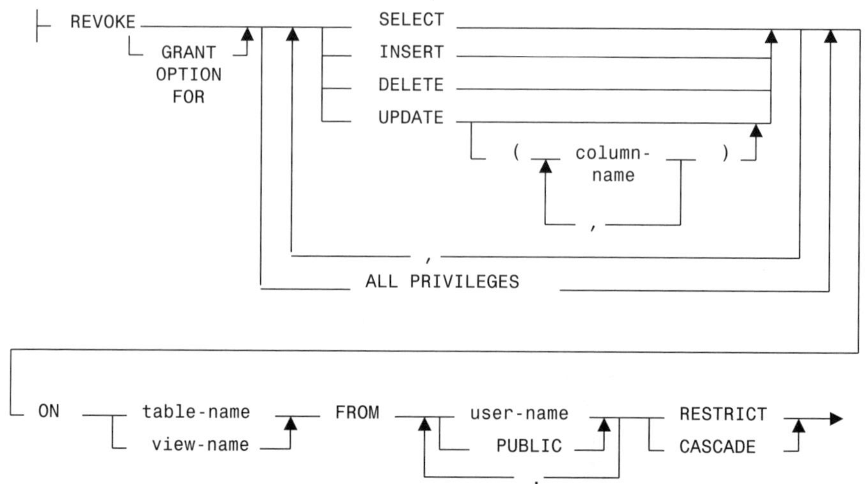

Figure 6.4 Syntax of the SQL REVOKE statement.

because of a7, which will not be abandoned because a6 still supports it. The reason for the failure will rather be authorization a5 (and therefore also a8), which would not have any supporting authorization after a4 is revoked. User B will receive an error message, and the privilege dependency graph will not change.

The CASCADE option takes a different approach in dealing with abandoned authorizations. If a privilege is revoked with the CASCADE option, abandoned authorizations will be revoked as well. Also, if views have been defined that need a privilege that is being revoked, the view should be dropped as well. In our example, revoking a4 with CASCADE option will succeed. Authorization a7 will not be revoked, because it is not abandoned, but a5 and a8 will be revoked as well.

Instead of revoking the privilege only, revoking the GRANT OPTION for a privilege is also possible. If this is done with the RESTRICT option set and if other users have received the privilege, the statement will fail. In case of a cascading revoke, those privileges will be revoked and the statement will succeed.

6.1.3 Structural Limitations

For expressing and enforcing security constraints in SQL databases there are two main concepts: structuring the granularity of the security objects by means of database views, and granting and revoking access privileges by the owners of view or base tables. At a first glance, one might get the impression that views and delegation of rights under the discretion of the users themselves are almost ideal for enforcing access restrictions. However, there are pros and cons and also some constraints and structural limitations, which will be discussed in this section.

- *View assets and drawbacks*: Views can give users a "personalized" interpretation of the database. As a benefit, the structure of the database may look simpler to the users, and user queries may become simpler because multiple-table queries may be expressed against a single view only. However, overall performance of the DBMS may decrease due to the views. This is because first views must be materialized during run time and second queries against the views must be translated into queries against the base tables. Additionally, there are update restrictions through views. Many views are read only.
- *Limitations due to different interpretations*: As discussed earlier, the semantics of the REVOKE statement are not always clear. As a result different DBMS vendors have implemented it differently. For example, Oracle DBMS implements a cascading revoke. There is no CASCADE keyword, but whenever a REVOKE statement is issued, the system will check for abandoned authorizations and revoke them as well. A restricted revoke is not supported. Revoking privileges in DB2 of IBM is similar to the approach taken in Oracle. In DB2 REVOKE works in a cascading way and no RESTRICT option is available. Microsoft SQL Server did not support the GRANT OPTION before release 6.5. Later Microsoft implemented the GRANT OPTION as well as a REVOKE statement that more or less complies with the standard.
- *Limitations due to problems not addressed*: There are certainly problems that are not addressed in the ANSI/ISO SQL standard. At least two, the lack of a noncascading revoke and the lack of negative authorizations, are of particular

interest. Organizational changes make it often necessary to change a security scheme. As an example, suppose there is a change in the tasks a user has to perform because of job promotion. This change may imply a change in his privileges. New authorizations may be granted to him, and existing ones revoked. A cascading revoke would revoke all authorizations granted by him in the past as well. However, these may probably still be appropriate and have to be reissued, which can be a time-consuming and thus an expensive task. What is missing in the standard is a noncascading revoke. There should be a possibility that abandoned authorizations that directly depend on the revoked one should be reconnected to the privilege dependency graph.

In the case in which a privilege is granted with the GRANT OPTION set, how can one be sure that a given user will never get the privilege? There is no means to specify that a particular user should never be able to perform a certain operation in the future. In the literature, this problem is also be known as the safety problem. What is missing are negative authorizations expressing an explicit denial of access. In Section 6.3, the case of IRO-DB is described, which includes an authorization scheme supporting positive as well as negative authorizations and the management of potential conflicts among them.

- *Limitations due to missing information flow control*: These problems exist in any system protected by discretionary access controls and may also comprise serious security flaws for databases. Consider the case of an authorized user having access to a security object. This user may act as "information flow channel" by passing information on to nonauthorized users. This may happen deliberately by copying the information into a security object he "owns" and granting access to the object to the nonauthorized user; it may happen by mistake or fault; or it may even happen through a Trojan horse attack. A Trojan horse would act under all authorizations of the user and, without the knowledge of the user, may simply issue GRANT commands and thereby pass privileges on to the nonauthorized users.

6.2 Multilevel Secure Databases

Mandatory access control policies address a higher level of threat than discretionary ones because, in addition to controlling the access to data, they control the flow of data. As a consequence, MAC techniques overcome the limitations of DAC-based protection as described earlier but—as it will be elaborated in this case—unfortunately introduce new ones. In this section, we will discuss the case of multilevel secure (MLS) relational databases combining relational database technology with MAC.

Mandatory security (see Section 5.2) requires that security objects and subjects are assigned to certain security levels coming from a hierarchy of sensitivity levels (e.g., top_secret > secret > confidential > unclassified) and the enforcement of two rules. The first rule protects the information of the database from unauthorized disclosure and requires for a subject s in order to successfully read data item d that clear(s) ≥ class(d). The clearance of the subject must dominate the classification of the object the user wishes to read. The second rule protects data from contamination or unauthorized modification by restricting the information flow from the higher security levels to lower ones. Subject s is allowed to write data item d if clear(s) ≤ class(d).

A few final sentences on MAC policies are in order. In many discussions, confusion has arisen about the fact that in mandatory systems it is not sufficient to only control the access to data. Why is it necessary to include the two rules and strong controls over who can write which data in systems with high security requirements? The reason is that a system with high security needs must protect itself against attacks from unauthorized as well as from authorized users. As described in DAC-based protection, there are several ways authorized users may disclose sensitive information to others. The simplest technique to disclose information by an authorized user is to retrieve it from the database, to copy it into an "owned" object, and to make the copy available to others. To prevent this, it is necessary to control the ability of the authorized user to make a copy (which implies the writing of data). In particular, once a transaction has successfully completed a read attempt, the protection system must ensure that there is no write to a lower security level (write-down) that is caused by a user authorized to execute a read transaction. As the read and write checks are both mandatory controls, a MAC system successfully protects against the attempt to copy information and to make the copy available to unauthorized users. By not allowing higher classified subjects to write-down on lower classified data, information flow among subjects with different clearances can efficiently be controlled.

In this case study the most general case of an MLS relational database will be considered, in which an individual attribute value is subject to security label assignment. As an example, consider Table 6.1 containing the `Project` table from the previous case and the existence of four sensitivity levels, denoted by TS, S, C, U (where TS > S > C > U) is assumed. For simplicity reasons, we have slightly changed the attributes in the table.

Consider the two different instances of relation `Project` as given in Table 6.1. `Project,TS` corresponds to the view of a subject s with clear(s) = TS. Because of the read access rule of Bell-LaPadula, users cleared at U would see the instances of `Project,U`. In this case the read access rule would automatically filter out data that dominate U. This may either be on a whole tuple level or on an attribute level. In

Table 6.1 Snapshot of MLS Database *Project Management*

Project, TS					
Title	Subject	Start	End	Budget	Manager
Alpha, U	Development, C	2007, U	2008, C	1.100, S	124, U
Beta, S	Research, TS	1999, S	2000, S	900, TS	125, S
Celsius, U	Production, U	2005, U	2006, U	2.200, U	123, U

Project, U					
Title	Subject	Start	End	Budget	Manager
Alpha, U	—	2007, U	—	—	124, U
Celsius, U	Production, U	2005, U	2006, U	2.200, U	123, U

the example, the whole project “Beta” is not visible for the user, and for alpha the information on the subject, the end date, and the budget is missing. The semantics of the NULL values are not defined (i.e., the user might think that this information has not yet been entered in the database, or the project will never have a budget, or it is not yet decided about the budget, or even the budget is secret and he does not have access). There are many different interpretations of the missing values possible.

6.2.1 Polyinstantiation and Side Effects

As an interesting optional feature, mandatory security and the Bell-LaPadula paradigm may lead to polyinstantiated databases. These are databases containing relations that may appear different to users with different clearances. This is due to the following two reasons: first, not all clearances may authorize all subjects to all data, and, second, the support of MAC may lead to polyinstantiation of attributes or tuples. The concept of polyinstantiation and resulting side effects will be discussed next.

Reconsider the example in Table 6.1 and a subject s with clear (s) = U issuing an insert operation where the user wishes to insert the tuple <Beta, Production, 2002, –, 800, 124> into the database. Because of the key integrity property (attribute Title is key in project), a standard relational DBMS would not allow this operation. (Although not seen by user s, Beta as a key already exists in relation `Project`.) However, from a security point of view, the insert must not be rejected because otherwise a covert signaling channel occurs from which s may conclude that sensitive information he is not authorized to access may exist. Such a signaling channel is interpreted as information flow between subjects cleared at different security levels, and this is not possible. With respect to security, much more may happen than just inferring the presence of a tuple. The success or failure of the service request, for example, can be used repeatedly to communicate one bit of information (0: failure, 1: success) to the lower level. Therefore, the problem is not only the inferring of a classified tuple; moreover, any information visible at the higher level can be sent through this covert signaling channel to the lower level. The outcome of the insert is shown in Table 6.2 and resulted in a polyinstantiated tuple in `Project`.

Polyinstantiation may also occur if a subject cleared for the U-level would wish to update project “Alpha” by replacing the null-values with certain data items. As another example of polyinstantiation, consider that subject s with clear(s)=S wants to update <Celsius, Production, 2005, 2006, 2.200, 123>. In systems supporting

Table 6.2 Polyinstantiation in MLS Databases

Project, TS					
Title	Subject	Start	End	Budget	Manager
Alpha, U	Development, C	2007, U	2008, C	1.100, S	124, U
Beta, S	Research, TS	1999, S	2000, S	900, TS	125, S
Beta, U	Production, U	2002, U	—, U	800, U	124, U
Celsius, U	Production, U	2005, U	2006, U	2.200, U	123, U

MAC, such an update is not allowed because of the write-access rule of BLP. This is necessary because an undesired information flow might occur between subjects cleared at the S-level to subjects cleared at the U-level. Thus, if an S-level subject wishes to update the tuple, the update again must result into polyinstantiation.

The theory of most data models is built around the concept that a fact of reality is represented in the database only once. Because of polyinstantiation, this fundamental property is no longer true for MLS databases, thus making the adaptation of the theory of the relational data model for MLS databases necessary. For example, in Table 6.2, attribute `Title` no longer holds as a key in `Project`. In general, a key in an MLS database consists of the key of the corresponding single level table, its classification, and all the other classifications of the attributes in the MLS table.

There is another side effect of polyinstantiation. As we have seen in the previous example, polyinstantiation may occur on several different occasions (e.g., because of a user with low clearance trying to insert a tuple that already exists with higher classification, because of a user wanting to change values in a lower classified tuple), but it may also occur because of a deliberate action in the form of a cover story, in which lower cleared users should not be supported with the proper values of a certain fact. The use of cover stories may be intended, or they may result as a side effect.

6.2.2 Structural Limitations

MLS databases and mandatory access control policies address a higher level of threat than discretionary ones, because, in addition to controlling the access to data, they control the flow of data among users as well. As a consequence, MAC techniques overcome the limitations of DAC-based protection but unfortunately introduce additional new ones. From the viewpoint of the level of security they support, MAC-based policies are very strong. However, the side effects and structural limitations they introduce make their practical relevance and use for civil applications at least questionable.

- *Granularity of security object*: It is not yet agreed about what should be the granularity of labeled data. Proposals range from protecting whole databases to protecting files, relations, attributes, or even certain attribute values. Commercially available products label on the level of whole databases, relations, or individual tuples. In any case, careful labeling is necessary to avoid inconsistent or incomplete label assignments. Additionally, databases usually contain a large collection of data and serve many users, and labeled data is not available in many civil applications. This is the reason manual security labeling is necessary, though it may result in an almost endless process for large databases. Therefore, supporting techniques, such as guidelines and design aids for multilevel databases, tools that help in determining the relevant security objects, and tools that suggest clearances and classifications, are missing.
- *Consequences of polyinstantiation*: As a consequence of polyinstantiation, the key of the corresponding single-level schemes no longer hold. This makes it necessary to revise the key. In relational databases, the key plays a central role. It is used as a unique identifier for each tuple. In addition, it influences

data consistency and normalization, and the DBMS internally uses key attributes to build index structures. For these reasons, it is very important to choose small (in terms of storage space) and efficient keys. For integrity reasons in MLS relational databases, all the classifications of all the data items must be part of the key. This results in very clumsy keys, which is very inconvenient and may cause dramatic effects on the overall performance of the database.

- *Semantic ambiguity*: It was already pointed out that in MLS databases, the semantics of NULL values are not clearly defined. But there are additional semantic ambiguities when it comes to queries. To illustrate, we will use a very simple example (see Table 6.3) consisting of one table. The table has only three attributes, and there are only two tuples in the table.

The table `Starship` probably is an example of one of the simplest database one can imagine. Now consider the simplest query a user can issue:

```
SELECT * FROM Starship
```

Where does Enterprise go? There are at least six different interpretations of a possible result of this query:

1. Enterprise goes to Mars, but the user should not talk to others. The organization trusts the user very much, and the user is cleared at security level *S*; however, there are also other users who have lower clearances.
2. Enterprise goes to the Moon, but instead of going back to Earth it continues on a secret mission to Mars.
3. Enterprise goes to Mars, which is *S*, but on its way back to Earth it stops on the Moon, and the mission becomes *U*.
4. Enterprise goes to the Moon and comes back to Earth. The next mission goes to Mars, which is *S*.
5. Enterprise goes to Mars on a secret mission. The next mission goes to the Moon, which is *U*.
6. There might even be two starships named Enterprise. One goes to the Moon; the other one goes on a secret mission to Mars.

For databases, it is essential that query results are very always clear and well defined. In real-world business scenarios, data accuracy is very important because many decisions are backed by data and query results from cooperate databases. The

Table 6.3 Query ambiguity in MLS databases.

Starship, S		
Name	Destination	Objective
Enterprise, U	Moon, U	Exploration, U
Enterprise, S	Mars, S	Garbage dumping, S

discussion here proves that the potential of MAC as an access control policy for those applications may be very limited.

However, we also need to be fair and the previous statement needs to be put into perspective slightly: All these examples have discussed the most general case in which for the granularity of the security object an individual attribute value was chosen. This may not always be the best solution for any application case. For certain applications a much coarser granularity (e.g., labeling on a relation level, file level, or even on a database level) may be sufficient or more appropriate. In such a case, the side effects and structural limitations of MLS databases are much less. This also explains why MLS technology is included in many of the commercial DMBS products. For example, Oracle Labeled Security was introduced in Oracle8i, which replaced Trusted Oracle MLS DBMS some time ago. Today it is included in the current version of Oracle Database 10g and linked with the Oracle Identity Management infrastructure. Other most noticeable commercial solutions including MLS technology are Informix Online/Secure, Sybase Secure SQL Server, or IBM's DB2 for z/OS.

6.3 Role-Based Access Control in Database Federations

Many organizations maintain information spread over several independent, possibly heterogeneous databases. While the databases may have been working sufficiently during the past years, in new applications often the need to integrate already existing databases into a database federation becomes evident. This may be because certain applications may wish to have federated views onto the information available, covering a more global aspect. Even database federations between different organizations or companies may be desirable in order to enable intercompany data sharing. It is important to note that in a federated database (FDBS), the participating component databases (CDBSs) shall keep their autonomy because existing local applications shall be still supported.

In FDBS, security plays an augmented role. In general, CDBS providers will only offer their local data to the federation if secrecy and integrity were still warranted. A federated security system has to be at least as strong as each of the local systems and on the other hand as transparent as possible to FDBS users.

In this case, the major design choices for discretionary security in FDBS will be discussed. We assume an object-oriented data model at the federation level, integrating heterogeneous (object-oriented and relational) and autonomous CDBSs.

6.3.1 Taxonomy of Design Choices

First of all, some choices concerning the granularity of security objects, security subjects, and access types have to be made. In general, a higher granularity makes it easier to fine-tune the security system so that it better suits particular requirements but burdens the usage and maintenance of the system.

The granularity of a *security object* increases depending on up to which level of the hierarchy access controls are supported. The highest possible level of granularity is the database level, which would allow protecting the database as a whole but would not offer any flexibility. The next level is up to the class level. This allows protecting a class in admitting access to all or none of its objects as a whole. A finer granularity is the class attribute/method level. In this case the construction of vertical views is possible, including or excluding particular attributes or methods from being accessed. The object level would be restricting access to a particular number of objects. The most general case is the object attribute/method level. Access is restricted to a number of attributes and methods for a particular set of objects.

There are two kinds of *security subjects*: local ones, which are able to use only one CDBS, and federated ones, which may retrieve data from several component databases. In many cases it is useful to support "sets" of security subjects like groups or roles. They may relieve the administrational expenditure and are apt to model the organizational structure of a company or reflect the functional and social position of a user. If roles are nested a generalization hierarchy of roles can be designed. In this case the effect of propagated authorization has to be considered as well as policies to resolve conflicts in the case of multiple superior roles. For a deeper study of role-based access control, see Section 5.4.

In general, the following five atomic *access types* can be identified: create, delete, read, write, and execute. Most of the other access types are combinations of them. Method access is a feature of the object-oriented paradigm and can be treated in two different ways: either they correspond to access types, meaning the internal actions of the methods (e.g., what other methods they call and what attributes they access) are the responsibility of the method implementation and no special authorization is required for that, or they correspond to security objects demanding a special kind of access type (e.g., execute). Furthermore, assumptions on propagated authorization have to be made, stating that, for instance, the right to create a database implies the right to create classes and objects, according to the hierarchy of security object granules.

A fundamental assumption describing the basic attitude of an *authorization system* is whether the system should be open or closed (closure assumption). In an open system, everything is accessible unless explicitly (or implicitly due to propagated authorization) forbidden, whereas a closed system represents the inverse situation. If security is an important objective, a closed system should be used, while flexibility requires open systems. Another fundamental decision concerns another basic attitude of the authorization system: should it be positive, negative, or mixed? Positive systems give only permissions (often in connection with a closed closure policy), negative systems give only prohibitions (frequently used with an open policy), and mixed systems allow for permissions and prohibitions, which of course requires a policy to solve conflicting authorizations. In positive systems it is burdensome to exclude one particular security subject from a particular kind of access, whereas the inverse situation is true for negative systems. Mixed systems are more flexible than strictly positive or negative ones, although they may be harder to maintain.

In the case in which there are large numbers of security objects and/or security subjects, the administrational expenditure may be significantly reduced if authorizations must not explicitly be stored or granted but may be automatically *inferred and implied*. Authorization may be propagated, for instance, from roles to individual users, along a role hierarchy, along a class hierarchy, within complex objects, or along the hierarchy of security object granules (e.g., from classes to their objects). In any case, the side effects of propagating authorization have to be thoroughly considered. For instance, a security subject that is authorized to fully access a particular class should not have propagated the same access to subclasses of that class, since they may contain more specific content.

Conflicts occur if two access rules determining contradicting access types between one particular security subject and one particular security object exist. These may be explicit rule conflicts because of different federation and component policies, for instance, or implicit conflicts due to propagated authorization. Examples are mixed authorization (conflicting permissions and prohibitions), multiple security object granules (e.g., conflicting access rules for a class and an object of that class), multiple security subject granules (e.g., conflicting group rights and user rights), multiple superior roles (combining two roles with conflicting access rights), and multiple roles at a time (a user may play more than one role at a time). Several superior policies for conflict resolution are feasible, like preferring the more restrictive rule (for security reasons), preferring the more permissive rule (for flexibility reasons), preferring the federated respectively component rule, depending on autonomy a.s.o. Furthermore, access rules could be time-stamped or could have explicit priorities, allowing a decision to be made about which rule should be applied.

6.3.2 Alternatives Chosen in IRO-DB

IRO-DB is a database federation supporting interoperable access between object-oriented and relational databases. The security model assumes a closed world, meaning that each request is prohibited unless a positive authorization can be inferred from the authorization base. It supports a mixed approach in the sense that positive and negative authorizations (permissions and prohibitions) may be granted to authorization subjects. Implications of authorizations are derived to reduce the number of explicit authorizations. Due to the principle of ownership, implied, positive, and negative authorizations conflicts may arise, which are solved within a conflict resolution schema The security system implements a combination of ownership and administration paradigm in which delegation of rights is under the discretion of roles "owning" an object, but managing roles, users, and authorization rules is under the responsibility of a central administration principal.

Implied authorizations are used to reduce the number of explicit authorizations in the authorization base. It means that an authorization of a certain type defined for a role on a certain database object may imply further authorizations along any combination of the three constituencies in an authorization rule—namely, the subject, type, and object. This is realized by introducing rules for implicit authorizations, which can be computed from the set of explicit authorizations granted to a subject and defined in the authorization base.

For resolving conflicts resulting from competing authorizations, a resolution schema consisting of three policies is used (see Figure 6.5).

The conflict resolution schema implements the following policy: Ownership takes the highest priority. Negative authorizations override positive ones, and explicit authorizations override implicitly stated authorizations. For the deeper

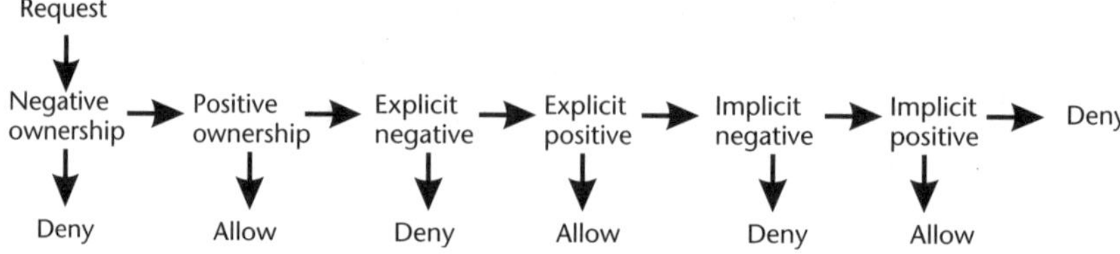

Figure 6.5 Resolution strategy for conflicting authorizations.

details of the IRO-DB security model, we refer the interested reader to the references given at the end of this chapter.

6.4 Conclusions

In this chapter we have seen that different access control models might be the choice for different types of data-centric applications. In most commercial or business environments, the discretionary access control model will probably be most suitable. However, we have seen that in systems protected by that kind of security mechanism, certain limitations and flaws may exist. This is the reason that for data-centric applications aiming at the support of a higher degree of protection, mandatory access controls may be appropriate. In order to get evaluated at the higher levels, the support of mandatory protection and the use of security labels is also demanded by most of the standards and evaluation criteria (e.g., the Common Criteria, ISO/IEC 15408) used to evaluate the security of a system. However, we have also seen that in systems protected by that kind of security mechanism, undesired side effects may occur, which make them less suitable for the use in commercial or business environments. Role-based access controls may be the choice in applications in which access should not be regulated on an individual user basis but in terms of organizational roles and job functions. Roles are a good abstraction mechanism, but in large-scale real-world applications it has turned out that it may become quite difficult to develop a proper role structure for a large enterprise. Overall, it is not possible to suggest "the" access control model for any data-centric applications. The model of choice will always depend on the respective circumstances and the types of threats the security mechanism has to address.

At the end of this chapter we will point the interested reader to some further literature. SQL has been standardized by both ANSI and ISO. The first source of reference for the security in relational databases of course is the recent standard, which unfortunately is not freely available but can be purchased from both organizations. As the semantics of the REVOKE statement are not clearly defined, the next source of reference is the documentation of the DBMS in use. There are also some books dealing with database security issues. The most comprehensive one is [1], but textbooks with more general content include chapters on database security (e.g., [2, 3]. A classic paper and a starting point for reading on statistical databases and tracker attacks is [4]. There are many papers dealing with database security research published in conference proceedings of different conference series or in academic journals. For the interested reader, a good starting point may be the proceedings of the IFIP TC 11/WG 11.3 Annual Conferences on Data and Applications Security, formerly called Database Security: Status and Prospects. A survey-style article is [5]. Among others, a pioneering work on the theory of MLS databases is [6]. Semantic ambiguities in MLS database queries are discussed in [7]. The standard reference to role-based access control is [8], published by the NIST. The access control subsystem of IRO-DB is further described in [9].

References

[1] Castano, S., et al., *Database Security,* Reading, MA: Addison-Wesley, 1995.

[2] Date, C. J., *Introduction to Database Systems*, 8th edition, Reading, MA: Addison-Wesley, 2003.

[3] Elmasri, R., and S. B. Navathe, *Fundamentals of Database Systems*, 5th edition, Boston, MA: Addison Wesley, 2006.

[4] Denning, D., P. J. Denning, and M. D. Schwartz, "The Tracker: A Threat to Statistical Database Security," *ACM Trans. Database Systems*, Vol. 4, No. 1, pp. 76–96.

[5] Pernul, G., "Database Security," in *Advances in Computers*, Vol. 41, M. C. Yovits (ed.), Academic Press, 1994, pp. 1–72.

[6] Jajodia, S., and R. S. Sandhu, "Towards a Multilevel Secure Relational Data Model," *Proceedings of the 1991 ACM SIGMOD International Conference on Management of Data*, Denver, CO, 1991.

[7] Smith, K., and M. Winslett, "Entity Modeling in the MLS Relational Model," *Proceedings of the 18th International Conference on Very Large Data Bases*, Vancouver, Canada, August 2002.

[8] Ferraiolo, D. F., R. D. Kuhn, and R. Chandramouli, *Role-Based Access Control*, 2nd Ed., Norwood, MA: Artech House, 2007.

[9] Essmayr, W., et al., "Authorization and Access Control in IRO-DB," *Proceedings IEEE Int. Conf. on Data Engineering (ICDE'96)*, New Orleans, LA, 1996.

CHAPTER 7

Modern Cryptology

Bart Preneel

7.1 Introduction

Cryptology is the science that studies mathematical techniques that provide secrecy, authenticity, and related properties for digital information [1]. Cryptology also enables us to create trust relationships over open networks; more in general, cryptographic protocols allow mutually distrusting parties to achieving a common goal while protecting their own interests. Cryptology is a fundamental enabler for security, privacy, and dependability in an online society. Cryptographic techniques can be found at the core of computer and network security, digital identification and digital signatures, digital content management systems, and so on. Their applications vary from e-business, m-business, e-health, e-voting, and online payment systems to wireless protocols and ambient intelligence.

The science of cryptology is almost as old as writing itself; for an historical perspective, the reader is referred to D. Kahn [2] and S. Sing [3]. Until the beginning of the twentieth century, the use of cryptology was restricted to kings, generals, and diplomats. This all changed with the advent of wireless communication, both for military and business purposes, around the mid-1910s. While the first generations of widely used devices were purely mechanical or electromechanical, electronic devices were introduced in the 1960s. For the next decades, cryptology was still restricted to hardware encryption on telecommunication networks and in computer systems; the main users were still governments and industry, in particular the financial and military sector. Around 1990 general purpose hardware became fast enough to allow for software implementations of cryptology, and, with the explosion of the Internet, cryptography became quickly a tool used by mass-market software; this includes protocols such as Transport Layer Security (TLS) and Secure Shell (SSH) at the transport layer, IPsec at the network layer [4], and more recently the WLAN protocols at the data link layer. Simultaneously, the success of the GSM network and the deployment of smart cards in the banking sector brought cryptography in the hands of every citizen. An increasing number of European governments is issuing electronic identity cards. In the next decade cryptographic hardware will be integrated into every general-purpose processor, while an ever-increasing number of small processors (hundreds or thousand of devices per user) will bring cryptology *everywhere*.

Without cryptology, securing our information infrastructure would not be possible. While cryptology as an art is very old, it has developed as a science in the last 60 years; most of the open research dates from the last 30 years. A significant number of successes has been obtained, and it is clear that cryptology should no longer be the weakest link in our modern security systems. Nevertheless, as a science and engineering discipline, cryptology is still facing some challenging problems.

This chapter intends to present the state of the art and to offer a perspective on open research issues in the area of cryptographic algorithms. The chapter will cover the principles underlying the design of cryptographic algorithms and protocols. First it discusses algorithms for confidentiality protection, (i.e., the protection against passive eavesdroppers) and for authentication (i.e., the protection against active eavesdroppers), who try to modify information. The chapter concludes with some comments on research challenges.

7.2 Encryption for Secrecy Protection

The basic idea in cryptography is to apply a complex mathematical transformation to protect the information. When the sender (usually called Alice) wants to convey a message to a recipient (Bob), the sender will apply to the *plaintext* P the mathematical transformation $E()$. This transformation $E()$ is called the encryption algorithm; the result of this transformation is called the *ciphertext* or $C = E(P)$. Bob, the recipient, will decrypt C by applying the inverse transformation $D = E^{-1}$, and in this way he recovers Por $P = D(C)$. For a secure algorithm E, the ciphertext C does not make sense to an outsider, such as Eve, who is tapping the connection and who can obtain C but not any partial information on the corresponding plaintext P.

This approach works only when Bob can keep the transformation D secret. While this secrecy is acceptable in a person-to-person exchange, it is not feasible for large-scale use. Bob needs a software or hardware implementation of D: either he has to program it himself or he has to trust someone to write the program for him. Moreover, he will need a different transformation and thus a different program for each correspondent, which is not very practical. Bob and Alice always have to face the risk that somehow Eve will obtain D (or E)—for example, by breaking into the computer system or by bribing the author of the software or the system manager.

This problem can be solved by introducing into the encryption algorithm $E()$, a secret parameter, the key K. Typically such a key is a binary string of 40 to a few thousand bits. A corresponding key K^* is used for the decryption algorithm D. One has thus $C = E_K(P)$ and $P = D_{K^*}(C)$. (See also Figure 7.1 which assumes that $K^* = K$.) The transformation strongly depends on the keys: if one uses a wrong key $K^{*\prime} \neq K^*$, then a random plaintext P' and not the plaintext P is obtained. Now it is possible to publish the encryption algorithm $E()$ and the decryption algorithm $D()$; the security of the system relies only on the secrecy of two short keys, which implies that $E()$ and $D()$ can be evaluated publicly and distributed on a commercial basis. Think of the analogy of a mechanical lock: everyone knows how a lock works, but to open a particular lock, one needs to have a particular key or know the secret combination.[1] In cryptography the assumption that the algorithm should remain secure even if it is known to the opponent is known as *Kerckhoffs's principle*. Kerckhoffs was a nineteenth century Dutch cryptographer who formulated this principle.

1 Note however that Matt Blaze has demonstrated in [5] that many modern locks are easy to attack and that their security relies to a large extent on *security through obscurity* (i.e., the security of locks relies on the fact that the methods to design and attack locks are not published).

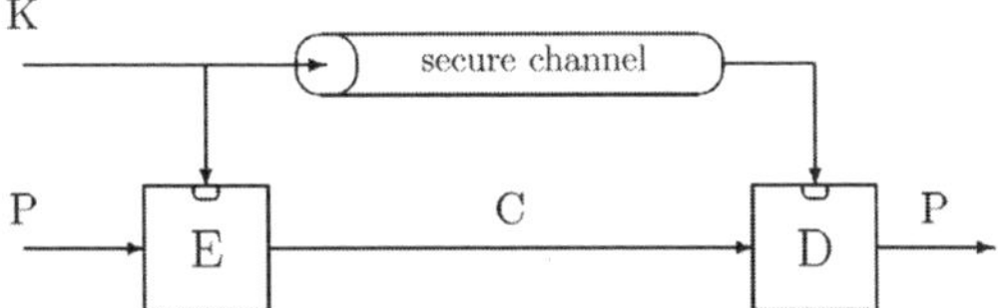

Figure 7.1 Model for conventional or symmetric encryption.

A simple example of an encryption algorithm is the Caesar cipher, named after the Roman emperor who used it. The plaintext is encrypted letter by letter; the ciphertext is obtained by shifting the letters over a fixed number of positions in the alphabet. The secret key indicates the number of positions. It is claimed that Caesar always used the value of 3 such that AN EXAMPLE would be encrypted to DQ HADPSOH. Another example is the name of the computer HAL from S. Kubrick's *A Space Odyssey* (2001), which was obtained by replacing the letters of IBM by their predecessor in the alphabet. This transformation corresponds to a shift over 25 positions or minus 1 position. It is clear that such a system is not secure, since it is easy to try the 26 values of the key and to identify the correct plaintext based on the redundancy of the plaintext.

The *simple substitution* cipher replaces a letter by any other letter in the alphabet. For example, the key could be

```
ABCDEFGHIJKLMNOPQRSTUVWXYZ
MZNJSOAXFQGYKHLUCTDVWBIPER
```

which means that an A is mapped to an M, a B to a Z, and so on; hence, THEEVENING would be encrypted as VXSSBSHFHA. For an alphabet of n letters (in English $n = 26$), there are $n!$ substitutions, which implies that there are $n!$ values for the secret key. Note that even for $n = 26$ trying all keys is not possible since $26! = 403291461126605635584000000 = 4 \cdot 10^{26}$. Even if a fast computer could try one billion (10^9) keys per second, it would take one billion years to try all the keys. However, it is easy to break this scheme by frequency analysis: in a standard English text, the character E accounts for 12 out of every 100 characters, if spaces are omitted from the plaintext. Hence it is straightforward to deduce that the most common ciphertext character, in this example S corresponds to an E. Consequently, the key space has been reduced by a factor of twenty-six. It is easy to continue this analysis based on lower frequency letters and based on frequent combinations of two (e.g., TH) and three (e.g., THE) letters that are called digrams and trigrams, respectively. In spite of the large key length, simple substitution is a very weak cipher, even if the cryptanalyst only has access to the ciphertext. In practice, the cryptanalyst may also know part of the plaintext (e.g., a standard opening such as DEARSIR).

A second technique applied in cryptology is a *transposition cipher* in which symbols are moved around. For example, the following mapping could be obtained:

```
TRANS      ONI S
POSIT  →   ROTIT
IONS       OSANP
```

Here the key would indicate where the letters are moved. If letters are grouped in blocks of n (in this example $n = 15$), there are $n!$ different transpositions. Again solving this cipher is rather easy (e.g., by exploiting digrams and trigrams or by fragments of known plaintexts). In spite of these weaknesses, modern ciphers designed for electronic computers are still based on a combination of several transpositions and substitutions (cf. Section 7.2.1.3).

A large number of improved ciphers has been invented. In the fifteenth and sixteenth centuries, *polyalphabetic substitution* was introduced, which uses $t > 1$ different alphabets. For each letter, one of these alphabets is selected based on some simple rule. The complexity of these ciphers was limited by the number of operations that an operator could carry out by hand. None of these manual ciphers is considered to be secure today. With the invention of telegraph and radio communications, more complex ciphers have been developed. The most advanced schemes were based on mechanical or electromechanical systems with rotors, such as the famous Enigma machine used by Germany in World War II and the Hagelin machines. Rotor machines were used between the 1920s and the 1950s. In spite of the increased complexity, most of these schemes were not sufficiently secure in their times. One of the weak points was users who did not follow the correct procedures. The analysis of the Lorenz cipher resulted in the development of Colossus, one of the first electronic computers.

A problem not yet been addressed is how Alice and Bob can exchange the secret key. The easy answer is that cryptography does not solve this problem; cryptography only moves the problem and at the same time simplifies the problem. In this case the secrecy of a (large) plaintext has been reduced to that of a *short key*, which can be exchanged beforehand. The problem of exchanging keys is studied in more detail in an area of cryptography that is called *key management*, which will not be discussed in detail in this chapter. See [1] for an overview of key management techniques.

The branch of science that studies the encryption of information is called *cryptography*. A related branch tries to break encryption algorithms by recovering the plaintext without knowing the key or by deriving the key from the ciphertext and parts of the plaintext; it is called *cryptanalysis*. The term *cryptology* covers both aspects. For more extensive introductions to cryptography, the reader is referred to [1,6–9].

So far we have assumed that the key for decryption K_D is equal to the encryption key K_E, or that it is easy to derive K_D from K_E. These types of algorithms are called *conventional* or *symmetric* ciphers. In *public-key* or *asymmetric* ciphers, K_D and K_E are always different; moreover, it should be difficult to compute K_D from K_E. This separation has the advantage that one can make K_E public; it has important implications for the key management problem. The remainder of this section discusses symmetric algorithms and public key algorithms.

7.2.1 Symmetric Encryption

This section introduces three types of symmetric encryption algorithms: the one-time pad also known as the Vernam scheme, additive stream ciphers, and block ciphers.

7.2.1.1 The One-Time Pad or the Vernam Scheme

In 1917 G. S. Vernam invented a simple encryption algorithm for telegraphic messages [10]. The encryption operation consists of adding a random key bit by bit to the plaintext. The decryption operation subtracts the same key from the ciphertext to recover the plaintext (see Figure 7.2). In practice, Vernam stored the keys on paper tapes.

The Vernam scheme can be formally described as follows. The *i*th bit of the plaintext, ciphertext, and key stream are denoted with p_i, c_i, and k_i, respectively. The encryption operation can then be written as $c_i = p_i \oplus k_i$. Here $\oplus$ denotes addition modulo 2 or exclusive or. The decryption operation is identical to the encryption or the cipher is an involution. Indeed $p_i = c_i \oplus k_i = (p_i \oplus k_i) \oplus k_i = p_i \oplus (k_i \oplus k_i = p_i \oplus 0 = p_i$. Vernam proposed use of a perfectly random key sequence, (i.e., the bit sequence k_i, $i = 1, 2, \ldots$ should consist of a uniformly and identically distributed sequence of bits).

In 1949 C. Shannon, the father of information theory, published his mathematical proof that shows that from observing the ciphertext, the opponent cannot obtain any new information on the plaintext, no matter how much computing power he has [11]. Shannon called this property *perfect secrecy*. The main disadvantage of the Vernam scheme is that the secret key is exactly as long as the message. C. Shannon also showed that the secret key cannot be shorter if one wants perfect secrecy. Until the late 1980s, the Vernam algorithm was used by diplomats and spies and even for the red-telephone system between Washington and Moscow.

Spies used to carry key pads with random characters: in this case p_i, c_i, and k_i are elements from $\mathbb{Z}_{26}$ representing the letters A through Z. The encryption operation is $c_i = (p_i + k_i) \bmod 26$ and the decryption operation is $p_i = (c_i - k_i) \bmod 26$. The keys were written on sheets of paper contained on a pad. The security of the scheme relies on the fact that every page of the pad is used only once, which explains the name one-time pad. Note that during World War II the possession of pads with random characters was sufficient to be convicted as a spy.

The one-time pad was also used for Soviet diplomatic communications. Under the codeword Venona, U.S. cryptologists attempted to break this system in 1943, which seems impossible as it offers perfectly secrecy. However, after two years, it was discovered that the Soviets used their pads twice (i.e., two messages were encrypted using the same key stream). This error was due to time pressure in the production of the pads. If c and c' are ciphertexts generated with the same pad, one finds that

$$c \oplus c' = (p \oplus k) \oplus (p' \oplus k) = (p \oplus p') \oplus (k \oplus k) = p \oplus p'.$$

This analysis implies that one can deduce from the ciphertext the sum of the corresponding plaintexts. Note that if the plaintexts are written in natural language, their

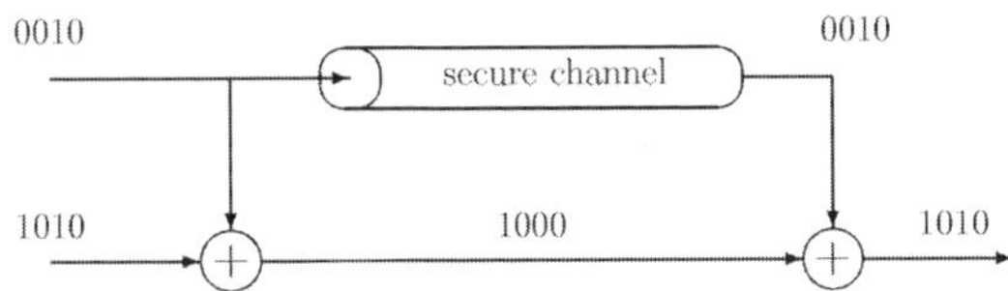

Figure 7.2 The Vernam scheme or one-time pad.

sum $p \oplus p'$ is not uniformly distributed, so it is possible to detect that the correct c and c' have been matched. Indeed, if c and c' would have been encrypted with different keys k and k', the sum $c \oplus c'$ would be equal to $(p \oplus p') \oplus (k \oplus k')$, which is uniformly distributed. By guessing or predicting parts of plaintexts, a clever cryptanalyst can derive most of p and p' from the sum $p \oplus p'$. In practice, the problem was more complex since the plaintexts were encoded with a secret method before encryption and cover names were used to denote individuals. Between 1943 and 1980, approximately 3,000 decryptions out of 25,000 messages were obtained. Some of these plaintexts contained highly sensitive information on Soviet spies. The Venona successes were only made public in 1995 and teach an important lesson: in using a cryptosystem, errors can be fatal even if the cryptosystem itself is perfectly secure. More details on Venona can be found in [12]. Section 7.4.3 discusses another weakness that can occur when implementing the Vernam scheme.

7.2.1.2 Additive Stream Ciphers

Additive stream ciphers are ciphers for which the encryption consists of a modulo 2 addition of a key stream to the plaintext. These ciphers try to mimic the Vernam scheme by replacing the perfectly random key stream by a pseudo-random key stream that is generated from a short key. Here pseudo-random means that the key stream looks random to an observer who has limited computing power. In practice one generates the bit sequence k_i with a keyed finite state machine (see Figure 7.3). Such a machine stretches a short secret key K into a much longer key stream sequence k_i. The sequence k_i is eventually periodic. One important but insufficient design criterion for the finite state machine is that the period has to be large (2^{80} is a typical lower bound) because a repeating key stream leads to a very weak scheme (cf. the Venona project). The values k_i should have a distribution that is close to uniform; another condition is that there should be no correlations between output bits. Note that cryptanalytic attacks may exploit correlations of less than one in 1 million.

Formally, the sequence k_i can be parameterized with a security parameter. For the security of the stream cipher, one requires that the sequence satisfies every polynomial time statistical test for randomness. In this definition, polynomial time means that the complexity of these tests can be described as a polynomial function of the security parameter. Another desirable property is that no polynomial time machine can predict the next bit of the sequence based on the previous outputs with

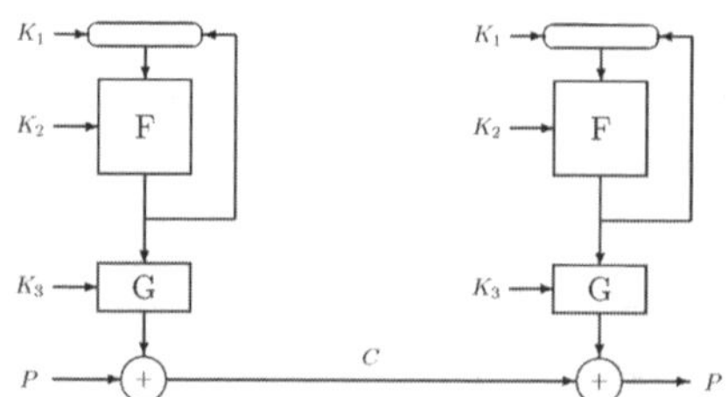

Figure 7.3 An additive stream cipher consisting of a keyed finite state machine. The initial state depends on K_1, the next state function F depends on K_2, and the output function G depends on K_3. The three keys K_1, K_2, and K_3 are derived from the user key K; this operation is not shown.

a probability that is significantly better than one-half. An important and perhaps surprising result in theoretical cryptology by A. Yao shows that these two conditions are in fact equivalent [13].

Stream ciphers have been popular in the twentieth century: they operate on the plaintext character by character, which is convenient and allows for a simple and thus inexpensive implementation. Most of the rotor machines are additive stream ciphers. Between 1960 and 1990, stream ciphers based on linear feedback shift registers (LFSRs) were very popular. (See, for example, the book by Rueppel [14].) However, most of these algorithms were trade secrets; every organization used its own cipher, and no standards were published. The most widely used LFSR-based stream ciphers are A5/1 and A5/2, which are implemented in hardware in the GSM phones. There are currently more than 1 billion GSM subscribers. The GSM algorithms were kept secret, but they leaked out and were shown to be rather weak [15]. In the last 15 years it has become clear that most LFSR-based stream ciphers are much less secure than expected. (See, for example, [16, 17].)

RC4, designed in 1987 by Ron Rivest, is based on completely different principles. RC4 is designed for 8-bit microprocessors and was initially kept as a trade secret. It was posted on the Internet in 1994 and is currently widely used in browsers (TLS protocol). While several statistical weaknesses have been identified in RC4 [18], the algorithm still seems to resist attacks that recover the key.

In the last 15 years, a large number of very fast stream ciphers has been proposed that are software oriented, suited for 32-bit processors, and that intend to offer a high level of security. However, for the time being, weaknesses have been identified in most proposals, and no single scheme has emerged as a standard or a de facto standard. Nevertheless, stream ciphers can be very valuable for encryption with very few hardware gates or for high-speed encryption. Developing strong stream ciphers is clearly an important research topic for the years ahead.

7.2.1.3 Block Ciphers

Block ciphers take a different approach to encryption: the plaintext is divided into larger words of n bits, called blocks; typical values for n are 64 and 128. Every block is enciphered in the same way, using a keyed one-way permutation (i.e., a permutation on the set of n-bit strings controlled by a secret key). The simplest way to encrypt a plaintext using a block cipher is as follows: divide the plaintext into n-bit blocks P_i, and encrypt these block by block. The decryption also operates on individual blocks:

$$C_i = E_K(P_i) \quad \text{and} \quad P_i = D_K(C_i).$$

This way of using a block cipher is called the electronic codebook (ECB) mode. Note that the encryption operation does *not* depend on the location in the ciphertext, as is the case for additive stream ciphers.

Consider the following attack on a block cipher, the so-called tabulation attack: the cryptanalyst collects ciphertext blocks and their corresponding plaintext blocks, which is possible as part of the plaintext is often predictable; these blocks are used to build a large table. With such a table, one can deduce information on other plaintexts encrypted under the same key. In order to preclude this attack, the value of n has to be quite large (e.g., 64 or 128). Moreover, the plaintext should not contain any repetitions or other patterns, as these will be leaked to the ciphertext.

This last problem shows that even if n is large, the ECB mode is not suited to encrypt structured plaintexts, such as text and images. This mode should only be used in exceptional cases where the plaintext is random, such for as the encryption of cryptographic keys. There is, however, an easy way to randomize the plaintext by using the block cipher in a different way. The most popular mode of operation for a block cipher is the cipher block chaining (CBC) mode (see Figure 7.4). In this mode the different blocks are coupled by adding modulo 2 to a plaintext block the previous ciphertext block:

$$C_i = E_K(P_i \oplus C_{i-1}) \quad \text{and} \quad P_i = D_K(C)_i \oplus C_{i-1}).$$

Note that this randomizes the plaintext and hides patterns. To enable the encryption of the first plaintext block ($i = 1$), one defines C_0 as the initial value IV, which should be randomly chosen and transmitted securely to the recipient. By varying this value, one can ensure that the same plaintext is encrypted into a different ciphertext under the same key. The CBC mode allows for random access on decryption: if necessary, one can decrypt only a small part of the ciphertext. A security proof the CBC mode has been provided by Bellare et al. [19]; it holds as long as the opponent can only obtain ciphertext corresponding to chosen plaintexts. Note that the CBC mode is insecure if the opponent can choose ciphertexts and obtain the corresponding plaintexts.

One can also use a block cipher to generate a key stream that can be used in an additive stream cipher: in the output feedback (OFB) mode, the block cipher is applied iteratively by feeding back the n-bit output to the input; in the counter (CTR) mode one encrypts successive values of a counter. An alternative stream cipher mode is the cipher feedback (CFB) mode; this mode is slower but it has better synchronization properties. The modes of operation have been standardized in FIPS 81 [20] (see also [21], which adds the CTR mode) and ISO/IEC 10116 [22].

Block ciphers form a very flexible building block. They have played an important role in the past 25 years because of the publication of two U.S. government Federal Information Processing Standards (FIPS) for the protection of sensitive but unclassified government information. An important aspect of these standards is that they can be used without paying a license fee.

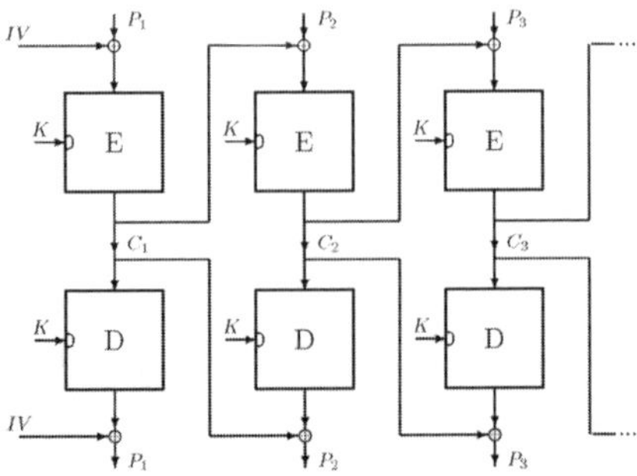

Figure 7.4 The CBC mode of a block cipher.

The first standardized block cipher is the Data Encryption Standard (DES) of FIPS 46 [23] published in 1977. This block cipher was developed by IBM together with the National Security Agency (NSA) in response to a call by the U.S. government. DES represents a remarkable effort to provide a standard for government and commercial use; its impact on both practice and research can hardly be overestimated. For example, DES is widely used in the financial industry. DES has a block size of 64 bits and a key length of 56 bits. (More precisely, the key length is 64 bits, but 8 of these are parity bits.) The 56-bit key length was a compromise: it would allow the U.S. government to find the key by brute force (i.e., by trying all $2^{56} \approx 7 \cdot 10^{16}$ keys one by one) but would put a key search beyond limits for an average opponent. However, as hardware got faster, this key length was sufficient only 10 to 15 years; hence, DES reached the end of its lifetime in 1987–1992 (see Section 7.4.3 for details). The block length of 64 bits is no longer adequate either because there exist matching ciphertext attacks on the modes of operation of an n-bit block cipher that require about $2^{n/2}$ ciphertext blocks [24]. For $n = 64$, these attacks require 4 billion ciphertexts, and with a high-speed encryption device this number is reached in less than a minute. The DES design was oriented toward mid-1970s hardware. For example, it uses a 16-round Feistel structure (see Figure 7.5) which implies that the hardware for encryption and decryption is identical. Each round consists of nonlinear substitutions from 6 bits to 4 bits, followed by some bit permutations or transpositions. The performance of DES in software is suboptimal; for example, DES runs at 40 cycles/byte on a Pentium III, which corresponds to 200 Mbps for a clock frequency of 1 GHz.

In 1978, one year after the publication of the DES standard, an improved variant of DES was proposed. Triple-DES consists of three iterations of DES: $E_{K1}(D_{K2}(E_{K3}(x)))$. Only in 1999, this variant was included into the third revision of FIPS 46 [23]. The choice of a decryption for the middle operation is motivated by backward compatibility: indeed, choosing $K_1 = K_2 = K_3$ results in single DES. Three-key triple DES has a 168-bit key, but the security level corresponds to a key of approximately 100 bits. Initially two key triple DES was proposed, with $K_3 = K_2$: its security level is about 80 to 90 bits. On first examination, the double-DES key length of 112 bits appears sufficient; however, it has been shown that the security level of double-DES is approximately 70 bits. For an overview of these attacks, see [1]. The migration from DES to triple-DES in the financial sector was started in 1986, but it is progressing slowly and has taken 20 years to complete. Triple-DES has the disadvantage that it is rather slow (115 cycles/byte on a Pentium III) and that the block length is still limited to 64 bits.

In 1997 the U.S. government decided to replace DES by the Advanced Encryption Standard (AES). AES is a block cipher with a 128-bit block length and key lengths of 128, 192, and 256 bits. An open call for algorithms was issued; 15 candidates were submitted by the deadline of June 1998. After the first round, five finalists remained and in October 2000 it was announced that the Rijndael algorithm, designed by the Belgian cryptographers Vincent Rijmen and Joan Daemen, was the winner. The FIPS standard was published in November 2001 [25]. It may not be a coincidence that the U.S. Department of Commerce Bureau of Export Administration (BXA) relaxed export restrictions for U.S. companies in September 2000. Note that otherwise it would have been illegal to export AES software from the U.S. to Belgium. In 2003, the U.S. government announced that it would also allow the use of AES for secret data, and even for top secret data; the latter application requires key lengths of 192 or 256 bits. Rijndael is a rather elegant and mathematical design. Among the five finalists, it offered the best combination of security, performance, efficiency, implementability, and flexibility. AES allows for efficient implementations on 32-bit architectures (15 cycles/byte on a Pentium III), but also for compact implementations on 8-bit smart cards. Moreover, hardware implementations of AES offer good tradeoffs between size and speed. AES consists of a substitution-permutation (SP) network with 10 rounds for a 128-bit key, 12 rounds for a 192-bit key, and 14 rounds for a 256-bit key (see Figure 7.5). Each round consists of nonlinear substitutions (from 8 bits to 8 bits), followed by some affine transformations, which move around the information. Note that Rijndael also supports 192-bit and 256-bit block lengths, but these have not been included in the AES standard. For a complete description of the AES and its design, see [26].

There exist many other block ciphers; a limited number of these has been included in products, such as Camellia, members of the CAST family, FEAL, Gost, IDEA, Kasumi, and Skipjack. For more details, the reader is referred to the cryptographic literature.

7.2.2 Public Key Encryption

The main problem left unsolved by symmetric cryptography is the key distribution problem. Especially in a large network it is not feasible to distribute keys between all user pairs. In a network with t users, there are $t\ (t - 1)/2$ such pairs; hence, even for 1,000 users approximately half a million keys are needed. An alternative is to manage all keys in a central entity that shares a secret key with every user. How-

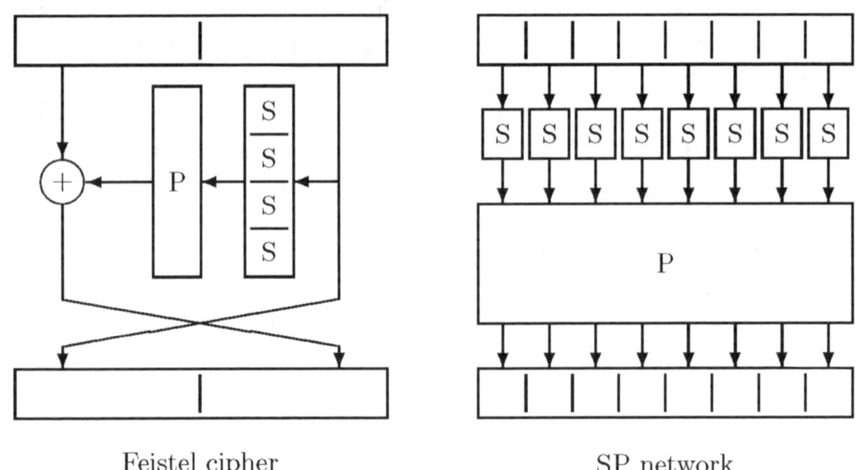

Figure 7.5 One round of a Feistel cipher (left) and of a SP network. Here S represents a substitution and P a permutation, which can be a bit permutation or an affine mapping. In a Feistel cipher, a complex operation on the right part is added to the left part; it has the advantage that the decryption operation is equal to the encryption operation, which simplifies hardware implementations. The substitution operation does not need to be invertible here. In an SP network, all the bits are updated in every round, which guarantees faster diffusion of information; the substitution and permutation operations need to be invertible.

ever, this entity then becomes a single point of failure and an attractive target of attack. A much more elegant solution to the key management problem is offered by public key cryptography invented in 1976, independently by W. Diffie and M. Hellman [27] and by R. Merkle [28].

7.2.2.1 Public Key Agreement

A public key agreement protocol allows two parties who have never met to agree on a secret key by way of a public conversation. Diffie and Hellman showed how to achieve this goal using the concept of *commutative one-way functions*. A *one-way function* is a function that is easy to compute but hard to invert. For example, in a block cipher, the ciphertext has to be a one-way function of the plaintext and the key: it is easy to compute the ciphertext from the plaintext and the key, but given the plaintext and the ciphertext it should be hard to recover the key; otherwise, the block cipher would not be secure. Similarly it can be shown that the existence of pseudo-random string generators, as used in additive stream ciphers, implies the existence of one-way functions. A *commutative one-way function* is a one-way function for which the result is the same independent of the order of the evaluation: for a function $f(\cdot, \cdot)$ with two arguments $f(f(z, x), y) = f(f(z, y), x)$.

The candidate commutative one-way function proposed by Diffie and Hellman is $f(a, x) = a^x \bmod p$; here p is a large prime number (large means 1,024 bits or more), $x \in [1, p-1]$, and a is a generator mod p, which means that $a^0, a, a^2, a^3, \ldots, a^{p-2} \bmod p$ run through all values between 1 and $p - 1$. For technical reasons, it is required that p is a safe prime, which means that $(p - 1)/2$ is a prime number as well. The Diffie-Hellman protocol works as follows (see also Figure 7.6).

- Alice and Bob agree on a prime number p and a generator a mod p.
- Alice picks a value x_A uniformly at random in the internal $[1, p - 1]$, computes $y_A = a^{x_A} \bmod p$, and sends this to Bob.
- Bob picks a value x_B uniformly at random in the interval $[1, p - 1]$, computes $y_B = a^{x_B} \bmod p$, and sends this to Alice.
- On receipt of y_B, Alice checks that $1 \le y_B \le p - 2$ and computes $k_{AB} = y_B^{x_A} \bmod p = a^{x_A x_B} \bmod p$.
- On receipt of y_A, Bob checks that $1 \le y_A \le p - 2$ and computes $k_{BA} = y_A^{x_B} \bmod p = a^{x_B x_A} \bmod p$.

2 *AliceBob*

$x_A \in_R [1, p-1], a^{x_A} \bmod p$

$a^{x_A} \bmod p$

$x_B \in_R [1, p-1], a^{x_B} \bmod p$

$a^{x_B} \bmod p$

$k_{AB} = (a^{x_B})^{x_A} \bmod p k_{AB} = (a^{x_A})^{x_B} \bmod p$

Figure 7.6 The Diffie-Hellman protocol.

- Alice and Bob compute the secret key as $h(k_{AB}) = h(k_{BA})$, with $h()$ a hash function or MDC (see Section 7.3.1).

It is easy to see that the commutativity implies that $k_{AB} = k_{BA}$; hence, Alice and Bob obtain a common value. Eve, who is eavesdropping the communication, only observes $y_A = a^{x_A} \bmod p$ and $y_B = a^{x_B} \bmod p$; there is no obvious way for her to obtain $k_{AB} = (a^{x_B})^{x_A} \bmod p$. If Eve could compute discrete logarithms (i.e., derive x_A from y_A and/or x_B from y_B), she could of course also derive k_{AB}. However, if p is large, this problem is believed to be difficult (cf. infra). Eve could try to find another way to compute k_{AB} from y_A and y_B. So far, no efficient algorithm has been found to solve this problem, which is stated as the Diffie-Hellman assumption: it is hard to solve the Diffie-Hellman problem—to deduce $(a^{x_B})^{x_A} \bmod p$ from $a^{x_A} \bmod p$ and $a^{x_B} \bmod p$. If the Diffie-Hellman assumption holds, the Diffie-Hellman protocol results in a common secret between Alice and Bob after a public conversation. It is clear from this discussion that the Diffie-Hellman problem cannot be harder than the discrete logarithm problem. It is known that for some prime numbers the two problems are equivalent.

It is very important to check that $y_A, y_B \notin \{0, 1, p - 1\}$: if not, Eve could modify y_A and y_B to one of these values and ensure in this way that $k_{AB} \in \{0, 1, p - 1\}$. However, the Diffie-Hellman protocol has another problem: how does Alice know that she is talking to Bob or vice versa? In the famous man-in-the-middle-attack, Eve sets up a conversation with Alice, which results in the key k_{AE}, and with Bob, which results in the key k_{BE}. Eve now shares a key with both Alice and Bob; she can decrypt all messages received from Alice, read them, and re-encrypt them for Bob and vice versa. Alice or Bob are unable to detect this attack; they believe that they share a common secret known only to the two of them. This attack shows that the common secret can only be established between two parties if there is an authentic channel, (i.e., a channel on which the information can be linked to the sender and the information cannot be modified). The conclusion is that the authenticity of the values y_A and y_B has to be established, by linking them to Alice and Bob, respectively. One way to achieve this goal is to read these values or hash values of these values (see Section 7.3.1) over the phone; this solution works if Alice and Bob know each other's voices or if they trust the phone system to connect them to the right person.

The Diffie-Hellman problem has another limitation: Alice and Bob can agree on a secret key, but Alice cannot use this protocol directly to tell Bob to meet her tonight at 9 pm. Alice can of course use the common key k_{AB} to encrypt this message using the AES algorithm in the CBC mode. We will explain in the next section how public key encryption can overcome this limitation.

7.2.2.2 Public Key Encryption

The key idea behind public key encryption is the concept of *trapdoor one-way functions* [27]. Trapdoor one-way functions are one-way functions with an additional property: given some extra information, the trapdoor, it becomes possible to invert the one-way function.

With such functions, Bob can send a secret message to Alice without the need for prior arrangement of a secret key. Alice chooses a trapdoor one-way function with public parameter P_A (i.e., Alice's public key) and with secret parameter S_A (i.e., Alice's secret key). Alice makes her public key widely available. For example, she can put it on her home page, but it can also be included in special directories. Anyone who wants to send some confidential information to Alice computes the ciphertext as the image of the plaintext under the trapdoor one-way function using the parameter P_A. On receipt of this ciphertext, Alice recovers the plaintext by using her trapdoor information S_A (see Figure 7.7). An attacker, who does not know S_A, sees only the image of the plaintext under a one-way function and will not be able to recover the plaintext. The conditions which a public-key encryption algorithm has to be satisfy are:

- The generation of a key pair (P_A, S_A) has to be easy.
- Encryption and decryption have to be easy operations.
- It should be hard to compute the public key P_A from the corresponding secret key S_A.
- $D_{S_A}(E_{P_A}(P)) = P$.

Note that if a person wants to send a message to Alice, that individual has to know Alice's public key P_A and has to be sure that this key really belongs to Alice and not to Eve, since only the owner of the corresponding secret key will be able to decrypt the ciphertext. Public keys do not need a secure channel for their distribution, but hey do need an authentic channel. As the keys for encryption and decryption are different, and Alice and Bob have different information, public key algorithms are also known as *asymmetric algorithms*.

Designing a secure public key encryption algorithm is apparently a difficult problem. From the large number of proposals, only a few have survived. The most popular algorithm is the RSA algorithm [29], which was named after its inventors R. L. Rivest, A. Shamir, and L. Adleman. RSA was published in 1978; the patent on RSA expired in 2000. The security of RSA is based on the fact that it is relatively easy to find two large prime numbers and to multiply these while factoring their product is not feasible with the current algorithms and computers. The RSA algorithm can be described as follows:

Key generation: Find two prime numbers p and q with at least one having 150 digits and compute their product, the modulus $n = p \cdot q$. Compute the Carmichael function $\lambda(n)$, which is defined as the least common multiple of

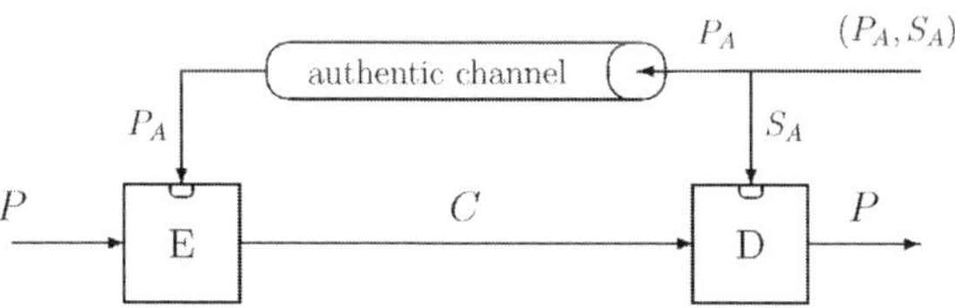

Figure 7.7 Model for public key or asymmetric encryption.

$p - 1$ and $q - 1$. In other words, $\lambda(n)$ is the smallest integer that is a multiple of both $p - 1$ and $q - 1$. Choose an encryption exponent e, which is at least 32 to 64 bits long and which is relatively prime to $\lambda(n)$ (i.e., has no common divisors with $\lambda(n)$). Compute the decryption exponent as $d = e^{-1} \bmod \lambda(n)$ using Euclid's algorithm. The public key consists of the pair (e, n), and the secret key consists of the decryption exponent d or the pair (p, q).

Encryption: Represent the plaintext as an integer in the interval $[0, n - 1]$ and compute the ciphertext as $C = P^e \bmod n$.

Decryption: $P = C^d \bmod n$.

The prime factors p and q or the secret decryption exponent d are the trapdoor that allows the inversion of the function $f(x) = x^e \bmod n$. Indeed, it can be shown that $f(x)^{\mathrm{d}} \bmod n = x^{\mathrm{ed}} \bmod n = x$. Note that the RSA function is the dual of the Diffie-Hellman function $f(x) = \alpha^x \bmod p$, which has a fixed base and a variable exponent.

Without explaining the mathematical background of the algorithm, it can be seen that the security of the RSA algorithm depends on the factoring problem. Indeed, if an attacker can factor n, he can find $l\ (n)$, derive d from e, and decrypt any message. However, to decrypt a ciphertext it is sufficient to extract modular eth roots. Note that it is not known whether it is possible to extract eth roots without knowing p and q. The RSA problem is the extraction of random modular eth roots, since this corresponds to the decryption of arbitrary ciphertexts. Cryptographers believe that the RSA problem is hard; this assumption is known as the RSA assumption. It is easy to see that the RSA problem cannot be harder than factoring the modulus. Some indication has been found that the two problems may not be equivalent.

For special arguments, the RSA problem is easy. For example, –1, 0, and 1 are always fixed points for the RSA encryption function, and, for small arguments, $P^e < n$ and extracting modular eth root simplifies to extracting natural eth roots, which is an easy problem. However, the RSA assumption states that extracting *random* modular eth roots is hard, which means that the challenge ciphertext needs to be uniformly distributed. Such a uniform distribution can be achieved by transforming the plaintext with a randomizing transform. A large number of such transforms is known, and many of these are ad hoc so there is no reason to believe that they should be effective. In 1993, Bellare and Rogaway published a new transform under the name optimal asymmetric encryption (OAEP) together with a security proof [30]. This proof shows that an algorithm that can decrypt a challenge ciphertext without knowing the secret key can be transformed into an algorithm that computes a random modular eth root. The proof is in the random oracle model, which means that the hash functions used in the OAEP construction are assumed to be perfectly random. However, seven years later Shoup pointed out that the proof was wrong [31]; the error has been corrected for by Fujisaki et al. in [32], but the resulting reduction is not meaningful—the coupling between the two problems is not very tight in this new proof (except when e is small). Currently the cryptographic community believes that the best way to use RSA is the RSA-KEM mode [33], which is a *hybrid* mode in which RSA is only used to transfer a session key while the plaintext is encrypted using a symmetric algorithm with this key. It is interesting to note that it has taken more than 20 years before cryptographers understood how RSA should be used properly for encryption.

7.2.2.3 The Factoring and the Discrete Logarithm Problem

The more complex properties of public key cryptography seem to require some "high level" mathematical structure; most public key algorithms are based on problems from algebraic number theory. While these number theoretic problems are believed to be difficult, it should be noted that there is no mathematical proof that shows that these problems are hard. Moreover, since the invention of public key cryptography, significant progress has been made in solving concrete instances. This evolution is due to a combination of more sophisticated algorithms with progress in hardware and parallel processing. Table 7.1 summarizes the progress made in factoring over the past 40 years. It is believed that the discrete logarithm problem mod$p$ is about as difficult as the factoring problem for the same size of modulus. This equivalence only holds if p satisfies certain conditions; a sufficient condition is that p is a safe prime as defined earlier.

The best known algorithm to factor an RSA modulus n is the general number field sieve. It has been used in all factoring records since 1996 and has a heuristic asymptotic complexity

$$O\left(\exp\left[(1.923 + o(1)) \cdot (\ln n)^{1/3} \cdot (\ln \ln n)^{2/3}\right]\right).$$

Note that this asymptotic expression should be used with care; extrapolations can only be made in a relatively small range due to the $o(1)$ term. Lenstra and Verheul provide an interesting study on the selection of RSA key sizes [34]. Currently, it is believed that factoring a 1,024-bit RSA modulus (308-digit) modulus requires 2^{80} steps. With special hardware proposed by Shamir and Tromer [35], the following

Table 7.1 Progress of Factorization Records for Products of Two Random Prime Numbers.

Year	*# Digits*	*# Bits*	*Computation*
1964	20	66	
1974	45	150	0.001 MY
1983	50	166	
1984	71	236	0.1 MY
1991	100	332	7 MY
1992	110	365	75 MY
1993	120	399	835 MY
1994	129	429	5000 MY
1996	130	432	1000 MY
1999	140	465	2000 MY
1999	155	512	8400 MY
2003	174	578	
2005	200	663	

Note: One MIPS year (MY) is the equivalent of a computation during one full year at a sustained speed of one million instructions per second, which corresponds roughly to the speed of a VAX 11/780.

cost estimates have been provided: with an investment of US$10 million, a 1,024-bit modulus can be factored in one year (the initial R&D cost is US$20 million). A 768-bit modulus can be factored for US$5,000 in 95 days, and a 512-bit modulus can be factored with a US$10,000 device in 10 minutes. Note that these cost estimates do not include the linear algebra step at the end; while this step takes additional time and effort, it should not pose any unsurmountable problems. Nevertheless, these estimates show that for long-term security, an RSA modulus of 2,048 bits is recommended.

7.2.2.4 Basing Public Key Cryptology on Other Problems

There has been a large number of proposals for other public key encryption algorithms. Many of these have been broken, the most notable example being the class of knapsack systems. The most important alternative to RSA is the ElGamal scheme, which extends the Diffie-Hellman scheme to public key encryption. In particular, the group of integers mod$p$ can also be replaced by a group defined by an elliptic curve over a finite field, as proposed by Miller and Koblitz in the mid 1980s. Elliptic curve cryptosystems allow for shorter key sizes (i.e., a 1,024-bit RSA key corresponds to a 170-bit elliptic curve key), but the operations on an elliptic curve are more complex (see [36–38]). Other alternatives are schemes based on hyperelliptic curves, multivariate polynomials over finite fields, lattice-based systems such as NTRU, and systems based on braid groups[2]; while these systems have particular advantages, it is believed that they are not yet mature enough for deployment. It is a little worrying that our digital economy relies to a large extent on the claimed difficult of a few problems in algebraic number theory.

7.2.2.5 Applying Public Key Cryptography

The main advantage of public key algorithms is the simplified key management; deployment of cryptology on the Internet largely relies on public key mechanisms (e.g., TLS, IPsec, and SSH [4]). An important question is how authentic copies of the public keys can be distributed; this problem will be briefly discussed in Section 7.3.2. The main disadvantages are the larger keys (typically 64 to 512 bytes) and the slow performance: both in software and hardware public key encryption algorithms are two to three orders of magnitude slower than symmetric algorithms. For example, a 1,024-bit exponentiation with a 32-bit exponent takes 360 μs on a 1-GHz. Pentium III; this corresponds to 2,800 cycles/byte; a decryption with a 1,024-bit exponent takes 9.8 ms or 76,000 cycles/byte. This speed should be compared to 15 cycles/byte for AES. Because of the large difference in performance, the large block length that influences error propagation, and the security reasons indicated earlier, one always employs *hybrid* systems: the public key encryption scheme is used to establish a secret key, which is then used in a fast symmetric algorithm.

2 Braid groups are noncommutative groups derived from geometric arrangements of strands; in a noncommutative group $a \cdot b$ is in general not equal to $b \cdot a$.

7.3 Hashing and Signatures for Authentication

Information authentication includes two main aspects:

- *Data origin authentication*, or who has originated the information;
- *Data integrity*, or has the information been modified.

Other aspects that can be important are the timeliness of the information, the sequence of messages, and the destination of information. These aspects can be accounted for by using sequence numbers and time stamps in the messages and by including addressing information in the data. In data communications, the implicit authentication created by recognition of the handwriting, signature, or voice disappears. Electronic information becomes much more vulnerable to falsification as the physical coupling between information and its bearer is lost.

Until the mid 1980s, it was widely believed that encryption with a symmetric algorithm of a plaintext was sufficient for protecting its authenticity. If a certain ciphertext resulted after decryption in a *meaningful* plaintext, it had to be created by someone who knew the key, and therefore it must be authentic. However, a few counterexamples are sufficient to refute this claim. If a block cipher is used in ECB mode, an attacker can easily reorder the blocks. For any additive stream cipher, including the Vernam scheme, an opponent can always modify a plaintext bit even without knowing whether a zero has been changed to a one or vice versa. The concept of meaningful information implicitly assumes that the information contains redundancy, which allows a distinction of genuine information from arbitrary plaintext. However, one can envisage applications where the plaintext contains very little or no redundancy (e.g., the encryption of keys). The separation between secrecy and authentication has also been clarified by public key cryptography: anyone who knows Alice's public key can send her a confidential message, and therefore Alice has no idea who has actually sent this message.

Two different levels of information authentication can be distinguished. If two parties trust each other and want to protect themselves against malicious outsiders, the term *conventional message authentication* is used. In this setting, both parties are at equal footing (e.g., they share the same secret key). If, however, a dispute arises between them, a third party will not be able to resolve it. For example, a judge may not be able to tell whether a message was created by Alice or by Bob. If protection between two mutually distrustful parties is required, which is often the case in a commercial relationships, an electronic equivalent of a manual signature is needed. In cryptographic terms this is called a *digital signature*.

7.3.1. Symmetric Authentication

The underlying idea is similar to that for encryption, where the secrecy of a large amount of information is replaced by the secrecy of a short key. In the case of authentication, one replaces the authenticity of the information by the protection of a short string, which is a unique fingerprint of the information. Such a fingerprint is computed as a hash result, which can also be interpreted as adding a special form of redundancy to the information. This process consists of two components. First, the information is compressed into a string of fixed length, with a cryptographic hash function. Then the resulting string, the hash result, is protected as follows:

- The hash result is communicated over an authentic channel, (e.g., it can be read over the phone). It is then sufficient to use a hash function without a secret parameter, which is also known as a manipulation detection code (MDC).
- The hash function uses a secret parameter (the key) and is then called a message authentication code (MAC) algorithm.

7.3.1.1 MDCs

If an additional authentic channel is available, MDCs can provide authenticity without requiring secret keys. Moreover an MDC is a flexible primitive, which can be used for a variety of other cryptographic applications. An MDC has to satisfy the following conditions:

- Preimage resistance: it should be hard to find an input with a given hash result;
- Second preimage resistance: it should be hard to find a second input with the same hash result as a given input;
- Collision resistance: it should be hard to find two different inputs with the same hash result.

An MDC satisfying these three conditions is called a *collision-resistant* hash function.

For a strong hash function with an n-bit result, solving the first two problems requires about 2^n evaluations of the hash function. This implies that $n = 90 \ldots 100$ is sufficient (cf. Section 7.4.3); larger values of n are required if one can attack multiple targets in parallel. However, finding collisions is substantially easier than finding preimages or second preimages. With high probability, a set of hash results corresponding to $2^{n/2}$ inputs contains a collision, which implies that collision-resistant hash functions need a hash result of 160 to 256 bits. This last property is also known as the *birthday paradox* based on the following observation: within a group of 23 persons the probability that there are two persons with the same birthday is about 50 percent. The reason is that a group of this size contains 253 different pairs of persons, which is rather larger compared to the 365 days in a year. The birthday paradox plays an essential role in the security of many cryptographic primitives. Note that not all applications need collision-resistant hash functions; sometimes preimage resistance or second preimage resistance is sufficient.

The most efficient hash functions are dedicated hash function designs. The hash functions MD4 and MD5 with a 128-bit hash result are no longer recommended. For the devastating attacks by Wang et al. see [39]. The most popular hash function today is SHA-1, but it has been pointed out by Wang et al. [40] that a shortcut collision attack exists that requires effort 2^{63} rather than 2^{80}. RIPEMD-160 is an alternative; both SHA-1 and RIPEMD-160 offer a 160-bit result. Recent additions to the SHA-family include SHA-256, SHA-384, and SHA-512 (see FIPS 180-2 [41]).

The ISO standard on dedicated hash functions (ISO/IEC 10118-3) contains RIPEMD-128, RIPEMD-160, SHA-1, SHA-256, SHA-384, SHA-512, and Whirlpool [42]. Part 2 of this standard specifies hash functions based on a block cipher, while part 4 specifies hash functions based on modular arithmetic.

7.3.1.2 MAC Algorithms

MAC algorithms have been used since the 1970s for electronic transactions in the banking environment. They require the establishment of a secret key between the communicating parties. The MAC value corresponding to a message is a complex function of every bit of the message and every bit of the key; it should be infeasible to derive the key from observing a number of text/MAC pairs or to compute or predict a MAC without knowing the secret key.

A MAC algorithm is used as follows (cf. Figure 7.8): Alice computes for her message P the value $\text{MAC}_K(P)$ and appends this MAC to the message (here MAC is an abbreviation of MAC result). Bob recomputes the value of $\text{MAC}_K(P)$ based on the received message P, and verifies whether it matches the received MAC result. If the answer is positive, he accepts the message as authentic (i.e., as a genuine message from Alice). Eve, the active eavesdropper, can modify the message P to P', but she is not able to compute the corresponding value $\text{MAC}_K(P')$ since she is not privy to the secret key K. For a secure MAC algorithm, the best Eve can do is guess the MAC result. In that case, Bob can detect the modification with high probability: for an n-bit MAC result Eve's probability of success is only $1/2^n$. The value of n lies typically between 32 and 96. Note that if separate techniques for encryption and authentication are combined, the keys for encryption and authentication need to be independent. Moreover, the preferred option is to apply the MAC algorithm to the ciphertext since this order of operations protects the encryption algorithm against chosen cipher text attacks.

A popular way to compute a MAC is to encrypt the message with a block cipher using the CBC mode, which is another use of a block cipher, and to keep only part of the bits of the last block as the MAC. However, Knudsen has shown

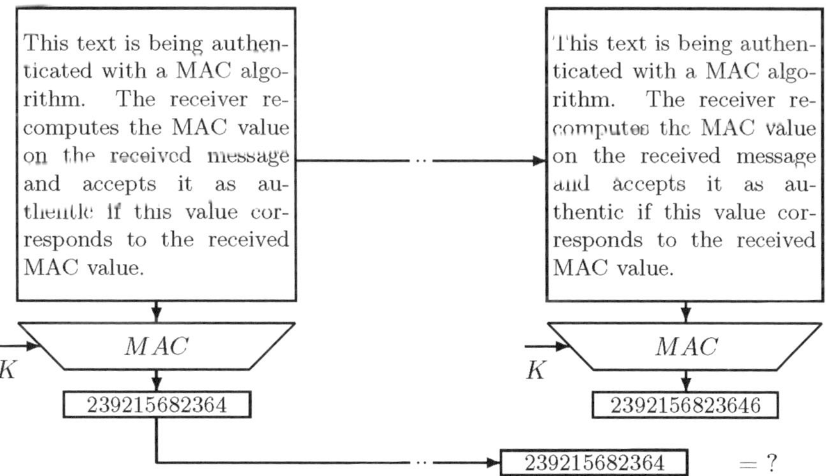

Figure 7.8 Using a message authentication code for data authentication.

that this approach is less secure than previously believed [43]. The recommended approach to use CBC-MAC consists of super-encrypting the final block with a different key, which may be derived from the first key. This scheme is known as EMAC; a security proof for EMAC has been provided by Petrank and Rackoff in [44]. Almost all CBC-MAC variants are vulnerable to a birthday-type attack that requires only $2^{n/2}$ known text-MAC pairs [45]. Another popular MAC algorithm is HMAC, which derives a MAC algorithm from a hash function such as SHA-1 [41]; an alternative for HMAC is MDx-MAC [45]. A large number of MAC algorithms has been standardized in ISO/IEC 9797 [46].

For data authentication, the equivalent of the Vernam scheme exists, which implies that a MAC algorithm can be designed that is unconditionally secure in the sense that the security of the MAC algorithm is independent of the computing power of the opponent. The requirement is again that the secret key is used only once. The basic idea of this approach is due to G. J. Simmons [8], who defined authentication codes, and Carter and Wegman [47,48], who used the term universal hash functions. The first ideas date back to the 1970s. It turns out that these algorithms can be computationally extremely efficient since the properties required from this primitive are combinatorial rather than cryptographic. The combinatorial property that is required is that if one takes the average over a key, the function values or pairs of function values need to be distributed almost uniformly. This property is much easier to achieve than cryptographic properties, which require that it should be hard to recover the key from input-output pairs, which is a much stronger requirement than a close to uniform distribution. Recent constructions are therefore one order of magnitude faster than other cryptographic primitives, such as encryption algorithms and hash functions, and achieve speeds up to 1–2 cycles/byte on a Pentium III for messages longer than 256 bytes (e.g., UMAC [49] and Poly 1305-AES [50]). A simple example is the polynomial hash function (see [51]). The key consists of two n-bit words denoted with K_1 and K_2. The plaintext P is divided into t n-bit words, denoted with P_1 through P_t. The MAC value, which consists of a single n-bit word, is computed based on a simple polynomial evaluation:

$$\mathrm{MAC}_{K_1K_2}(P) = K_1 + \sum_{i=1}^{t} P_i \cdot (K_2)^i,$$

in which addition and multiplication are to be computed in the finite field with 2^n elements. It can be proved that the probability of creating another valid message/MAC pair is upper bounded by $t/2^n$. A practical choice is $n = 64$, which results in a 128-bit key. For messages up to one megabyte, the success probability of a forgery is less than $1/2^{47}$. Note that K_2 can be reused; however, for every message a new key K_1 is required. This key could be generated from a short initial key using an additive stream cipher, but then the unconditional security is lost. However, it can be argued that it is easier to understand the security of this scheme than that of a computationally secure MAC algorithm. An even better way to use universal hash functions is to apply a pseudo-random function to its output concatenated with a value that occurs only once (e.g., a serial number or a large random number).

7.3.2 Digital Signatures

A digital signature is the electronic equivalent of a manual signature on a document. It provides a strong binding between the document and a person, and in case of a dispute, a third party can decide whether or not the signature is valid based on public information. Of course a digital signature will not bind a person and a document, but will bind a public key and a document. Additional measures are then required to bind the person to his or her key. Note that for a MAC algorithm, both Alice and Bob can compute the MAC result; hence, a third party cannot distinguish between them. Block ciphers and even one-way functions can be used to construct digital signatures, but the most elegant and efficient constructions for digital signatures rely on public key cryptography.

We now explain how the RSA algorithm can be used to create digital signatures with message recovery. The RSA mapping is a bijection—more specifically, a trapdoor one-way permutation. If Alice wants to sign some information P intended for Bob, she adds some redundancy to the information, resulting in $\tilde{P}$, and decrypts the resulting text with her secret key. This operation can only be carried out by Alice. On receipt of the signature, Bob encrypts it using Alice's public key and verifies that the information $\tilde{P}$ has the prescribed redundancy. If so, he accepts the signature on P as valid. Such a digital signature, which is a signature with message recovery, requires the following condition on the public key system: $E_{P_A}(D_{S_A}(\tilde{P})) = \tilde{P}$. Anyone who knows Alice's public key can verify the signature and recover the message from the signature.

Note that if the redundancy is left out, any person can pick a random ciphertext C^* and claim that Alice has signed $P^* = C^{*e} \bmod n$. It is not clear that P^* is a meaningful message, which will require some extra tricks, but it shows why redundancy is essential. A provably secure way to add the redundancy is PSS-R [52]; however, in practice other constructions are widely used, and most of them combine a hash function with a digital signature scheme (Figure 7.9).

If Alice wants to sign very long messages, digital signature schemes with message recovery result in signatures that are as long as the message. Moreover, signing with a public key system is a relatively slow operation. In order to solve these problems, Alice does not sign the information itself but rather the hash result of the information that is computed with an MDC (see also Figure 7.10). This approach corresponds to the use of an MDC to replace the authenticity of a large text by that of a short hash value (cf. Section 7.3.1). The signature now consists of a single block that is appended to the information. This type of signature scheme is sometimes called a digital signature with appendix. In order to verify such a signature, Bob

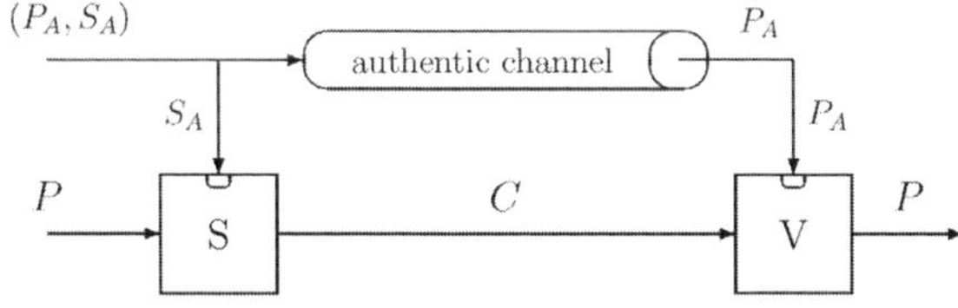

Figure 7.9 A digital signature scheme with message recovery based on a trapdoor one-way permutation; S and V denote the signing and verification operation, respectively.

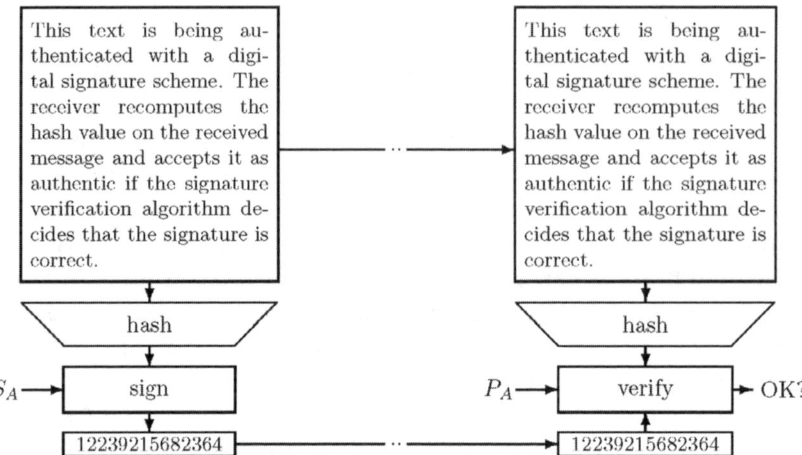

Figure 7.10 A digital signature scheme with appendix.

recomputes the MDC of the message and encrypts the signature with Alice's public key. If both operations give the same result, Bob accepts the signature as valid. MDCs used in this way need to be collision resistant: if Alice can find two different messages (P, P') with the same hash result, she can sign P and later claim to have signed P' (P and P' will have the same signature!).

Note that there exist other signature schemes with appendix, such as the DSA from FIPS 186 [53], which are not derived immediately from a public key encryption scheme. For these schemes it is possible to define a signing operation using the secret key and a verification operation using the public key without referring to decryption and encryption operations. The security of the DSA is based on the discrete logarithm problem similar to the Diffie-Hellman scheme. There also exists an elliptic curve–based variant of DSA called ECDSA. There are many more digital signature schemes than public key encryption schemes. Other digital signature schemes include ESIGN, Fiat-Shamir, Guillou-Quisquater, and Schnorr.

7.3.2.1 Certificates

Digital signatures move the problem of authenticating data to the problem of authenticating a link between an entity and its public key. This problem can be simplified using digital *certificates*. A certificate is a digital signature of a third party on an entity's name, its public key, and additional data, such algorithms and parameters, key usage, and begin and end of validity period. It corresponds roughly to an electronic identity. The verification of a certificate requires the public key of the trusted third party, and in this way the problem of authenticating data will be replaced by authenticating a single public key. In principle, authenticating this key is very easy: it can be published in a newspaper, listed on a webpage, or added in a browser. The infrastructure that provides public keys is called a public key infrastructure (PKI). Deploying such an infrastructure in practice is rather complex, since it is necessary to consider revocation of certificates, upgrades of keys, multiple trusted parties, and integration of the public key management with the application (see, for example [54]).

7.4 Analysis and Design of Cryptographic Algorithms

In this section we compare several approaches in cryptography. Next we describe the typical phases in the life of a cryptographic algorithm. Finally we discuss why, in spite of the progress in cryptology, weak cryptographic algorithms are still in use, and which problems cryptographers are facing in the next decades.

7.4.1 Different Approaches in Cryptography

Modern cryptology follows several approaches: the information theoretic approach, the complexity theoretic approach, the bounded-storage approach, the quantum cryptology approach, and the system-based approach. These approaches differ in the assumptions about the capabilities of an opponent, in the definition of a cryptanalytic success, and in the notion of security.

From the viewpoint of the cryptographer, the most desirable are unconditionally secure algorithms. This approach is also known as the *information theoretic* approach. It was developed in the seminal work of C. Shannon in 1943 and published a few years later [11]. This approach offers a perfect solution, since the security can be proved independent of the computational power of the opponent, and the security will not erode over time. However, few such schemes exist that are secure in this model; examples are the Vernam scheme (Section 7.2.1) and the polynomial MAC algorithm (Section 7.3.1.2). While they are computationally extremely efficient, the cost in terms of key material may be prohibitively large. For most applications, users have to live with schemes that offer only conditional security.

A second approach is to reduce the security of a cryptographic algorithm to that of other well-known difficult problems or to that of other cryptographic primitives. The *complexity theoretic* approach starts from an abstract model for computation and assumes that the opponent has limited computing power within this model [55]. The most common models used are Turing machines and RAM models. This approach was started in cryptology by Yao [13] in 1982. It has many positive sides and has certainly contributed toward moving cryptology from an art to a science:

- It forces the formulation of exact definitions and the clear statement of security properties and assumptions, which may seem trivial, but it has taken the cryptographic community a long time to define what secure encryption, secure message authentication, secure digital signatures, and secure authenticated encryption are. It turns out that there are many variants for these definitions, depending on the power of the opponent and the goals, and it takes a substantial effort to establish the relationship between them. For more complex primitives, such as e-payment, e-voting, interactions with more than two parties, and in general multiparty computation, establishing correct definitions is even more complex.
- The complexity theoretic approach results in formal reductions. It can be formally proven that if a particular object exists, another object exists as well. For example, one-way functions imply that digital signature schemes, or the

existence of pseudo-random permutations, which corresponds to a secure block cipher, sufficient to prove that the are CBC mode with secret and random initial values, is a "secure" mode for encryption, assuming that the opponent can choose plaintexts and observe the corresponding ciphertexts. The terms secure have very specific definitions as explained previously. These reductions often work by contradiction. For example, it can be shown that if an opponent can find a weakness in CBC encryption using a chosen plaintexs attack, the implication is that there is a weakness in the underlying block cipher, which allows a distinction of this block cipher from a pseudo-random permutation. Once the proofs are written down, any person can verify them, which is very important, as it turns out that some of these proofs have very subtle errors.

The complexity theoretic approach also has some limitations:

- Many cryptographic applications need building blocks, such as one-way functions, one-way permutations, collision-resistant compression functions, and pseudo-random functions, which cannot be reduced to other primitives. In terms of the existence of such primitives, complexity theory has only very weak results: it is not known whether one-way functions exist. In nonuniform complexity, which corresponds to Boolean circuits, the best result proved thus far is that there exist functions that are twice as hard to invert as to compute, which is far too weak to be of any use in cryptography [56]. To some extent cryptographers need to rely on number theoretic assumptions, such as the assumptions underlying the security of the Diffie-Hellman protocol and RSA. For the others, we rely on unproven assumptions on functions, such as RC4, DES, AES, RIPEMD-160, and SHA-1, which means that even when there are strong reductions, the foundations of the reductions are still weak.
- Sometimes the resulting scheme is much more expensive in terms of computation or memory than a scheme without security proof; however, in the last decade a substantial effort has been made to improve the efficiency of the constructions for which there exists a security proof .
- The security proof may not be very efficient. Some reductions are only asymptotic or are very weak. For example, for RSA-OAEP, if the security property is violated (i.e., a random ciphertext can be decrypted), the security assumption, in this case the computation of a modular eth root is infeasible, is only violated with a very small probability. Many reductions require the random oracle assumption, which states that the hash function used in the scheme behaves as a perfectly random function. It has been shown that this approach can result in security proofs for insecure schemes; however, these reductions still have some value as a heuristic in the sense that these schemes are typically better than schemes for which no reduction in the random oracle model is known.

The term concrete complexity has been coined by Bellare and Rogaway to denote security proofs with concrete reductions focusing on efficiency (i.e., without asymptotics and hidden constants). For an overview of complexity theoretic results in cryptology, the reader is referred to the work of Goldwasser and Bellare [57] and Goldreich [58].

There has also been some interest in the bounded-storage model, in which it is assumed that the storage capacity of an adversary is limited (see, for example, [59]). This model can be considered to be part of the information theoretic approach if one imposes that only limited information is available to the adversary or part of the complexity theoretic approach if space rather than time is considered to be limited.

Quantum cryptology does not work based on any computational assumptions; rather, it starts from the assumption that quantum physics provides a complete model of the physical world. It relies on concepts such as the Heisenberg uncertainty principle and quantum entanglement. Quantum cryptography provides a means for two parties to exchange a common secret key over an authenticated channel. The first experimental results were obtained in the early 1990s, and 10 years later several companies offer encryption products based on quantum cryptology. While this offers a fascinating approach, quantum cryptology needs authentic channels, which seems to make them not very useful to secure large open networks such as the Internet. Indeed, without public key cryptography, the only way to achieve an authentic channel is by establishing prior secrets manually; this seems rather impractical for a large network. Moreover, most implementations today have practical limitations. They are not yet compatible with packet-switched systems; they typically allow for low speeds only, although high-speed links have been demonstrated recently; and the communication distances are limited. For high-speed communication, one still needs to stretch the short keys with a classic stream cipher; of course, the resulting scheme then depends on standard cryptographic assumptions, which to some extent defeats the purpose of using quantum cryptology in the first place. Moreover, it can be expected that the current implementations are vulnerable to side-channel attacks as discussed later.

In view of the limitations of these approaches, modern cryptology still has to rely on the *system-based* or *practical* approach. This approach tries to produce practical solutions for basic building blocks, such as one-way functions, pseudo-random bit generators or stream ciphers, pseudo-random functions, and pseudo-random permutations. The security estimates are based on the best algorithm known to break the system and on realistic estimates of the necessary computing power or dedicated hardware to carry out the algorithm. By trial and error procedures, several *cryptanalytic principles* have emerged, and it is the goal of the designer to avoid attacks based on these principles. The second aspect is to design *building blocks with provable properties*, and to assemble such basic building blocks to design cryptographic primitives. The complexity theoretic approach has also improved our understanding of the requirements to be imposed on these building blocks. In this way, this approach is also evolving toward a more scientific approach. Nevertheless, it seems likely that for the next decades, cryptographers still need to rely the security of some concrete functions such as AES, SHA-512, and their successors.

7.4.2 Life Cycle of a Cryptographic Algorithm

This section discusses the life cycle of a cryptographic algorithm, evaluates the impact of open competitions, and compares the use of public versus secret algorithms.

A cryptographic algorithm usually starts with a new idea of a cryptographer, and a first step should be an evaluation of the security properties and of the cryptographic algorithm in which the cryptographer tries to determine whether or not the scheme is secure for the intended applications. If the scheme is unconditionally secure, he has to write the proofs and convince himself that the model is correct and matches the application. For computational security, it is again important to write down security proofs and check these for subtle flaws. Moreover, a cryptographer has to assess whether the assumptions behind the proofs are realistic. For the system-based approach, it is important to prove partial results and write down arguments that should convince others of the security of the algorithm. Often such cryptographic algorithms have security parameters, such as the number of steps and the size of the key. It is then important to give lower bounds for these parameters and to indicate the value of the parameters that corresponds to a certain security level.

The next step is the publication of the algorithm at a conference, in a journal, or in an Internet request for comment (RFC), which hopefully results in an *independent* evaluation of the algorithm. Often more or less subtle flaws are then discovered by other researchers. These flaws can vary form small errors in proofs to complete security breaks. Depending on the outcome, this evaluation can lead to a small fix of the scheme or to abandoning the idea altogether. Sometimes such weaknesses can be found in real time when the author is presenting his ideas at a conference, but often evaluating a cryptographic algorithm is a time-consuming task; for example, the design effort of the DES, has been more than seventeen man-years and the open academic evaluation since that time has taken a much larger effort. Cryptanalysis is quite destructive; in this respect, it differs from usual scientific activities, even when proponents of competing theories criticize each other.

Few algorithms survive the evaluation stage; ideally, this stage should last for several years. The survivors can be integrated into products and find their way to the market. Sometimes they are standardized by organizations such as the National Institute of Standards and Technology (NIST), IEEE, IETF, ISO, or ISO/IEC.

As will explained, even if no new security weaknesses are found, the security of a cryptographic algorithm degrades over time; if the algorithm is not parameterized, the moment will come when it has to be taken out of service or when its parameters need to be upgraded, and often such upgrades are not planned for.

7.4.3 Insecure Versus Secure Algorithms

Today secure cryptographic algorithms that offer good performance at an acceptable cost are publicly available. Nevertheless, one finds in many applications insecure cryptographic algorithms. This section explains some of the technical reasons for this problem. We will not focus on other explanations such as plain incompetence (a "do-it-yourself" attitude), political reasons (national security and law enforcement), and economic reasons (cost of upgrade is too high compared to economic benefits).

7.4.3.1 Brute Force Attacks Become Easier Over Time

Brute force attacks are attacks that exist against any cryptographic algorithm that is conditionally secure, no matter how it works internally. These attacks depend only on the size of the external parameters of the algorithm, such as the block length of a block cipher or the key length of any encryption or MAC algorithm. It is the task of the designer to choose these parameters such that brute force attacks are infeasible.

A typical brute force attack against an encryption or MAC algorithm is an exhaustive key search key search that is equivalent to breaking into a safe by trying all the combinations of the lock. The lock should be designed such that a brute force attack is not feasible in a reasonable amount of time. This attack requires only a few known plaintext/ciphertext or plaintext/MAC pairs, which can always be obtained in practice. It can be precluded by increasing the key length. Note that adding one bit to the key doubles the time for exhaustive key search. It should also be guaranteed that the key is selected uniformly at random in the key space.

On a standard PC, trying a single key for a typical algorithm requires between 0.1 and 10 μs depending on the complexity of the algorithm. For example, a 40-bit key can be recovered in 1–100 days. If a local area network (LAN) with 100 machines can be used, the key can be recovered in 20 minutes to a day. For a 56-bit key, such as DES, a key search requires a few months if several thousand machines are available, as has been demonstrated in the first half of 1997. However, if dedicated hardware is used, a different picture emerges. M. Wiener has designed in 1993 a US$1 million hardware DES key search machine that can recover a 56-bit DES key in about 3 hours [60]. If such a machine were built in 2007, it would be about 400 times faster, and thus recovering a 56-bit key would take about half a minute on average. In 1998, a US$250,000 machine called Deep Crack was built that finds a 56-bit DES key in about 4 days [61]; the design, which required 50 percent of the cost, has been made available for free.

These numbers makes a person wonder how the U.S. government could publish in 1977 a block cipher with a 56-bit key. However, it should be taken into account that a variant of Moore's law formulated in 1967 [62] states that computers will double their speed every 18 months, which implies that in 1977, recovering a 56-bit key with a US$10 million hardware machine would take about 20 days; such a machine was clearly only feasible for very large organizations, including the U.S. government. This discussion also explains the initial controversy over the DES key length.

Experts believe that Moore's law will be holding for at least another 15 years, which means that if data needs to be protected for 15 years, against an opponent with a budget of US$10 million, a key length of at least 90 bits needed, which corresponds to the security level of three key triple-DES (Section 7.2.1.3). However, as the cost of increasing the key size is quite low, it is advisable to design new algorithms with variable key size from 128 to 256 bits. Indeed, searching for the smallest AES key (128 bits) is a factor 2^{72} times more expensive than finding a DES key. Even with a key search machine of US$$10^{12}$, it would take in 2024 one million years to recover a 128-bit key. This calculation shows that if symmetric algorithms with 128-bit keys are used, it may no longer necessary to worry about exhaustive key search for the next decades.

7.4.3.2 Shortcut Attacks Become More Efficient

Many algorithms are less secure than suggested by the size of their external parameters. It is often possible to find more effective attacks than trying all keys (see, for example, the discussion on two key and three-key triple-DES in Section 7.2.1.3).

Assessing the strength of an algorithm requires cryptanalytic skills and experience as well as hard work. During the last 15 years, powerful new tools have been developed, including differential cryptanalysis [63], which analyzes the propagation of differences through cryptographic algorithms; linear cryptanalysis [64], which is based on the propagation of bit correlations; fast correlation attacks on stream ciphers [17]; and algebraic attacks on stream ciphers [16]. For example, the FEAL block cipher with eight rounds was published in 1987 and can now be broken with only 10 chosen plaintexts. Similar progress has been made in the area of public key cryptology, for example, with attacks based on lattice reduction [65] and factoring methods discussed in Section 7.2.2.2.

7.4.3.3 New Attack Models

The largest threats however originate from new attack models. One of these models is a quantum computer. Another new threat is the exploitation of side-channel attacks and attacks based on fault injection.

Feynman realized in 1982 that computers based on the principles of quantum physics would be able to achieve exponential parallelism: having n components in the quantum computer would allow it to perform 2^n computations in parallel. Note that in a classic computer, having n computers can speed up the computation with a factor up to n. A few years later, Deutsch realized that a general-purpose quantum computer could be developed on which any physical process could be modeled, at least in principle. However, the first interesting application of quantum computers outside physics was proposed by Shor in 1994 [66], who showed that quantum computers were perfectly suitable to number theoretic problems, such as factoring and discrete logarithm problems. This result implies that if a large quantum computer could be built, the most popular public key systems would be completely insecure. Building quantum computers, however, is a huge challenge. A quantum computer maintains state in a set of qubits. A qubit can hold a one, or a zero, or a superposition of one and zero; this superposition is the essence of the exponential parallelism, but it can be disturbed by outside influences, called decoherence. The first quantum computer with 2 qubits was built in 1998. Currently the record is a quantum computer with 7 bits, 3 of these used for quantum error correction. This allowed the factorization of the integer 15 in 2002 [67]. Experts are divided on the question whether sufficiently powerful quantum computers can be built in the next 15 to 20 years, but no one seems to expect a breakthrough in the next 5 to 10 years. For symmetric cryptography, quantum computers are less of a threat, since they can reduce the time to search a $2n$-bit key to the time to search an n-bit key, using Grover's algorithm [68]. Hence doubling the key length (from 128 to 256 bits) offers an adequate protection. Collision search on a quantum computer reduces from $2^{n/2}$ steps to $2^{n/3}$ steps [68], so it is sufficient to increase the number of bits of a hash result from 256 to 384.

Cryptographic algorithms have to be implemented in hardware or software, but it should not be overlooked that software runs on hardware. The opponent can try to make life easier by obtaining information from the hardware implementation, rather than trying to break the cryptographic algorithm using fast computers or clever mathematics. Side-channel attacks have been known for a long time in the classified community; these attacks exploit information on the time to perform a computation [69], on the power consumption [70], or on the electromagnetic radiation [71,72] to extract information on the plaintext or even the secrets used in the computation. A very simple side-channel attack on the Vernam scheme (see Section 7.2.1) exploits the fact that if there is a logical zero on the line, this can be the result of $0 \oplus 0$ or $1 \oplus 1$. If the device implementing the scheme is not properly designed, the two cases may result in different electrical signals, which immediately leaks information on half of the plaintext bits.

Protecting implementations against side-channel attacks is notoriously difficult and requires a combination of countermeasures at the hardware level—such as adding noise, special logic, decoupling power source—at the algorithmic level—for example, blinding and randomization—and at the protocol level—for example, frequent key updates (see [73]). While many countermeasures have been published, many fielded systems are still vulnerable. This vulnerability is in part due to cost reasons and delays in upgrades and also due to the development of ever more sophisticated attacks. Developing efficient implementations that offer a high security level against side-channel attacks is an important research challenge for the coming years.

The most powerful attacks induce errors in the computations (e.g., by varying clock frequency or power level or by applying light flashes). Such attacks can be devastating because small changes in inputs during a cryptographic calculation typically reveal the secret key material [74]. Protecting against these attacks is nontrivial since it requires continuous verification of all calculations, which should also include a check on the verifications—and even this countermeasure may not be sufficient as has been pointed out in [75]. It is also clear that pure cryptographic measures will never be sufficient to protect against such attacks.

7.5 Conclusions

This chapter covers only part of the issues in modern cryptology since the discussion is restricted to an overview of cryptographic algorithms. Other problems solved in cryptography include identification, timestamping, sharing of secrets, and electronic cash. Many interesting problems are studied under the concept of secure multiparty computation, such as electronic elections and the generation and verification of digital signatures in a distributed way. An important aspect is the underlying key management infrastructure, which ensures that secret and public keys can be established and maintained throughout the system in a secure way—which is where cryptography meets the constraints of the real world.

This chapter has demonstrated that while cryptology has made significant progress as a science, there are some interesting research challenges ahead both from a theoretic and a practical viewpoint. Progress needs to be made in theoretical foundations, such as how to prove that a problem is hard, the development of new cryptographic algorithms (such as new public key cryptosystems that do not depend on algebraic number theory and light-weight or low-footprint secret-key cryptosystems), and secure and efficient implementations of cryptographic algorithms. Hopefully this will allow the cryptographic community to build schemes that offer long-term security at a reasonable cost.

References

[1] Menezes, A. J., P. C. van Oorschot, and S. Vanstone, *Handbook of Applied Cryptography*, CRC Press, 1997.

[2] Kahn, D., *The Codebreakers: The Story of Secret Writing*, New York: MacMillan, 1967.

[3] Singh, S., *The Code Book: The Science of Secrecy from Ancient Egypt to Quantum Cryptography*, Boca Raton, FL: Anchor, 2000.

[4] Stallings, W., *Cryptography and Network Security*, 3rd edition, New York: Prentice Hall, 2003.

[5] Blaze, M., "Rights Amplification in Master-Keyed Mechanical Locks," *IEEE Security & Privacy*, Vol. 1, No. 2, 2003, pp. 24–32.

[6] Koblitz, N., *A Course in Number Theory and Cryptography*, Berlin: Springer-Verlag, 1987.

[7] *State of the Art in Applied Cryptography, Lecture Notes in Computer Science 1528*, B. Preneel and V. Rijmen (eds.), Berlin: Springer-Verlag, 1998.

[8] *Contemporary Cryptology: The Science of Information Integrity*, G. J. Simmons (ed.), New York: IEEE Press, 1991.

[9] Stinson, D., *Cryptography: Theory and Practice*, 3rd edition, Boca Raton, FL: CRC Press, 2005.

[10] Vernam, G. S., "Cipher Printing Telegraph System for Secret Wire and Radio Telegraph Communications," *Journal American Institute of Electrical Engineers*, Vol. XLV, 1926, pp. 109–115.

[11] Shannon, C. E., "Communication Theory of Secrecy Systems," *Bell System Technical Journal*, Vol. 28, No. 4, 1949, pp. 656–715.

[12] Haynes, J. E., and H. Klehr, *Venona: Decoding Soviet Espionage in America*, New Haven: Yale University Press, 1999.

[13] Yao, A.C., "Theory and Applications of Trapdoor Functions." In *Proc. 23rd IEEE Symposium on Foundations of Computer Science*, Washington DC: IEEE Computer Society, 1982, pp. 80–91.

[14] Rueppel, R. A., *Analysis and Design of Stream Ciphers*, Berlin: Springer-Verlag, 1986.

[15] Biryukov, A., A. Shamir, and D. Wagner, "Real Time Cryptanalysis of A5/1 on a PC." In *Fast Software Encryption, Lecture Notes in Computer Science 1978*, B. Schneier (ed.), Berlin: Springer-Verlag, 2002, pp. 1–18.

[16] Courtois, N., and W. Meier, "Algebraic Attacks on Stream Ciphers with Linear Feedback." In *Advances in Cryptology, Proc. Eurocrypt'03, Lecture Notes in Computer Science 2656*, E. Biham (ed.). pp. 345–359, Berlin: Springer-Verlag, 2003.

[17] Meier, W., and O. Staffelbach, "Fast Correlation Attacks on Stream Ciphers," *Journal of Cryptology*, Vol. 1, 1989, pp. 159–176.

[18] Fluhrer, S., I. Mantin, and A. Shamir, "Weakness in the Key Scheduling Algorithm of RC4." In *Selected Areas in Cryptography, Lecture Notes in Computer Science 2259*, S. Vaudenay and A. Youssef (eds.), Berlin: Springer-Verlag, 2001, pp. 1–24.

[19] Bellare, M., et al., "A Concrete Security Treatment of Symmetric Encryption." In *Proc. 3th Annual Symposium on Foundations of Computer Science, FOCS'97*, Washington DC: IEEE Computer Society, 1997, pp. 394–403.

[20] FIPS 81, *DES Modes of Operation*, Federal Information Processing Standard, NBS, US Dept. of commerce, December 1980.

[21] NIST, *SP 800-38A Recommendation for Block Cipher Modes of Operation—Methods and Techniques*, December 2001.

[22] ISO/IEC 10116, *Information Technology—Security Techniques—Modes of Operation of an* n-*Bit Block Cipher Algorithm*, 1997.

[23] FIPS 46, *Data Encryption Standard*, Federal Information Processing Standard, NBS, U.S. Dept. of Commerce, January 1977 (revised as FIPS 46-1, 1988; FIPS 46-2, 1993, FIPS 46-3, 1999).

[24] Knudsen, L. R. "Block Ciphers: Analysis, Design and Applications," PhD thesis, Aarhus University, Denmark, 1994.

[25] FIPS 197, *Advanced Encryption Standard*, Federal Information Processing Standard, NIST, U.S. Dept. of Commerce, November 26, 2001.

[26] Daemen, J., and V. Rijmen, *The Design of Rijndael, AES, The Advanced Encryption Standard*, Berlin: Springer-Verlag, 2001.

[27] Diffe, W., and M. E. Hellman, "New Directions in Cryptography," *IEEE Transactions on Information Theory*, Vol. IT-22, No. 6, 1976, pp. 644–654.

[28] Merkle, R., *Secrecy, Authentication, and Public Key Systems*, UMI Research Press, 1979.

[29] Rivest, R. L., A. Shamir, and L Adleman. "A Method for Obtaining Digital Signatures and Public-Key Cryptosystems," *Communications ACM*, Vol. 21, No. 2, 1978, pp. 120–126.

[30] Bellare, M., and P. Rogaway, "Random Oracles Are Practical: A Paradigm for Designing Efficient Protocols." In *Proc. ACM Conference on Computer and Communications Security*, New York: ACM Press, 1993, pp. 62–73.

[31] Shoup, V., "OAEP Reconsidered." In *Advances in Cryptology, Proc. Crypto'01, Lecture Notes in Computer Science 2139*, J. Kilian (ed.), Berlin: Springer-Verlag, 2001, pp. 239–259.

[32] Fujisaki, E., et al., "RSA-OAEP Is Secure Under the RSA Assumption." In *Advances in Cryptology, Proc. Crypto'01, Lecture Notes in Computer Science 2139*, J. Kilian (ed.), Berlin: Springer-Verlag, 2001, pp. 260–274.

[33] NESSIE, http://www.cryptonessie.org.

[34] Lenstra, A. K., and E. R. Verheul, "Selecting Cryptographic Key Sizes," *Journal of Cryptology*, Vol. 14, 2001, pp. 255–293.

[35] Shamir, A., and E. Tromer, "Factoring Large Numbers with the TWIRL Device." In *Advances in Cryptology, Proc. Crypto'03, Lecture Notes in Computer Science 2729*, D. Boneh (ed.), Berlin: Springer-Verlag, 2003, pp. 1–26.

[36] Avanzi, R. M., et al., *Handbook of Elliptic and Hyperelliptic Curve Cryptography*, H. Cohen and G. Frey (eds.), Boca Raton, FL: Chapman & Hall/CRC, 2005.

[37] Blake, F., I. G. Seroussi, and N. P. Smart, *Elliptic Curves in Cryptography*, Cambridge: Cambridge University Press, 1999.

[38] Hankerson, D., A. Menezes, and S. Vanstone, *Guide to Elliptic Curve Cryptography*, Berlin: Springer-Verlag, 2004.

[39] Wang X., and H. Yu, "How to Break MD5 and Other Hash Functions." In *Advances in Cryptology, Proc. Eurocrypt'05, Lecture Notes in Computer Science 3494*, R. Cramer (ed.), Berlin: Springer-Verlag, 2005, pp. 19–35.

[40] Wang, X., Y. L. Lin, and H. Yu, "Finding Collisions in the Ful SHA-1." In *Advances in Cryptology, Proc. Crypto'05, Lecture Notes in Computer Science 3621*, V. Shoup (ed.), Berlin: Springer Verlag, 2005, pp. 17–36.

[41] FIPS 180, *Secure Hash Standard*, Federal Information Processing Standard (FIPS), Publication 180, National Institute of Standards and Technology, US Department of Commerce, Washington D.C., May 11, 1993 (revised as FIPS 180-1, 1995; FIPS 180-2, 2003).

[42] ISO/IEC 10118, *Information Technology—Security Techniques—Hash-Functions Part 1: General, 2000. Part 2: Hash-Functions Using an* n*-Bit Block Cipher Algorithm, 2000. Part 3: Dedicated Hash-Functions, 2003. Part 4: Hash-Functions Using Modular Arithmetic*, 1998.

[43] Knudsen, L. R. "Chosen-Text Attack on CBC-MAC," *Electronics Letters*, Vol. 33, No. 1, 1997, pp. 48–49.

[44] Petrank E., and C. Rackoff, "CBC MAC for Real-Time Data Sources," *Journal of Cryptology*, Vol. 13, 2000, pp. 315–338.

[45] Preneel, B., and P. van Oorschot, "MDx-MAC and Building Fast MACs from Hash Functions." In *Proc. Crypto'95, Lecture Notes in Computer Science 963*, D. Coppersmith (ed.), Berlin: Springer-Verlag, 1995, pp. 1–14.

[46] ISO/IEC 9797, *Information Technology—Security Techniques—Message Authentication Codes (MACs). Part 1: Mechanisms Using a Block Cipher, 1999. Part 2: Mechanisms Using a Dedicated Hash-Function*, 2002.

[47] Carter, J. L., and M. N. Wegman, "Universal Classes of Hash Functions," *Journal of Computer and System Sciences*, Vol. 18, 1979, pp. 143–154.

[48] Wegman, M. N., and J. L. Carter, "New Hash Functions and Their Use in Authentication and Set Equality," *Journal of Computer and System Sciences*, Vol. 22, No. 3, 1981, pp. 265–279.

[49] Black, J., et al., "UMAC: Fast and Secure Message Authentication." In *Advances in Cryptology, Proc. Crypto'99, Lecture Notes in Computer Science 1666*, M. Wiener (ed.), Berlin: Springer-Verlag, 1999, pp. 216–233.

[50] Bernstein, D. J. "The Poly1305–AES Message-Authentication Code." In *Fast Software Encryption, Lecture Notes in Computer Science 3557*, H. Gilbert and H. Handschuh (eds.), Berlin: Springer-Verlag, 2005, pp. 32–49.

[51] Kabatianskii, G. A., T. Johansson, and B. Smeets, "On the Cardinality of Systematic A-Codes via Error Correcting Codes," *IEEE Transactions on Information Theory*, Vol. IT-42, No. 2, 1996, pp. 566–578.

[52] Bellare, M., and P. Rogaway, "The Exact Security of Digital Signatures: How to Sign with RSA and Rabin." In *Advances in Cryptology, Proc. Eurocrypt'96, Lecture Notes in Computer Science 1070*, U. Maurer (ed.), Berlin: Springer-Verlag, 1996, pp. 399–414.

[53] FIPS 186, *Digital Signature Standard*, Federal Information Processing Standard, NIST, US Dept. of Commerce, May 1994 (revised as FIPS 186-1, 1998; FIPS 186-2, 2000; change notice published in 2001).

[54] Adams, C., and S. Lloyd, *Understanding PKI: Concepts, Standards, and Deployment Considerations*, 2nd edition, Addison-Wesley, 2003.

[55] Garey, M. R., and D. S. Johnson, *Computers and Intractability: A Guide to the Theory of NP-Completeness*, W. H. Freeman and Company, 1979.

[56] Hiltgen, A. P. L., "Construction of Feebly-One-Way Families of Permutations." In *Proc. Auscrypt'92, Lecture Notes in Computer Science 718*, J. Seberry and Y. Zheng (eds.), Berlin: Springer-Verlag, 1993, pp. 422–434.

[57] Goldwasser, S., and M. Bellare, *Lecture Notes on Cryptography*, http://www.cs.ucsd.edu/users/mihir/papers/gb.html.

[58] Goldreich, O., *Foundations of Cryptography: Volume 1, Basic Tools*, Cambridge: Cambridge University Press, 2001.

[59] Dziembowski, S., and U. Maurer, "Optimal Randomizer Efficiency in the Bounded-Storage Model," *Journal of Cryptology*, Vol. 17, 2004, pp. 5–26.

[60] Wiener, M. J., "Efficient DES Key Search," *Presented at the Rump Session of Crypto'93.* Reprinted in *Practical Cryptography for Data Internetworks*, W. Stallings (ed.), Washington DC: IEEE Computer Society, 1996, pp. 31–79.

[61] EFF, *Cracking DES: Secrets of Encryption Research, Wiretap Politics & Chip Design*, O'Reilly, 1998.

[62] Schaller, R. R., "Moore's Law: Past, Present, and Future," *IEEE Spectrum*, Vol. 34, No. 6, 1997, pp. 53–59.

[63] Biham, E., and A. Shamir, *Differential Cryptanalysis of the Data Encryption Standard*, Berlin: Springer-Verlag, 1993.

[64] Matsui, M., "The First Experimental Cryptanalysis of the Data Encryption Standard." In *Advance in Cryptology, Proc. Crypto'94, Lecture Notes in Computer Science 839*, Y. Desmedt (ed.), Berlin: Springer-Verlag, 1994, pp. 1–11.

[65] Joux, A., and J. Stern, "Lattice Reduction: A Toolbox for the Cryptanalyst," *Journal of Cryptology*, Vol. 11, 1998, pp. 161–185.

[66] Shor, P. W., "Algorithms for Quantum Computation: Discrete Logarithms and Factoring." In *Proc. 35th Annual Symposium on Foundations of Computer Science*, S. Goldwasser (ed.), Washington DC: IEEE Computer Society, 1994, pp. 124–134.

[67] Vandersypen, L. M. K., et al., "Experimental Realization of Shor's Quantum Factoring Algorithm Using Nuclear Magnetic Resonance," *Nature*, Vol. 414, 2001, pp. 883–887.

[68] Grover, L. K., "A Fast Quantum Mechanical Algorithm for Database Search." In *Proc. 28th Annual ACM Symposium on Theory of Computing*, New York: ACM Press, 1996, pp. 212–219.

[69] Kocher, P., "Timing Attacks on Implementations of Diffie-Hellman, RSA, DSS, and Other Systems." In *Advances in Cryptology, Proc. Crypto'96, Lecture Notes in Computer Science 1109*, N. Koblitz (ed.), Berlin: Springer-Verlag, 1996, pp. 104–113.

[70] Kocher, P., J. Jaffe, and B. Jun, "Differential Power Analysis," In *Advances in Cryptology, Proc. Crypto'99, Lecture Notes in Computer Science 1666*, M. Wiener (ed.), Berlin: Springer-Verlag, 1999, pp. 388–397.

[71] Gandolfi, K., C. Mourtel, and F. Olivier, "Electromagnetic Analysis: Concrete Results." In *Proc. Cryptographic Hardware and Embedded Systems—CHES 2001, Lecture Notes in Computer Science 2162*, C. K. Koç, D. Naccache, and C. Paar (eds.), Berlin: Springer-Verlag, 2001, pp. 251–261.

[72] Quisquater, J.-J., and D. Samide, "ElectroMagnetic Analysis (EMA): Measures and Countermeasurs for Smart Cards." In *Smart Card Programming and Security, International Conference on Research in Smart Cards, E-smart 2001, Lecture Notes in Computer Science 2140*, I. Attali and T. Jensen (eds.), Berlin: Springer-Verlag, 2001, pp. 200–210.

[73] Borst, J., B. Preneel, and V. Rijmen, "Cryptography on Smart Cards," *Journal of Computer Networks*, Vol. 36, 2001, pp. 423–435.

[74] Boneh, D., R. A. DeMillo, and R. J. Lipton, "On the Importance of Eliminating Errors in Cryptographic Computations," *Journal of Cryptology*, Vol. 14, 2001, pp. 101–119.

[75] Yen, S.-M., and M. Joye, "Checking Before Output May Not Be Enough Against Fault Based Cryptanalysis," *IEEE Transactions on Computers*, Vol. 49, No. 9, 2000, pp. 967–970.

CHAPTER 8

Network Security

Sokratis Katsikas and Natalia Miloslavskaya

Many periods in the history of mankind have been named after significant technological discoveries that occurred within them or after the dominant technology of the time. The stone age, the bronze age, and the iron age constitute early examples of such periods that were named after the dominant material for building tools and the respective technology. More recently, one may distinguish the steam age, the electricity age, and the silicon age. Our times might, in the future, be named the "information age." If this happens, the name will be due to the realization of the Information Society, which affects all human activities and influences the way in which we live, work, entertain ourselves, engage in entrepreneurial activities, perform transactions, care for our health, communicate with each other, and so on. In this environment, networks have permeated almost all aspects of our everyday life, changing it, improving it, and making it easier, while at the same time increasing its dependence on information and communication technologies (ICT). This is the main reason why the security of networks is the most frequently cited reason that influences the further expansion of e-services. Indeed, the more our society depends on ICT, the more significance will be attributed to securing these technologies. In this scene, where almost all organizations base their operation on processing information, new dependencies, new vulnerabilities, and new risks appear.

8.1 Network Security Architectures

The term *computer network* refers to an interconnected collection of autonomous computers. Two computer systems are interconnected if they can exchange data, whereas they are autonomous if no master/slave relation between them is established. A computer network is often confused with a *distributed system*, which is composed of a collection of processes, spatially apart, that communicate through the exchange of messages. In addition, the time delay in transmitting a message in a distributed system is not negligible compared to the time lapse between events in a simple process. The fundamental difference between a computer network and a distributed system is in the degree of transparency when a user exploits multiple autonomous computer systems that comprise a distributed system to perform a task: a distributed system would exhibit a far higher degree of transparency than that of a computer network. As transparency mainly refers to software rather than to hardware, a distributed system can be considered a special case of a computer network, if the underlying software exhibits high cohesion and transparency.

Computer networks constitute a complicated collection of cooperating hardware and software. In order to facilitate their understanding, the scientific community has developed common means to model them. A *reference model* is meant to be one that explains the way in which the individual components of a system cooperate and contains descriptions of the interfaces among these components. A common practice when designing reference models is to break down the overall functionality of the system into layers, so that its complexity is reduced.

8.1.1 ISO/OSI Network Security Architecture

The International Standards Organization (ISO) has created the Open Systems Interconnection (OSI) standard reference model that defines a layered architecture for computer networks. This model, defined in the ISO 7498 standard [1] and adopted in the X.800 International Telecommunication Union (ITU) Recommendation comprises seven layers—namely, from bottom to top, the physical layer, the data link layer, the network layer, the transport layer, the session layer, the presentation layer, and the application layer.

Every layer (N) uses the services offered by the layer immediately below (N-1) and enhances them, to produce a more integrated service to the layer immediately above (N+1). Naturally, as every layer uses the services of the layer immediately below, it also uses the services of all layers below. Layer N is implemented by two or more layer N communicating *entities* that exchange data and commands via a layer N *protocol*. The entities activate the services of layer N-1 to exchange layer N *protocol data units* (PDUs) and subsequently offer the integrated service to layer N+1 through internal interfaces.

Each layer may offer two kinds of services: *connection-oriented services* and *connectionless services*. The former require the establishment of a session between the two communicating entities, its maintenance throughout the communication, and its termination on the end of the communication. The latter do not require establishing a session. In this case, every message exchanged between the communicating entities is transmitted autonomously and independently of other messages, and the message transmission sequence is not necessarily maintained.

8.1.2 ISO/OSI Network Security Services

ISO 7498-2 [2] defines a set of services that the network must support so that its security is preserved. These are categorized in the following classes: *authentication services, access control services, confidentiality services, integrity services, and non-repudiation services.*

Authentication services are further categorized in two subclasses. *Data origin authentication services* aim to allow data transmitters to verify their identity and are usually provided during the data transmission phase. *Peer entity authentication services* aim to allow entities to verify that a peer entity participating in a session is the one that it claims to be and are usually provided during the session establishment phase, even though their use during data transmission is possible.

Access control services aim to protect resources, files, data, and applications in the network from unauthorized access. These are the first ones that come to mind when we refer to computer or network security, and they are strongly related to user identification and authentication.

Confidentiality services aim at protecting data transmitted over a network from disclosure to unauthorized persons, processes, and entities. Four such services are defined in the standard: *connection-oriented confidentiality service*, *connectionless confidentiality service*, *selected field confidentiality service*, and *traffic confidentiality service*. The first three pertain to protecting the confidentiality of data transmitted within an established session or of a data unit in the absence of an established session or of a selected field within a data unit, either within an established session or not. The latter aims at ensuring that deducing information by monitoring the network traffic is not possible.

Integrity services aim to protect data transmitted over a network from unauthorized modification. Five such services are defined in the standard: *connection-oriented integrity service with recovery*, *connection-oriented integrity service without recovery*, *connection oriented selected field integrity service*, *connectionless integrity service*, and *connectionless selected field integrity service*. The first two differ in that the first allows both the detection of unauthorized data modification and subsequent attempts to recover, whereas the second only allows the detection of the event.

Finally, nonrepudiation services aim to ensure that an entity that sent a message or undertook some action cannot deny that it did so. They also aim to ensure that an entity that received a message cannot deny that it did so. Two such services are defined in the standard: *nonrepudiation service with proof of origin and nonrepudiation service with proof of delivery*. The former aims to deliver to the recipient proof of the message origin, whereas the latter aims to deliver to the sender proof of the message delivery.

ISO 7498-2 also recommends the placement of the security services in the network layers, as shown in Table 8.1. As it can easily be seen, the mapping of services to layers is not one to one. This means that the placement of network security services is a design decision. To come to this decision, several factors must be considered. These are related to the degree of flexibility that the security policy allows to the users to protect their data, to the degree of the divergence of the security characteristics that lower layer network elements have, to the number of network points that we need to protect, to the existence of a requirement to protect message headers as well as content, and so forth.

The security services discussed earlier are implemented through a set of *security mechanisms*. These are called *specific security mechanisms* and are cryptography, digital signatures, access control, data integrity, authentication exchange, traffic padding, routing control, and notarization. The mapping of mechanisms to services is depicted in Table 8.2.

In addition to the specific security mechanisms, the standard also defines a set of *pervasive security mechanisms* that are meant not to implement any specific service, but rather to provide overall security functionality. These mechanisms are *trusted functionality*, *security labels*, *event detection*, *security audit trail, and security recovery*.

Table 8.1 Security Services per OSI Layer

Layer	*Service*
7 Application	Authentication Access control Data integrity Data confidentiality Nonrepudiation
6 Presentation	Data confidentiality
5 Session	–
4 Transport	Authentication Access control Data integrity Data confidentiality
3 Network	Authentication Access control Data integrity Data confidentiality
2 Data Link	Authentication Access control Data integrity Data confidentiality
1 Physical	Data confidentiality

Table 8.2 Implementing Services with Mechanisms

	Mechanism							
Service	*CR*	*DS*	*AC*	*DI*	*AE*	*TP*	*RC*	*N*
Peer entity authentication	✓	✓			✓			
Data origin authentication	✓	✓						
Access control			✓					
Connection confidentiality	✓						✓	
Connectionless confidentiality	✓						✓	
Selective field confidentiality	✓							
Traffic confidentiality	✓					✓	✓	
Connection integrity with recovery	✓			✓				
Connection integrity without recovery	✓			✓				
Connection oriented selective field integrity	✓			✓				
Connectionless integrity	✓	✓		✓				
Connectionless selective field integrity	✓	✓		✓				
Nonrepudiation of origin		✓		✓				✓
Nonrepudiation of receipt		✓		✓				✓

8.1.3 Internet Security Architecture

The ISO/OSI reference model does not exactly map to the architecture of the Internet. This is mainly due to the fact that the conception and initial development of the Internet preceded that of the reference model. The differences are that, in the Internet model, the reference model layers 7, 6, and 5 collapse to one, called the *application layer*, and the reference model layers 2 and 1 again collapse to one, called the *network layer*. Furthermore, the reference model network layer is called *Internet layer* in the Internet model. Thus, the Internet architecture uses four layers—

namely, from bottom to top, the network layer, the Internet layer, the transport layer, and the application layer.

The network layer uses communication protocols that pertain to either local area network technologies or point-to-point connection technologies. Examples of typical protocols for local networks are the token bus, token ring, and, the most widely used one, the Ethernet protocol. On the other hand, the most popular protocol used for point-to-point connections is the PPP (point-to-point protocol) [3], commonly used to connect to the Internet through the public switched telephone network (PSTN). We will be looking at security at the network layer in Section 8.2.

The dominant protocol of the Internet layer is the Internet protocol (IP), which is used to interconnect hosts belonging to the same or different networks. It is a connectionless protocol, in the sense that no specific communication path is established between hosts, but rather each individual packet may follow a different route before it reaches its destination. This further means that packets may be lost, retransmitted, or received in an order different than the one in which they were transmitted. IP does not assume responsibility for handling such errors; this lies with higher layer protocols. It is, however, the responsibility of IP to address and route packets on one hand and to define the ways in which network entities handle packets. The current version of IP is IP version 6—IPv6 [4]. However, most installations continue to use the earlier version, IP version 4—IPv4 [5]. We will look at the security of the IP in Section 8.3. The Internet layer also uses *routing protocols*, whose role is to allow routing decisions to be taken. Examples of such protocols are the routing Internet protocol (RIP) [6], the open shortest path first (OSPF) protocol [7], the interior gateway routing protocol (IGRP) [8] and the border gateway protocol (BGP) [9]. Finally, the Internet layer uses support protocols that carry out functions supporting IP, such as error control, routing modifications, and address translation. Such protocols are the Internet control message protocol (ICMP) [10], the address resolution protocol (ARP) [11] and the reverse address resolution protocol (RARP) [12].

The transport layer uses two protocols: the transport control protocol (TCP) [13] and the user datagram protocol (UDP) [14]. The former is the most widely used transport layer protocol. TCP provides connection-oriented transport services that are bidirectional and reliable, ensuring that packets will be delivered in the correct order, double packets will be deleted, and lost packets will be retransmitted. UDP is similar to TCP, but intentionally unreliable, as delivery is based on the best attempt principle. UDP does not support error correction, retransmission or detection of lost packets, or packets out of sequence, whereas error detection is possible but not mandatory. We will look at transport layer security issues in Section 8.4.

Finally, a large variety of application protocols and services run over TCP and UDP at the application layer. Examples, of such well-known and widely used services are email service, the World Wide Web, the file transfer service, and the network management service. Such services are implemented through the simple mail transfer protocol (SMTP) [15], the hyper text transfer protocol (HTTP) [16], the file transfer protocol (FTP) [17], and the simple network management protocol (SNMP) [18]. We will look into some areas of application layer security in Section 8.5.

Internet standardization work is mainly carried out by the Internet Engineering Task Force (IETF). Research-oriented issues are taken up by the Internet Research Task Force (IRTF).

8.2 Security at the Network Layer

Imagine someone trying to connect to her company intranet when physically being in another country. There are at least two ways to do this: An obvious solution would be to use the PSTN or the integrated services digital network (ISDN) to directly connect (perhaps via a modem) to the company intranet's remote access server (RAS) and, through the PPP, to the destination server. This is a generally available and simple solution, even though its security is low, as all traffic between the client and the server is unprotected, and its cost is high, as the user is charged with the long-distance phone call costs. An alternative would be to use a virtual private network (VPN) channel or tunnel. This can be done by encapsulating a given network layer protocol (e.g., IP, IPX, AppleTalk) within PPP-encrypting PPP frames and encapsulating data using a tunneling protocol that typically is IP but could also be ATM or frame relay; this approach is known as layer 2 tunneling protocol (L2TP) [19]. Even though a VPN can also be established using a layer 3 protocol, L2TP is currently considered the best choice.

Coming back to our problem of connecting someone remotely to the company intranet through a VPN, we identify the following elements of the overall solution: a remote system or a dial-up client is a computer or router that can be either the founder or the recipient of layer 2 tunnel. An L2TP access concentrator (LAC) is the node acting as the one end of the tunnel (on the client's side) and an L2TP network server (LNS) is LAC's peer to the other end of the tunnel (server's side). A network access server (NAS) can serve as both an LAC and an LNS. Three layer 2 tunneling protocols are discussed in the next paragraphs.

8.2.1 Layer 2 Forwarding Protocol (L2F)

This was the first layer 2 forwarding protocol and was proposed by Cisco Systems. It focused mainly on two areas: how to encapsulate layer 2 frames (e.g., PPP frames) within L2F and how to manage the connection for establishing and terminating the layer 2 tunnel. The protocol, as described in RFC 2341 [20], uses the UDP port 1701 both as a source and as a destination port and does not protect the confidentiality of encapsulated data. Today, L2F has only historical value.

8.2.2 Point-to-Point Tunneling Protocol (PPTP)

PPTP is the result of the cooperation between Microsoft and a set of communication equipment manufacturers that formed the PPP Forum. Its specifications were sent to the IETF Point-to-Point Extensions (PPEXT) Working Group in 1996 and were subsequently recorded in RFC 2637 [21].

In terms of the problem addressed in the beginning of this section, the remote system can connect to the LNS either directly, if the remote system supports PPTP, or using the LAC of an ISP supporting inbound PPP connections, if the remote system does not support PPTP. In the first case, the situation is rather simple. The remote system first establishes a PPP connection to the ISP's LAC and then uses PPTP to transmit its encapsulated PPP frames to the LNS. The LAC simply forwards the packets that encapsulate the PPP frames. In the second case, the situation is more complicated, as the LAC must use PPTP to encapsulate PPP frames on behalf of the remote system. In effect, two connections are established in this case: a PPP connection between the remote system and the LAC, and a PPTP connection between the LAC and the LNS.

PPTP uses a complicated encapsulation scheme to route PPP frames through a TCP/IP network that connects the LAC to the LNS: The network layer data units are initially encapsulated using PPP. The resulting PPP frames are then encapsulated using a generic routing encapsulation (GRE) routing header and the IP header. An additional intermediate header is also added to the resulting packets before they are forwarded to the Internet access interface. In addition to the data channel, PPTP also establishes a TCP session on port 1723, which is used for control purposes. This makes the use of PPTP difficult, particularly over a network protected by a firewall.

The PPTP specification does not specify algorithms to be used for authentication and cryptography. Instead, the specification provides a negotiation framework for selecting specific algorithms. The negotiation process is not specified in PPTP, but it is based on the PPP negotiation options, like the PPP compression control protocol (CCP) [22], the challenge handshake authentication protocol (CHAP) [23], the PPP encryption control protocol (ECP) [24], or the PPP extensible authentication protocol (EAP) [25].

8.2.3 Layer 2 Tunneling Protocol (L2TP)

L2TP was founded on a joint proposal of Microsoft and Cisco to combine MS-PPTP and L2F. Originally specified in RFC 2661, its current version (L2TPv3) is specified in RFC 3931 [19].

As with L2F and PPTP, L2TP forwards encapsulated frames through a network, in a way as much as possible transparent to users and to applications. In contrast to the other protocols we have discussed, L2TP uses IPsec security associations to protect the confidentiality of data transmitted between an LAC and an LNS. Securing L2TP with IPsec is described in RFC 3193 [26].

L2TP can be used to create either *voluntary* or *compulsory* tunnels. In the voluntary tunnel case, the user creates a tunnel between himself and the LAC. In the compulsory tunnel case, a tunnel is created between the LAC and the LNS, either on the initiative of the LAC or the LNS.

8.3 Security at the Internet Layer

Security at the Internet layer is mainly looked on by the Internet Engineering Task Force IPsec Working Group (IETF IPsec WG). Before this group started working, several transport layer security protocols had been proposed. Security protocol 3 (SP3) is such a protocol that was developed by the National Security Agency (NSA) and the National Institute of Science and Technology (NIST), as part of the secure data network system (SDNS) suite of protocols. The network layer security protocol (NLSP) was developed by the ISO to provide security to the connectionless network protocol (CLNP). The integrated NLSP (I-NLSP) was developed by NIST to provide security services to the IPv4 and to the CLNP. The SWIPE protocol was initially developed by J. Ioannidis and M. Blaze. The similarities among these are more than their differences, as they all use encapsulation as the basic enabling technique. Following the establishment of the IETF IPsec WG, in 1992, the fact that the security architecture adopted for IPv6 could equally well be applied to IPv4 was made apparent. This architecture consists of an IP security protocol (IPSP) and an Internet key management protocol (IKMP) [27].

8.3.1 IP Security Protocol (IPSP)

The IPSP provides authentication of origin, connectionless data integrity, and connectionless confidentiality services. It may also provide nonrepudiation of origin services, as well as (optionally) protection from replay attacks, but it does not provide protection from traffic analysis. The authentication, data integrity, and nonrepudiation services and optional protection from replay attacks are provided through the authentication header (AH) mechanism [28], whereas the encapsulating security payload (ESP) mechanism [29] provides additional confidentiality service. Note that ESP can also be used to provide only integrity, without confidentiality. These mechanisms have been designed for either independent or joint use. However, as experience has shown that there are very few contexts in which the ESP cannot provide the requisite security services, the use of AH in IPSP implementations has been downgraded to optional. Both mechanisms use and rely on the concept of security associations (SAs).

8.3.1.1 Security Associations

An SA is basically an agreement between two or more parties on the security services they want to use in their communication and how these will be provided. This agreement boils down to a set of security parameters that comprise at least the crypto algorithm, the mode, and the keys for the AH mechanism; the crypto algorithm, the mode, and the keys for the ESP mechanism; the existence or not and the size of the initialization vector to be used for synchronizing the crypto algorithm used by the ESP mechanism; the lifetime of the keys and of the SA itself; the SA source address; and the sensitivity level of the data, when the servers require multilevel security in a mandatory access control setting. When an IP packet is received, it can only be processed if it is associated with a specific SA. This means that SAs must be uniquely identified. This is done by associating an SA with a suitable identifier, called security parameter index (SPI). The combination of an SPI and a destination address uniquely identifies an SA. SA identification is easier to do in a unicast rather than a multicast setting.

8.3.1.2 Authentication Header Mechanism

The AH mechanism adds authentication data to IP packets. These data are computed by using a crypto algorithm with the respective key. They are computed by the sender before the packet is sent and are verified by the receiver on reception. However, bearing in mind that some fields in the IP header may change during the packet's transmission (e.g., the hop-count field in the IPv6 header and the TTL field in the IPv4 header), the authentication data must be computed excluding these fields, leading to the conclusion that the sender must compile a temporary version of the IP header to compute the AH data.

The next step is to process the AH data by using a suitable crypto algorithm. This algorithm is agreed on during the setting up of the SA. Options include the use of HMAC-SHA1-96, AES-XCBC-MAC-96, or HMAC-MD5-96.

The protocol number assigned by the Internet Assigned Numbers Authority (IANA) to the IPSP AH is 51. Therefore, in IPv4, the IP header protocol field must contain the number 51, whereas the AH itself is placed between the IP header and the packet PDU. In IPv6, as the AH header is one of the header extensions, the header extension preceding the AH header must contain, in its next header field, the number 51. Note that the presence or absence of the AH does not, in any way, alter the behavior of the IP, nor of any other network or transport layer protocol.

8.3.1.3 Encapsulating Security Payload Mechanism

The encapsulating security payload (ESP) mechanism can be used to provide confidentiality, data origin authentication, connectionless integrity, an antireplay service, and (limited) traffic flow confidentiality. The set of services provided depends on options selected at the time of SA establishment and on the location of the implementation in a network topology. Confidentiality is provided by using encryption either over the full IP packet (in tunnel mode) or over the payload (in transport mode). Encryption may be provided by a variety of algorithms including DES, 3-DES, RC5, and CAST.

Transport Mode

In transport mode, the sender first selects the higher layer protocol data of the IP packet to be sent and encapsulates it into an ESP. Then, the sender determines the appropriate SA and applies the crypto algorithm defined therein to encrypt the payload data as well as the contents of some additional fields.

The ESP is then included as payload in the IP packet to be sent. As the protocol number assigned by the IANA to the IPSP ESP is 50, the IPv4 header protocol field is set to 50 and the payload type of the ESP header is set to the value corresponding to the higher layer protocol that is now encapsulated within ESP. In IPv6 the ESP header is always the last extension header.

On the other side, the recipient processes the IP header and part of the ESP cleartext to recover the SPI, which is in turn used to determine the appropriate SA. Finally, the encrypted part of ESP is decrypted using the algorithm, the key, and the IV form defined in the SA.

Tunnel Mode

In this mode, whole network layer PDUs (such as IP packets) may be encrypted and encapsulated in new IP packets. ESP in tunnel mode is primarily used by security gateways for packets not originating by the gateway itself, but that must be securely forwarded. Thus, ESP may be used to create a secure tunnel between two firewalls or to set up an IP tunnel or a VPN.

The sender determines the appropriate SA and applies the crypto algorithm defined therein to encrypt the whole IP packet. This is then encapsulated within an ESP and this is in turn included as the payload in a new IP packet. The protocol field in IPv4 or the next header field in IPv6 is set to 50, whereas the payload type field of the ESP is set to 4 (IPv4) or 6 (IPv6).

On the other side, the recipient processes the cleartext IP header and part of the ESP cleartext to recover the SPI. This is used as an index to a local table to determine the agreed SA parameters. The encrypted part of ESP is decrypted using the algorithm, the key, and the IV form defined in the SA. Finally, the recipient detaches the IP packet that had been placed within ESP, to transmit it further.

8.3.2 Internet Key Exchange Protocol

Establishing an SA requires sharing keys known only among the legitimate parties of the SA. This in turn requires a key distribution mechanism and a key management protocol. The IETF IPsec WG developed the Internet key exchange (IKE) protocol [30], which is based on two earlier protocols, namely the Internet Security Association key management protocol (ISAKMP) [31], developed by NSA and the OAKLEY protocol [32]. The support of IKE is optional for IPv4 implementations and compulsory for IPv6 implementations. However, there are a number of other key management protocols that have been proposed, including the modular key management protocol (MKMP) [33], the photuris protocol [34], the secure key exchange mechanism (SKEME) [35], and the simple key management for Internet protocol (SKIP) [36]. The current key establishment protocol for IPsec is the Internet key exchange protocol (IKE).

8.3.2.1 Internet Security Association Key Management Protocol

ISAKMP was designed to be used as common framework for the Internet, in which many security mechanisms with many options for each one of them can coexist. It is used to establish an SA between two parties and runs in two phases. In the first phase, a basic set of security properties is agreed on, which can then be used for protecting ISAKMP exchanges. This includes agreement on the authentication and key exchange methods to be used. The phase is concluded with the establishment of an ISAKMP SA. If a common basic set of security properties already exists, the phase may be omitted. During the second phase, the ISAKMP SA is used to allow the communicating entities to negotiate the security services that will be included in the SA for other security protocols or applications.

8.3.2.2 OAKLEY

OAKLEY was designed to allow sharing a secret key among authenticated entities as a component compatible with ISAKMP. It is based on an authenticated Diffie-Hellman key exchange to achieve perfect forward secrecy (PFS) for the shared secret keys. OAKLEY allows the communicating entities to agree on the authentication, encryption, and key exchange algorithms that they will use.

OAKLEY proceeds in three steps—namely, a cookie exchange step meant to provide protection against resource-clogging attacks using IP spoofed addresses, a

Diffie-Hellman key exchange step meant to share a secret among the communicating entities, and an authentication step meant to mutually authenticate the communicating entities and to verify the integrity of the values exchanged in the previous steps.

8.3.2.3 Simple Key Management for Internet Protocol

SKIP was designed to provide secure connectionless datagram distribution services, such as those provided by the IP. Its characteristic is that it does not require any previous communication to establish the authentication of the IP traffic and of the encryption keys, by using packet-oriented rather than session-oriented keys, which are transmitted together with the IP packets.

SKIP is based on the fact that, assuming that every participant has a certified Diffie-Hellman public key, every pair of participants silently shares a mutually authenticated key that can be computed on the sole knowledge of certified Diffie-Hellman public keys of the other participants. Therefore, it assumes the existence of an automatic mechanism for the distribution and recovery of Diffie-Hellman public key certificates.

8.3.2.4 Internet Key Exchange Protocol

IKE was introduced in 1998 [30]. The protocol uses part of Oakley and part of SKEME in conjunction with ISAKMP to obtain authenticated keying material for use with ISAKMP and for other security associations such as AH and ESP. In 2005, IKE version 2 was introduced in [37]. IKE performs mutual authentication between two parties and establishes an IKE SA that includes shared secret information that can be used to efficiently establish SAs for ESP and/or AH and a cryptographic suite to be used by the SAs to protect the traffic that they carry. An initiator proposes one or more cryptographic suites by listing supported algorithms that can be combined into suites in a mix-and-match fashion.

All IKE communications consist of exchanges, each consisting of a request and a response. The first request/response of an IKE session negotiates security parameters for the SA, sends nonces, and sends Diffie-Hellman values. The second request/response transmits identities, proves knowledge of the secrets corresponding to the two identities, and sets up an SA for the AH and/or ESP.

Subsequent exchanges either create additional SAs for the AH and/or ESP or delete an SA, report error conditions, or do other housekeeping. A request with no payloads (other than the empty encrypted payload required by the syntax) is commonly used as a check for liveness. These subsequent exchanges cannot be used until the initial exchanges have been completed.

8.4 Security at the Transport Layer

Security at the transport layer is mainly looked on by the Internet Engineering Task Force Transport Layer Security Working Group (IETF TLS WG). Before this group started working, several transport layer security protocols had been proposed. Security protocol 4 (SP4) is a protocol that was developed by the National

Security Agency (NSA) and the National Institute of Science and Technology (NIST), as part of the secure data network system (SDNS) suite of protocols. The transport layer security protocol (TLSP) was developed by ISO. Furthermore, M. Blaze and S. Bellovin of the AT&T Bell Labs developed the encrypted session manager (ESM) software, which is similar to the secure shell (SSH). Today, the most widely known and used transport layer security protocols are SSH [38], SSL [39], and TLS [40].

8.4.1 Secure Shell

SSH is a piece of software that can be used to connect an entity securely to a remote host, allowing the secure remote execution of commands and the secure transfer of files. SSH was created by T. Ylönen, of the Helsinki University of Technology, in Finland. It comes in two versions, a publicly available free version and a commercial version.

Both versions may use any transport layer protocol. When they are used over TCP/IP, port 22 has been assigned for the communication. SSH provides for both server and client authentication, data compression, data confidentiality, and data integrity services. SSH does not necessarily rely on the use of a public key infrastructure, but may employ public keys that are distributed out of band.

Originally, SSH was monolithic. However, following the work of IETF Secure Shell Working Group (SECSH WG), it was broken down in two parts: the SSH transport layer protocol and the SSH authentication protocol, both referred to commonly as SSH 2.0.

8.4.1.1 SSH Transport Layer Protocol

The SSH transport layer protocol provides for the cryptographic authentication of a host, as well as data confidentiality and integrity services, but it does not provide a user authentication service; the provision of this service is the responsibility of the SSH authentication protocol.

The protocol supports several ways to exchange private and public keys, hash functions, and message authentication techniques that have to be agreed on by the communicating parties when the connection is set up. To enable this, when two communicating parties use SSH to establish a TCP/IP connection, they first exchange identifying information, including the software they use and the protocol version that they support. They then start the key exchange process. During this process, the data compression, crypto, and message authentication algorithms to be used in all subsequent communication are selected among several options and agreed on.

When the SSH transport layer protocol has been completed, the client may request a service on behalf of the user, by sending an appropriate message to the server. If the server supports the service and the client is allowed to use it, the server replies with another message. From then on, all data are transferred using SSH_STREAM_DATA messages. When the service use is completed, either on the server or the client side, a SSH_STREAM_EOF message is sent to the other side. When either side wishes to terminate communication, it sends to the other a

SSH_STREAM_CLOSE message. On receipt of this message, the recipient side also sends a SSH_STREAM_CLOSE message.

8.4.1.2 SSH Authentication Protocol

The SSH authentication protocol has been designed to operate over the SSH transport layer protocol to provide the user authentication service. When this is required, the client first declares the name of the service and the name of the user wishing to use the service. The server replies by sending all the possible options for the authentication methods available for this service, and the client returns an authentication request. This dialogue is continued until the request is either granted or denied.

SSH supports several authentication methods, from traditional password authentication to challenge-response–based methods, to signature-based methods, to Kerberos-based methods.

8.4.2 The Secure Sockets Layer Protocol

A popular method for achieving interprocess communication in Internet programming is the use of BCD sockets. The development of the secure sockets layer protocol was based on the idea that such interprocess communication can be made secure by authenticating peer entities and by allowing them to exchange cryptographic keys that can later be used for exchanging authenticated and encrypted information.

SSL was created by Netscape Communications Corporation. The first version was only used internally, whereas version 2.0 was incorporated in Netscape Navigator v.1 and v.2. Because of limitations in both functionality and cryptographic security, SSL was upgraded to version 3.0, which is the current version.

Formally speaking, SSL does not exactly belong to the transport layer, as it lies between the transport and the application layers, relying on a TCP/IP transport service and providing a peer authentication service using cryptographic keys, a data confidentiality service using symmetric cryptography, and a message authentication service using MAC authentication. SSL does not provide a service to protect from traffic analysis.

An analysis of the security offered by SSL indicates that no serious vulnerabilities or security problems exist with it. Thus, SSL is thought to offer excellent protection against passive attacks, even though some sophisticated active attacks against it are possible. This is probably the reason why SSL is currently the most widely used security protocol in the Internet and the web. The price to pay is that the client-server communication is slowed down due to the encryption overhead.

SSL actually comprises several protocols, namely the SSL handshake protocol, the SSL record protocol, the SSL alert protocol, and the SSL change cipher spec protocol.

8.4.2.1 SSL Handshake Protocol

The SSL handshake protocol is an entity authentication and key exchange protocol that negotiates, initializes, and synchronizes the security parameters at the two ends of a connection. The protocol is executed in four main steps that involve a number of compulsory as well as several optional message exchanges.

A typical SSL handshake usually starts with the client and the server exchanging "hello" messages. In the second step of the protocol, the server processes the client's request and responds with either another "hello" or with an error message. The server may, at this step, optionally send additional messages to the client, if, for example, server authentication is desired. At the end of the second step, the server sends a message to the client indicating completion of this step of the protocol. In the third step, the client sends a message containing information to be used in order to establish the master and the session keys. The step is concluded with the client sending a message to the effect that the agreed cryptographic options have been set and another indicating the completion of the step. In step 4, it is the server's turn to confirm the adoption of the agreed cryptographic options and to indicate the completion of the step at its end.

Following the completion of the SSL handshake protocol, data are exchanged through the SSL record protocol.

8.4.2.2 SSL Record Protocol

The SSL record protocol provides authentication, confidentiality, and data integrity services, as well as protection against message replay, over a connection-oriented transport service, such as one provide by TCP/IP. It receives data from higher layer protocols, fragments them, compresses them, authenticates them, and encrypts them.

8.4.2.3 SSL Alert Protocol

The SSL alert protocol is used to transmit alerts over the SSL record protocol. These alerts are composed as a certain message type that includes information on the level of the alert and a description of the alert.

8.4.2.4 SSL Change Cipher Spec Protocol

The SSL change cipher spec protocol is used whenever a selected cryptographic option needs to be changed. Cryptographic options are normally selected during the execution of the SSL handshake protocol, but they can be modified at any time.

8.4.3 Transport Layer Security Protocol

TLS is the result of the work of the IETF Transport Layer Security Working Group (TLS WG). The main goal of the group was to develop a transport layer security specification that would closely follow the SSL, SSH, and PCT protocols. The result is not considerably different than SSL 3.0. Nevertheless, the differences are enough to make cooperation between TLS and SSL 3.0 a nontrivial task. However, TLS does provide a mechanism that can make a TLS implementation fully compatible with SSL 3.0.

Like SSL, TLS is on its own a layered protocol. At the highest level, handshaking is used to agree on a session state, which comprises a session identifier, a peer certificate, a compression method, a crypto spec, a master key, and a binary variable indicating whether or not this is a new session. All this is used subsequently by the TLS record protocol. The overall TLS handshaking process comprises three subprotocols—namely, the TLS change cipher spec protocol, the TLS alert protocol, and the TLS handshake protocol. Following the completion of a TLS handshake, the client and the server may start exchanging data messages.

At the lowest level, similarly to SSL, the TLS record protocol receives messages from higher layer protocols, fragments and optionally compresses the data, computes and attaches a MAC to every data segment, and finally encrypts and sends the result. When a record is received, it is decrypted, confirmed, decompressed, and put together again before being sent to a higher layer client.

8.5 Security at the Application Layer

Providing security at the application layer is a flexible option because the limits and the power of the offered protection can be adapted to the needs of every specific application. There are two approaches to providing security at the application layer: one may integrate security services within the application itself, or a generic security system can be created that is used for providing security services within different applications.

8.5.1 Secure Email

The cornerstone of the Internet mailing system is the simple mail transfer protocol (SMTP) [41]. The format of ASCII messages is described in [42], and the multipurpose Internet mail extensions (MIME) are described in [43]. None of these specifies any security services. Three schemes for providing such services through the use of the digital enveloping technique are the privacy enhanced mail (PEM) scheme [44–47], the pretty good privacy (PGP) system [48], and the secure multipurpose Internet mail extensions (S/MIME) specification [49].

8.5.1.1 Privacy Enhanced Mail

PEM is described in [44–47]. PEM supports message authentication, data integrity, and nonrepudiation of origin services, using a message integrity check (MIC), as well as a data confidentiality service implemented by encrypting the whole message. Access control and nonrepudiation of receipt services are not provided.

PEM messages comprise a header and a text portion. The header contains information related to the algorithms used, to the MIC and to the certificates. The text portion contains the message itself, possibly encrypted and encoded. A PEM message is thus constructed in four steps: a canonicalization step that transforms the original message to a standard format, a digital signature step that digitally signs the MIC, an optional encryption step, and an optional transmission encoding step that transforms the message to a form compatible to SMTP coding.

PEM uses a key management scheme based on certificates compatible with the ITU-T X.509 recommendation. The PEM hierarchy of certificates consists of three levels: the Internet policy registration authority (IPRA) that serves as the root certification authority, a number of policy certification authorities (PCAs) within IPRA's domain, and a number of certification authorities (CAs) within the each PCA's domain. This scheme does not support cross certification.

Originally PEM did not support binary attachments or MIME messages. It was later extended to do so, resulting in the MIME object security services (MOSS) specification, described in RFCs 1847–1848. In fact, there are some additional differences between MOSS and PEM, the most important of which is that MOSS does not require users to possess X.509 certificates but simply to have a pair of keys.

8.5.1.2 Pretty Good Privacy

PGP is a complete system, developed by P. Zimmermann. It provides message authentication, data integrity, and no-repudiation of origin services, as well as a data confidentiality service implemented by digital enveloping. It can also encrypt local files; it allows digital signatures to be detached from messages and stored separately and uses publicly available programs to compress messages.

In contrast to PEM, PGP does not rely on the existence of a fully operational certificate hierarchy; instead, it is based on the concept of trust among users. As in the social notion of trust, PGP applies transitivity to trust, creating a web of trust among its users. Every PGP user may select a public key from his own public key ring and associate with this a trust level. He can also collect certificates for this key, issued by several other users. Each PGP user also has a secret key ring containing his own private key, encrypted with a secret key that is produced by a user passphrase.

8.5.1.3 Secure Multipurpose Internet Mail Extensions

The S/MIME specification was developed by a working group from industry led by RSA Data Security.

As with the PEM specification, S/MIME is based on the ITU-T X.509 recommendation. However, to avoid the need for a fully operational certificate hierarchy, S/MIME is more flexible with the specification of CAs and certificate hierarchies than PEM.

S/MIME relies on the public key cryptography standards PKCS#7 and PKCS#10 to ensure crypto compatibility and interoperability among implementations. PKCS #7 specifies a standard syntax for crypto messages and defines bit encodings for digital signatures and digital envelopes. PKCS #10 defines the message syntax for certificate requests.

S/MIME differs from PGP in its message structure as well as in the way that keys and certificates are handled.

8.5.2 Web Transactions

The wide deployment and use of the Internet is largely due to the existence of the World Wide Web and HTTP. Originally, like the Internet, the web was not designed with security in mind, as it was thought that all information on the web would be publicly available. Clearly, today, this is no longer the case, as the web is being increasingly used for all kinds of transactions, including transactions of a financial nature as opposed to a mere source of information.

As a result, efforts to secure web transactions started in the mid 1990s. One possible approach is to establish a normal TCP/IP connection and run a "security enhanced version" of HTTP over it. This is the approach taken by secure HTTP (S-HTTP) [50].

Alternatively, it is technically possible to secure web transactions by first establishing a secure TCP/IP connection, possibly using SSL, and then run HTTP over the secure connection. This is the approach taken by the HTTPS uniform resource identifier (URI) scheme. HTTPS protects from eavesdropping and man-in-the-middle attacks, and authenticates web servers using public key certificates; it may also be used to authenticate clients. HTTPS is syntactically identical to the HTTP scheme. Using an HTTPS URL indicates that HTTP is to be used, but with a different default TCP port (443) and an additional encryption/authentication layer between the HTTP and TCP. Thus, strictly speaking, HTTPS is not a separate protocol. This approach seems to have become the current standard option.

8.5.3 Domain Name System

As the Internet domain name system (DNS) was not originally designed to be secure and it is crucial to the operation of Internet, it has been the subject of several attacks. An example of a spoofing attack against the DNS could be to relate a logical name to an IP address other than the legitimate one. Such an attack would be successful if the response to a request cannot be verified as coming from the authentic DNS server or if the integrity of this response cannot be guaranteed. In [51], several threats against DNS are described. To address these, a set of extensions to the DNS specification, collectively known as DNSSEC, were developed and are described in [52–54].

DNSSEC uses public key cryptography and digital signatures to provide data integrity and authentication services for information provided by the DNS and included in the DNS protocol. To achieve this, DNSSEC defines a number of new DNS resource records (RRs) and two new message header bits. With DNSSEC a zone administrator digitally signs a resource record set (RRSet) and publishes this digital signature, along with the zone administrator's public key, in the DNS. The DNSSEC client validates this, as well as the zone administrator's public key, to verify the integrity and authenticity of the response data.

8.5.4 Network Management

Network management systems support the network administrators in the carrying out of their administrative functions to ensure the efficient and reliable operation of the network. As such, the existence of security services within them is significant. A general architecture for network management is based on the client-server model and consists of three main elements: a set of managed resources; at least a network management station, usually referred to as manager; and a network management protocol, used by the manager and its agents to exchange network management–

related information. For network management purposes, resources are represented by objects in a database that is called the management information base (MIB). An object is in practice a variable that represents an aspect of the resource. The state of a resource can be determined or modified by examining or setting the value of the relevant variable, respectively. Network management can be compactly described in terms of a set of management functional units, known collectively as fault, configuration, accounting, performance, and security (FCAPS). The security management functional unit provides for entity authentication, access control, confidentiality, and data integrity services. Additionally, event detection, audit trail logs management, and alarm management are also included. Clearly, not all network management protocols are required to support all these services. However, the usual security requirements applicable to any other type of application apply to network management systems as well.

TCP/IP network management is based on the simple network management protocol (SNMP). ISO/OSI network management is based on the common management information protocol (CMIP) and the management of telecommunication networks is based on the ITU-T telecommunication network management (TMN) framework.

8.5.4.1 Simple Network Management Protocol

The first version of SNMP, known as SNMPv1, is specified in [55] and other related RFCs. The security it offers is poor, as authentication of origin is performed only by including the unencrypted name of a "community" as a password in the message. A simple access control service to the MIB is also provided.

The SNMP version 2, known as SNMPv2, is specified in [56] and other related RFCs. It was meant to improve the functionality of SNMPv1 in performance, security, and manager-to-manager communications. However, its security-related provisions were thought to be very complex and were not widely accepted. As a result, a variant of SNMPv2, known as the community-based SNMP version 2, or SNMP v2c, was specified in [57] and other related RFCs. A compromise that would offer increased security over SNMPv1 but without the complexity of SNMPv2 was found in another variant, the user-based SNMP version 2, or SNMP v2u, that was specified in [58] and other related RFCs.

The last version of SNMP is SNMP version 3, known as SNMPv3, that is specified in [18] and other related RFCs. As it is common practice for SNMPv3 implementations to support earlier versions of the protocol, their coexistence is specified in [59]. SNMPv3 provides authentication, data integrity, and confidentiality services, in addition to an access control service.

The SNMP architecture is defined by a set of distributed SNMP entities. Each SNMP entity comprises an SNMP engine that in turn comprises four subsystems—namely, the dispatcher, the message processing subsystem, the security subsystem and the access control subsystem, and a set of applications.

The security subsystem provides the authentication and confidentiality services. Each outgoing or incoming message is sent by the message processing subsystem to the security subsystem. Depending on the required services, the message is processed and possibly transformed and sent back to the message processing subsystem. The security subsystem may support several security models. However, the model currently defined is the user-based security model (USM).

USM has been designed to thwart threats such as information modification, masquerading, message flow modification, and disclosure. It does not address denial of service attacks or traffic analysis attacks. It provides authentication and confidentiality services to SNMP. To implement these services, the SNMP engine needs each user to own a private key and an authentication key. These are stored separately for each user and are not accessible through SNMP. Two communicating SNMP engines must share the same authentication key, which is used to produce and verify a MAC. The same requirement applies for encryption.

8.5.5 Distributed Authentication and Key Distribution Systems

Extensive effort has been invested in the past few years to develop distributed authentication and key distribution systems. Examples of such systems are Kerberos [60], developed by MIT, which is presented in more detail below, as it is the most widely deployed one; SESAME [61]; NetSP [62]; SPX [63]; and TESS [60].

8.5.5.1 Kerberos

Kerberos is organized into realms (domains). In each realm there exists a central, physically secure authentication server (AS) that shares a secret key K_p with each participant (principal) P. The system operates by providing principals with tickets that can be used to authenticate them, as well as with secret keys that can be used for secure communication. The AS authenticates users on login and provides them with a ticket-granting ticket (TGT). This can be used for issuing tickets by a ticket-granting server (TGS). These tickets can then be used as credentials when contacting other servers. The Kerberos protocol is based on key distribution protocols that were initially developed by Needham and Schroeder and were later modified to accommodate timestamping. The protocol can be summarized in six steps and involves three exchanges: one in steps 1 and 2 between the client C and the AS, one in steps 3 and 4 between C and the TGS, and one in steps 5 and 6 between C and a server S.

In step 1, C sends the user name and the name of a TGS to the AS. The AS creates the TGT and returns it to C. C asks the user to provide her password and, if this is correct, C creates the necessary key to decrypt the TGT. In step 3, C creates an authenticator and forwards it, along with the name of a server S and the appropriate ticket to the TGS. TGS decrypts the ticket, retrieves the key to decrypt the authenticator, and checks the validity of the timestamp. If both the ticket and the timestamp are valid, TGS issues the ticket and returns it to C. C can now decrypt the message and retrieve the new session key. C creates a new authenticator and forwards it, along with the ticket to S. S decrypts the ticket, retrieves the session key, decrypts the authenticator, and checks the validity of the timestamp. If both the ticket and the timestamp are valid, the service is provided.

8.5.6 Firewalls

Connecting to the Internet is no longer an option, but a necessity. Notwithstanding the benefits of this connectivity, it also allows the outside world to access assets of the local network, thus making it vulnerable to attacks. One may naturally try to protect the local network by protecting its individual assets. Another approach, complementary to the latter, is to protect the perimeter of the local network by introducing a firewall between the local network (intranet) and the outside world (Internet).

A firewall must satisfy three distinct design goals: First, all traffic between the intranet and the Internet must go through the firewall. Second, only authorized traffic must be able to pass through the firewall. Third, the firewall itself must be robust against attacks. These goals can be achieved by controlling the use of network services or the direction of the service traffic, or the access of users to services or the behavior of the services themselves.

There are three main firewall types—namely, packet-filtering routers, application-level gateways, and circuit-level gateways. The former applies a specified rule set to each incoming IP packet and consequently forwards or rejects the packet. Application-level gateways (alias proxy servers) function as traffic retransmitters at the application level. Users do not directly connect to remote servers when requesting a service; rather, they connect to the proxy server, which checks all necessary parameters and allows or disallows the service provision to the specific user. Circuit-level gateways do not allow end-to-end TCP connections; rather, they establish two TCP connections, one between themselves and the user requesting a service and another between themselves and the remote server. Once the connections have been set up, the gateway simply forwards traffic, without checking its content. Application-level gateways and circuit-level gateways are typically implemented via dedicated systems, called bastion hosts.

By combining these types, it is possible to implement more complex firewall configurations. Thus, it is possible to combine a packet-filtering router with an application-level gateway, or to combine a packet-filtering router with two application-level gateways, or to combine two packet-filtering routers with two application-level gateways. In such a way, the intranet is transformed to a *screened subnet*.

8.6 Security in Wireless Networks

Wireless technologies and mobile communications are becoming increasingly popular as alternatives to the traditional wired option, as they can effectively support the demand of users for continuous provision of services anywhere, anytime. However, wireless networks, due to their nature, are susceptible to more threats than their wired counterparts, as wireless networks are characterized by the absence of any form of access control over the area covered by the network. This means that any potential attacker may anonymously move into the area covered by the network and may launch a variety of attacks, some of which are not applicable to the traditional network case.

With respect to coverage, wireless networks are broadly categorized as wireless local area networks (WLANs) that follow the IEEE 802.11 standard [61], as wireless metropolitan area networks (WMANs) that follow the IEEE 802.16 standard [62], and as short-range local area networks that utilize the Bluetooth technology. We will look into the security of these categories in the next three subsections. We will also look into the security of mobile communications systems that conform to the universal mobile communications system (UMTS) standard.

8.6.1.1 IEEE 802.11 LAN Security

Wireless networks that follow the IEEE 802.11 standard consist of access points (APs) that act as bridges between the wireless and the wired network, a distribution system (DS) that connects APs, the wireless medium (which may be on the radio frequencies or the infrared range of the spectrum), and stations that exchange information through the network. The basic building block of every IEEE 802.11 network consists of a set of communicating stations and is called basic service set (BSS). Its limits are determined by the area covered by the network, called the basic service area (BSA). Wireless networks are categorized, with regard to their topology and architecture, as structured (infrastructure mode) or unstructured (ad hoc). The former requires the use of APs to which the users connect, whereas the latter does not. The structured network provides a better platform to deploy security policies.

Originally, IEEE 802.11 provided the following security services:

- Two different methods for authentication, namely open system and shared key;
- The wired equivalent privacy (WEP) for the protection of confidentiality;
- A WEP-encrypted integrity check value (ICV) for the protection of the integrity of messages.

However, as security weaknesses of WEP were revealed, a new standard, for the security of IEEE 802.11 networks, known as IEEE 802.11i was developed. This provides for improved mechanisms for network device authentication; key management algorithms; session key agreement methods; an improved encapsulation mechanism, known as the counter mode cipher block chaining message authentication code protocol (CCMP) and (optionally) the temporal key integrity protocol (TKIP).

IEEE 802.11i defines three methods for authentication—namely, the two methods already provided for by IEEE 802.11 and, additionally, a network association method based on the 802.1X framework or on pre-agreed secret keys, called robust security network association (RSNA). It supports three different crypto algorithms for the protection of confidentiality—namely, WEP, CCMP, and TKIP. WEP and TKIP use the RC4 crypto engine, whereas CCMP is based on the AES. Key distribution is achieved via a four-way handshake, using appropriate group key handshake protocols. Data origin authenticity is provided by CCMP or TKIP. IEEE 802.1X provides a port-based authentication service.

Even though RSNA networks seem to provide effective confidentiality protection when CCMP is used, as well as satisfactory authentication and key management services, as availability is not addressed, they are susceptible to several types of DoS attacks.

8.6.1.2 Bluetooth Security

Bluetooth is a short-range wireless technology that was developed to cover the networking needs of personal devices. The Bluetooth protocol suite [63] is structured in two layers—namely, the physical layer and the baseband layer. The physical layer is where modulation and demodulation, as well as radio signal processing, take place. Over the physical layer, the baseband layer is split in two sublayers. At this layer packets are formed, headers are created, checksums are computed, retransmission is handled, and encryption and decryption, if applicable, are carried out. The lower baseband layer protocols are implemented by the link controller (LC), whereas the upper baseband layer is implemented by the link manager (LM).

Bluetooth devices can form ad hoc networks to exchange data. Pairs of devices in a network may establish a trusted relationship using a shared secret *passkey*. In such a trusted relationship, devices can authenticate each other, and they also may exchange encrypted data. Authentication and key derivation are implemented by algorithms based on the SAFER+ and E22 algorithms. Encryption is implemented by the E0 algorithm.

Security flaws in Bluetooth were discovered as early as 2003. Some of them were the result of poor implementations, whereas others are due to weaknesses in the PIN mechanism used to establish device pairings.

8.6.1.3 IEEE 802.16 WMAN Security

The basic building block of IEEE 802.16 WMAN networks is the cell, which comprises one or more base stations (BSs) and multiple mobile stations (MSs). BSs manage the access medium for MSs and are also responsible for resource allocation, key management, and so on. Two functional areas are distinguished in a WMAN—namely, the access service network (ASN) that provides BSs with services for accessing the network, and the connectivity service network (CSN) that provides IP connectivity. Six different interfaces (R1–R6) between the network elements are defined.

Originally, IEEE 802.16 security services were limited. Confidentiality was provided through encryption of the MACPDUs (MPDUs) with DES-CBC using a 56-bit key. The integrity of MPDUs was not efficiently protected, and only MSs had to authenticate with BSs. These limitations were addressed with the IEEE 802.16e specification.

This standard allows for the use of symmetric methods, such as preagreed keys and subscriber identity module (SIM) cards for controlling access to the network, in addition to the asymmetric ones. In this way, the MS can authenticate to the BS either directly, using public certificates, or indirectly, by authenticating to an authentication server using the IEEE 802.1X framework. It also supports mutual authentication between BSs and MSs by introducing the privacy key management (PKMv2) protocol, thus thwarting man-in-the-middle attacks. The standard also revisits the algorithms used for the confidentiality and integrity of MPDUs, allowing the use of the AES-CCM and HMAC/CMAC algorithms. It increases the key lengths and provides protection from replay attacks. The MS can associate with the BS and obtain the necessary keys faster. Nevertheless, some vulnerabilities still exist.

8.6.1.4 Mobile Communication Networks Security

The most widely deployed mobile communication network today is the global system for mobile communication (GSM). GSM started operating in 1991 and belongs to the family of mobile communication networks known as second generation mobile communication networks (2G), the first generation being based on analog technology. The third generation of such networks (3G) is known as universal mobile telecommunication system (UMTS) [64]. Such systems are also in operation today, in several countries, in parallel to 2G systems. The fourth generation (4G) is expected to implement the full unification of wired and wireless systems of heterogeneous technologies under a common framework, based on the IP.

User authentication in GSM networks is based on symmetric cryptography. The public land mobile network (PLMN) operators provide their subscribers with a SIM card, containing a unique secret key that is used to authenticate the subscriber to the network. The same key is also stored in the PLMN's authentication center. Only the user equipment is authenticated to the network—not vice versa.

Confidentiality of the data transmitted through the radio link in GSM networks is achieved by encryption based on the A8 algorithm for the key generation and the A5 algorithm for encryption itself.

Privacy protection in GSM networks is achieved by using temporary mobile subscriber identities (TMSIs) that change every time the subscriber connects to the network and are transmitted in encrypted form. Location privacy is also achieved with the same mechanism.

The user authentication mechanism used in UMTS networks, known as authentication and key agreement (AKA), is designed along the same lines as those of the GSM user authentication mechanism. The same holds true for the encryption and privacy protection mechanisms. Therefore, the security mechanisms in both GSM and UMTS are similar, even though the technical details vary.

8.7 Network Vulnerabilities

Thousands of attacks are launched against networks everyday. However, only a few of them prove to be successful. This primarily happens because the absolutely necessary prerequisite for an attack to succeed is its ability to successfully exploit an existing vulnerability. There are three basic overlapping types of vulnerabilities: bugs or misconfiguration problems in software, design flaws, and cleartext traffic in the network.

Bug-free software seems to be very difficult to build. These are exploited in the server daemons, the client applications, the operating system, and the network stack. Well-known types of software bugs include buffer overflows that allow an attacker to execute unauthorized commands, unexpected combinations of layered code, invalid input that the software does not expect and is not prepared to handle, and race conditions among system resources. System configuration bugs include insecure default configurations that are not changed by administrators, inadvertent insecure settings, and maintenance of trust relationships among machines.

Even if a software implementation is completely correct according to the design, there still may be flaws in the design itself that may lead to a successful attack. Such flaws are known to exist in the design of several protocols and of operating systems.

Much of the traffic in a network travels in cleartext. This may be due to the design of the network itself or it may be an operational option. In either case, if the decision on which data to transmit in cleartext is based on a careful examination of the security risks that the network faces in conjunction with its operational requirements, the chances that an attack aiming to intercept data of some value to a potential attacker are slim. However, if the decision is taken on other grounds, such as on sacrificing security in favor of efficiency, a significant risk is being taken.

The number of vulnerabilities reported yearly rises to several thousands. According to their importance in terms of how many different systems they affect, vulnerabilities can be ranked as follows:

- Default installations of operating systems and applications;
- Accounts with no passwords or weak passwords;
- Nonexistent or incomplete backups;
- Large numbers of open ports;
- Not filtering packets for correct incoming and outgoing addresses;
- Nonexistent or incomplete logging;
- Vulnerable CGI programs.

8.8 Remote Attacks

An attack is any unauthorized action undertaken with the intent of hindering, damaging, incapacitating, or breaching the security of a system.

8.8.1 Types of Attacks

Attacks can be categorized in many ways. One way is to classify them according to the *passive* or *active* nature of the attacker's actions. Another is to classify attacks according to the physical or logical location of their launching. Thus, a *remote attack* is any attack that is initiated against a machine that the attacker does not currently have control over; this differentiates it from a *local attack*. Yet another is to classify attacks according to the number of the attacking machines and the number of attacked machines.

In a *one-to-one attack*, an attacker chooses a single victim to launch an attack against. If there is more than one attacker collaborating to attack the same target, a *many-to-one attack* is manifested. In a *one-to-many attack*, the attacker hits several remote hosts. Clearly, the complexity of such an attack is directly proportional to the number of simultaneously attacked hosts. This is why the *many-to-many attack*, whereby several concerted adversaries perform unauthorized actions toward multiple targets at the same time, is more frequently encountered (see Figure 8.1).

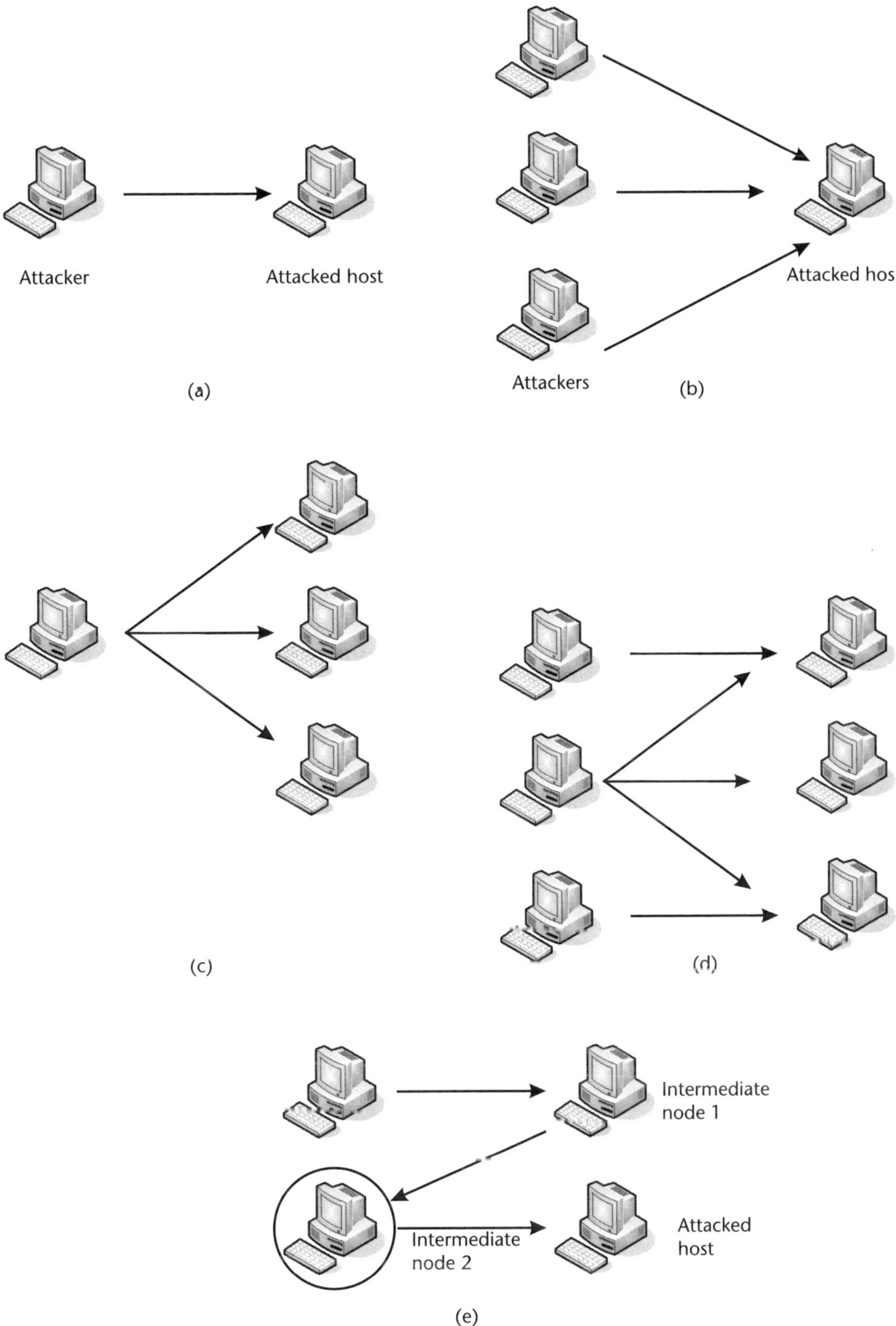

Figure 8.1 Models of attacks.

An attack may be launched either directly from the attacker's machine or, more frequently, as one of the goals of the attacker is also to cover his traces, by using one or more intermediate nodes. In this case, the attacker obtains control over one or more nodes, which he then uses to launch the attack.

8.8.2 Severity of Attacks

There are six levels of the severity of attacks. Attacks classified in the level-one category include mail bombing and DoS attacks. The latter may be distributed, constituting a distributed DoS (DDoS) attack. We will be looking at DoS and DDoS attacks in more detail later. Level-one attacks usually compromise the availability of systems and data and can also compromise the integrity of data. Attacks in levels two and three involve inside users gaining unauthorized access to information. Such attacks adversely affect the confidentiality and/or integrity of data, but the damage may be contained within the organization. Attacks in level four are usually launched by outsiders, who gain unauthorized read access to internal information. Such attacks adversely affect the confidentiality of data; the damage may not be contained within the organization. Attacks in levels five and six consist of conditions whereby remote users can read, write, and execute files. Such attacks adversely affect the confidentiality and/or integrity of data and the availability of systems and data, and the damage may not be contained within the organization.

8.8.3 Typical Attack Scenario

In terms of the sequence of actions taken to launch an attack, a typical scenario involves four distinct steps.

In the first step, the attacker performs an outside reconnaissance to find out as much as possible about the victim without actually giving himself away. This can be done in a legitimate way by finding publicly available information that resides in DNS tables, public web sites, and anonymous FTP sites. It can also be complemented by searching news articles and press releases about the target organization.

In the second step, the attacker advances to perform an inside reconnaissance to scan for more information in a more obtrusive manner. This can be done by visiting web pages and looking for CGI scripts, by performing a ping sweep to see which machines in the target network are alive, or by carrying out a UDP/TCP scan/strobe on target machines to see what services are available.

In the third step, the attacker actively engages in the attack itself by exploiting the identified vulnerabilities of the target. It is at this step that the attacker will take advantage of all the information collected in the previous two steps to gain access to the system.

In the fourth step, the attacker has succeeded in compromising the security of a machine in the target network and has gained access to the system. He is now able to install software that will ensure their continuing access, he may replace existing services with his own Trojan horses, or he may create his own user accounts.

In the final step, the attacker takes advantage of his accomplishments so far to carry out his ultimate purpose, that may be the acquisition of information, the misuse of system resources, or the malevolent modification of data.

8.8.4 Typical Attack Examples

A *DoS* is an attack aiming to disrupt the normal provision of services. This can be done by either flooding the network with a high volume of traffic, in which case a bandwidth attack takes place, or by flooding a computer with a high volume of connection requests, in which case a connectivity attack takes place. In both cases, all available system resources are consumed, and the system can no longer process legitimate user requests.

A *DDoS* attack is a DoS attack that uses many nodes against one or more targets. These nodes may be acting knowingly or unknowingly. The attack is usually launched by software agents installed within the conspiring nodes.

A *sniffing attack* is a passive attack aiming to eavesdrop information traveling along a network. This information is then stored on some media and archived for later viewing.

In a *spoofing attack* the attacker creates misleading context to lure the target into undertaking action that assists the attacker. The *IP spoofing attack* involves forging one's source address. Other forms of spoofing, such as *DNS spoofing*, occur when an attacker has compromised a DNS server and explicitly alters the hostname-IP address tables.

Web spoofing is a "man in the middle" type of attack in which the attacker creates a convincing but false copy of the entire web site, which the attacker controls, so that all network traffic between the victim's browser and the site goes through him.

Fingerprinting is a type of attack in which the attacker uses information like the format of responses, error messages, or spelling of key words (upper case or lower case) to deduce the version of the software running on the target.

According to the 2007 CSI Computer Crime and Security Survey, types of attacks or misuse detected in the last 12 months (up to June 2007) by percentage of respondents are the following: insider abuse of Internet access: 59 percent; viruses: 52 percent; laptop/mobile theft: 50 percent; phishing where respondent's organization was fraudulently represented as sender: 26 percent; instant messaging misuse: 25 percent; denial of service: 25 percent; unauthorized access to information: 25 percent; (ro)bots within the organization: 21 percent; and so on.

8.9 Anti-Intrusion Approaches

An *intrusion* is the act of somebody attempting to break into or misuse a system. The "filtering" of successful intrusions is graphically depicted by the narrowing of the successful intrusion attempt band (Figure 8.2). *Prevention* precludes or severely handicaps the likelihood of a particular intrusion's success. *Preemption* strikes offensively against likely threat agents prior to an intrusion attempt to lessen the

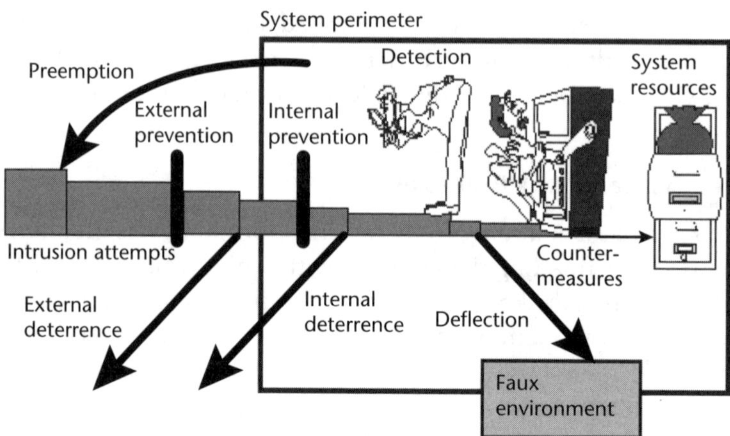

Figure 8.2 Anti-intrusion approaches.

likelihood of a particular intrusion occurring later. *Deterrence* averts the initiation or continuation of an intrusion attempt by increasing the necessary effort for an attack to succeed, increasing the risk associated with the attack, and/or devaluing the perceived gain that would come with success. *Deflection* leads an intruder to believe that he has succeeded in an intrusion attempt, whereas instead he has been attracted or shunted off to where harm is minimized. *Detection* discriminates intrusion attempts and intrusion preparation from normal activity and alerts the authorities. *Countermeasures* actively and autonomously counter intrusions as they are being attempted.

8.9.1 Intrusion Detection and Prevention Systems

An *intrusion detection system (IDS)* is a system for detecting intrusions. Intrusion detection is the art of detecting inappropriate, incorrect, or anomalous activity. An *intrusion prevention system* (IPS) is a computer security device that monitors network and/or system activities for malicious or unwanted behavior and can react, in real time, to block or prevent those activities. Intrusion prevention technology is considered by some to be an extension of intrusion detection (IDS) technology. IPSs are designed to sit inline with traffic flows and prevent attacks in real time.

The main architectural elements of almost all IDSs and IPSs are the *data-gathering module*, the *prefiltering module*, the *decision-making module*, the *communications module*, and the *human interface module*. Some systems may have an additional module, which automatically or semi-automatically applies countermeasures when an intrusion is detected. IDSs and IPSs gather and process data that may come from directly observing a system's operation, from monitoring network traffic, or from a system's files, such as audit log files. Data gathering may be done in a distributed manner, utilizing software agents residing in different machines in a network. The role of the prefiltering module is to filter gathered data, to discard those that are not relevant to security, and to forward the remaining ones to the decision-making module. Clearly, some decision making is involved in this module, particularly if event correlation and data fusion are considered. The decision-making module has the ultimate responsibility to decide whether or not an intrusion is indeed taking place and also to identify the exact nature of the intrusion. There are two complementary approaches to detecting intrusions—*misuse detection* and *anomaly detection*. Misuse detection techniques apply the knowledge accumulated about specific attacks and system vulnerabilities to detect attempts to exploit these vulnerabilities. Anomaly detection techniques assume that an intrusion can be detected by observing a deviation from normal or expected behavior of the system or its users. The model of normal or valid behavior is extracted from reference information collected by various means. It is also possible to combine the two approaches.

IDSs and IPSs can be arranged as either *centralized* or *distributed*. A distributed system consists of multiple IDSs or IPSs over a large network, all of which communicate with each other. They can also be classified as *host based* or *network based;* the former monitors a single host whereas the latter monitors an entire network.

8.10 Conclusions

Network security is an important aspect within the field of information and communications security today. On its own, network security deserves—and has found—treatment in much more voluminous texts than the one herein. We have only touched on the subject here, giving a general overview of the area, structured according to the layers of the OSI architecture, along with pointers to material that the interested user can look on to acquire more detailed information.

References

[1] ISO/IEC 7498, “Information Processing Systems—Open System Interconnection Model—Part 1: The Basic Model,” 1982.

[2] ISO/IEC 7498, “Information Processing Systems—Open System Interconnection Model Part 2: Security Architecture,” 1982.

[3] Simpson, W., “The Point-to-Point Protocol,” RFC 1661, July 1994, http://tools.ietf.org/html/rfc1661.

[4] Deering, S., and R. Hinden, “Internet Protocol, Version 6,” RFC 2460, December 1998, http://tools.ietf.org/html/rfc2460.

[5] Information Sciences Institute, University of Southern California, “Internet Protocol,” RFC 791, September 1981, http://tools.ietf.org/html/rfc791.

[6] Hedrick, C., “Routing Information Protocol,” RFC 1058, June 1988, http://tools.ietf.org/html/rfc1058.

[7] Coltun, R., D. Ferguson, and J. Moy, “OSPF for IPv6,” December 1999, http://tools.ietf.org/html/rfc2740.

[8] Hedrick, C., “An Introduction to IGRP,” August 1991, http://www.cisco.com/warp/public/103/5.html.

[9] Rekhter, Y., T. Li, and S. Hares, “A Border Gateway Protocol 4 (BGP-4),” RFC 4271, January 2006, http://tools.ietf.org/html/rfc4271.

[10] Postel, J., “Internet Control Message Protocol,” RFC 792, September 1981, http://tools.ietf.org/html/rfc792.

[11] Plummer, D., "An Ethernet Address Resolution Protocol," RFC 826, November 1982, http://tools.ietf.org/html/rfc826.

[12] Finlayson, R., et al., "A Reverse Address Resolution Protocol," RFC 903, June 1984, http://tools.ietf.org/html/rfc903.

[13] Information Sciences Institute, University of Southern California, "Transport Control Protocol," RFC 793, September 1981, http://tools.ietf.org/html/rfc793.

[14] Postel, J., "User Datagram Protocol," RFC 768, August 1980, http://tools.ietf.org/html/rfc768.

[15] Klensin, J., "Simple Mail Transfer Protocol," RFC 2821, April 2001, http://tools.ietf.org/html/rfc2821.

[16] Fielding, R., et al., "Hypertext Transfer Protocol—HTTP/1.1," RFC 2616, June 1999, http://tools.ietf.org/html/rfc2616.

[17] Postel, J., and J. Reynolds, "File Transfer Protocol," RFC 959, October 1985, http://tools.ietf.org/html/rfc959.

[18] Harrington, D., R. Presuhn, and B. Wijnen, "An Architecture for Describing Simple Network Management Protocol (SNMP) Management Frameworks," RFC 3411, December 2002, http://tools.ietf.org/html/rfc3411.

[19] Lau, J., M. Townsley, and I. Goyret, "Layer Two Tunneling Protocol (L2TP)," RFC 3931, March 2005, http://tools.ietf.org/html/rfc3931.

[20] Valencia, A., M. Littlewood, and T. Kolar, "Cisco Layer 2 Forwarding Protocol "L2F"," RFC 2341, May 1998, http://tools.ietf.org/html/rfc2341.

[21] Hamzeh, K., et al., "Point-To-Point Tunneling Protocol (PPTP)," RFC 2637, July 1999, http://tools.ietf.org/html/rfc2637.

[22] Rand, D., "The PPP Compression Control Protocol," RFC 1962, June 1996, http://tools.ietf.org/html/rfc1962.

[23] Simpson, W., "The PPP Challenge Handshake Authentication Protocol (CHAP)," RFC 1994, August 1996, http://tools.ietf.org/html/rfc1994.

[24] Meyer, G., "The PPP Encryption Control Protocol," RFC 1968, June 1996, http://tools.ietf.org/html/rfc1968.

[25] Aboba, B., et al., "Extensible Authentication Protocol (EAP)," RFC 3748, June 2004, http://tools.ietf.org/html/rfc3748.

[26] Patel, B., B. Aboba, and W. Dixon, "Securing L2TP Using IPsec," RFC 3193, November 2001, http://tools.ietf.org/html/rfc3193.

[27] Kent, S., and K. Seo, "Security Architecture for the Internet Protocol," December 2005, RFC 4301, http://tools.ietf.org/html/rfc4301.

[28] Kent, S., "IP Authentication Header," December 2005, RFC 4302, http://tools.ietf.org/html/rfc4302.

[29] Kent, S., "IP Encapsulating Security Payload," December 2005, RFC 4303, http://tools.ietf.org/html/rfc4303.

[30] Harkins, D., and D. Carrel, "The Internet Key Exchange," RFC 2409, November 1998, http://tools.ietf.org/html/rfc2409.

[31] Maughan, D., et al., "Internet Security Association and Key Management Protocol (ISAKMP)," RFC 2408, November 1998, http://tools.ietf.org/html/rfc2408.

[32] Orman, H., "The OAKLEY Key Determination Protocol," RFC 2412, November 1998, http://tools.ietf.org/html/rfc2412.

[33] Cheng, P., et al., "Modular Key Management Protocol," Internet Draft draft-cheng-modular-ikmp-00, November 1994, http://citeseer.ist.psu.edu/cache/papers/cs/3026/ftp:zSzzSzathos.rutgers.eduzSzinternet-draftszSzdraft-cheng-modular-ikmp-00.pdf/cheng94modular.pdf.

[34] Karn, P., and W. Simpson, "Photuris: Session-Key Management Protocol," RFC 2522, March 1999, http://tools.ietf.org/html/rfc2522.

[35] Krawczyk, H., "SKEME: A Versatile Secure Key Exchange Mechanism for the Internet," *Proc. IEEE 1996 Symposium on Network and Distributed Systems Security*, 1994, p. 114.

[36] Aziz, A., and M. Patterson, "Simple Key-Management for Internet Protocols (SKIP)," *Proc., INET '95*, 1995.

[37] Kaufman, C., "Internet Key Exchange (IKEv2) Protocol," RFC 4306, December 2005, http://tools.ietf.org/html/rfc4306.

[38] Ylonen, T., and C. Lonvick, "The Secure Shell (SSH) Protocol Architecture," RFC 4251, January 2006, http://tools.ietf.org/html/rfc4251.

[39] Freier, A., P. Karlton, and P. Kocher, "The SSL Protocol Version 3.0," Internet Draft draft-freier-ssl-version3-02.txt, November 1996, http://wp.netscape.com/eng/ssl3/draft302.txt.

[40] Dierks, T., and E. Rescorla, "The Transport Layer Security (TLS) Protocol Version 1.1," RFC 4346, April 2006, http://tools.ietf.org/html/rfc4346.

[41] Klensin, J., "Simple Mail Transfer Protocol," RFC 2821, April 2001, http://tools.ietf.org/html/rfc2821.

[42] Resnick, P., "Internet Message Format," RFC 2822, April 2001, http://tools.ietf.org/html/rfc2822.

[43] Borenstein, N., and N. Freed, "MIME (Multipurpose Internet Mail Extensions) Part One: Mechanisms for Specifying and Describing the Format of Internet Message Bodies," RFC 1521, September 1993, http://tools.ietf.org/html/rfc1521.

[44] Linn, J., "Privacy Enhancement for Internet Electronic Mail: Part I: Message Encryption and Authentication Techniques," RFC 1421, February 1993, http://tools.ietf.org/html/rfc1421.

[45] Kent, S., "Privacy Enhancement for Internet Electronic Mail: Part II: Certificate-Based Key Management," RFC 1422, February 1993, http://tools.ietf.org/html/rfc1422.

[46] Balenson, D., "Privacy Enhancement for Internet Electronic Mail: Part III: Algorithms, Modes and Identifiers," RFC 1423, February 1993, http://tools.ietf.org/html/rfc1423.

[47] Kaliski, B., "Privacy Enhancement for Internet Electronic Mail: Part IV: Key Certification and Related Services," RFC 1424, February 1993, http://tools.ietf.org/html/rfc1424.

[48] Garfinkel, S., *PGP: Pretty Good Privacy*. Cambridge, MA: O'Reilly & Associates, 1991.

[49] Galvin, J., et al., "Security Multiparts for MIME: Multipart/Signed and Multipart/Encrypted," RFC 1847, October 1995, http://tools.ietf.org/html/rfc1847.

[50] Rescorla, E., and A. Schiffman, "The Secure Hypertext Transfer Protocol," RFC 2660, August 1999, http://tools.ietf.org/html/rfc2660.

[51] Atkins, D., and R. Austein, "Threat Analysis of the Domain Name System (DNS)," RFC 3833, August 2004, http://tools.ietf.org/html/rfc3833.

[52] Arends, R., et al., "DNS Security Introduction and Requirements," RFC 4033, March 2005, http://tools.ietf.org/html/rfc4033.

[53] Arends, R., et al., "Resource Records for the DNS Security Extensions," RFC 4034, March 2005, http://tools.ietf.org/html/rfc4034.

[54] Arends, R., et al., "Protocol Modifications for the DNS Security Extensions," RFC 4035, March 2005, http://tools.ietf.org/html/rfc4035.

[55] Case, J., M. Fedor, and M. Schoffstall, "A Simple Network Management (SNMP) Protocol," RFC 1157, May 1990, http://tools.ietf.org/html/rfc1157.

[56] Case, J., K. McCloghrie, and M. Rose, "Introduction to Version 2 of the Internet-Standard Network Management Framework," RFC 1441, April 1993, http://tools.ietf.org/html/rfc1441.

[57] Case, J., et al., "Introduction to Community-Based SNMPv2," RFC 1901, January 1996, http://tools.ietf.org/html/rfc1901.

[58] McCloghrie, K., "An Administrative Infrastructure for SNMPv2," RFC 1909, February 1996, http://tools.ietf.org/html/rfc1909.

[59] Frye, R., et al., "Coexistence Between Version 1, Version 2 and Version 3 of the Internet-Standard Network Management Framework," RFC 3584, August 2003, http://tools.ietf.org/html/rfc3584.

[60] Oppliger, R., *Authentication Systems for Secure Networks*, Norwood, MA: Artech House, 1996.

[61] IEEE, "Standard for Information technology—Telecommunications and Information Exchange Between Systems—Local and Metropolitan Area Networks-Specific Requirements—Part 11: Wireless LAN Medium Access Control (MAC) and Physical Layer (PHY) Specifications," http://ieeexplore.ieee.org/xpl/freeabs_all.jsp?tp=&isnumber=4248377&arnumber=4248378&punumber=4248376.

[62] IEEE, "Standard for Local and Metropolitan Area Networks Part 16: Air Interface for Fixed Broadband Wireless Access Systems," http://ieeexplore.ieee.org/xpl/freeabs_all.jsp?tp=&isnumber=29691&arnumber=1350465&punumber=9349.

[63] IEEE, "Standard for Information Technology—Telecommunications and Information Exchange Between Systems—Local and Metropolitan Area Networks—Specific Requirements—Part 15.1: Wireless Medium Access Control (MAC) and Physical Layer (PHY) Specifications for Wireless Personal Area Networks (WPANs)," http://ieeexplore.ieee.org/xpl/freeabs_all.jsp?tp=&isnumber=32053&arnumber=1490827&punumber=9980.

[64] International Engineering Consortium, "UMTS Protocols and Protocol Testing," http://www.iec.org/online/tutorials/acrobat/umts.pdf, 2007.

CHAPTER 9

Standard Public Key and Privilege Management Infrastructures

Javier Lopez

9.1 Key Management and Authentication

The security services that are based on cryptographic mechanisms assume that cryptographic keys are distributed among parties before they establish a secure communication. Therefore, the aim of key management is to provide secure procedures for the use of cryptographic keys because even the most sophisticated security concepts are wasteful if key management is not properly addressed.

A crucial problem of key management is key distribution. More precisely, it is necessary to guarantee to users the origin and integrity of the keys and, in the case of symmetric encryption, also confidentiality. The keys can be distributed either manually or automatically. However, in large networks like the Internet, manual distribution is not possible, even less so when the number of users is high. Thus, there is a need to separate users and keys into management domains, where any domain specifies the management area or scope from a functional point of view, and not necessarily from a physical point of view. Any flexible organization of those domains originates interrelation structures that influence the security policy of the system and the protocols used for the automatic distribution of the keys.

Due to the lack of a direct relation among most of the users, the protocols rely on the existence of a third party that gets directly involved in the distribution of the keys. This means that every user will base his operations not only on the messages sent and received during the protocol run, but also on the trust put on that intermediary entity. For this reason, it is called *trusted third party* (TTP). The TTP helps not only in the distribution of the keys but also in the mutual authentication of the users given that, as mentioned, they will not know one another.

Key distribution requirements in symmetric cryptosystems are very different from those of public-key cryptosystems. As already known, in the first type, the two users, Alice and Bob, involved in the communication must have a copy of the secret key (*session key*) used to protect the channel, keeping it hidden from the rest of users. In the second type, the private key must be known only by its owner (say, Alice), while the associated public key should be made known to all those who wish to communicate securely with Alice. Clearly, in the public key distribution process, confidentiality is not necessary, though it is essential to guarantee the authenticity of the key so that no dishonest user can substitute it. Additionally, it is necessary that Alice and Bob are authenticated in such a way that Alice is certain of communicating with Bob, and vice versa.

In the case of cryptographic mechanisms based on symmetric schemes, the TTP is referred to as either *key distribution center* (KDC) or *authentication server*. On the other hand, and as we will see later, if the mechanisms used are based on asymmetric schemes, the TTP is known as *certification authority* (CA) or *key certification center*. Basically, both KDC and CA are responsible for ensuring that the security policy of the system is enforced and for guaranteeing its integrity. In addition, they must register and identify users as well as provide them with some type of credentials.

As previously stated, users have to place a certain degree of trust in these types of intermediaries. For instance, in those scenarios in which a KDC is used to generate and distribute a session key, whenever two users wish to establish a secure connection, the users necessarily place *unconditional trust* on it. The reason is that the KDC would have the capability to intercept any communication protected by that key. In other systems, users have to trust CAs, which instead of issuing keys issue a sort of user credential. In this case, the trust users place on the CA is a *functional trust* [1].

In the case of the KDC, this one manages a kind of database containing as many secret keys as users that are registered in the system, thus sharing one specific key with each of those users. In this way, any of them can communicate securely with the KDC, encrypting and authenticating the messages exchanged with it. Therefore, every time a new user is registered in the system, or when there is any suspicion that a user's secret key has been compromised, only one single point needs to be updated.

In the simplest distribution protocol [2], Alice indicates to the KDC that she wants to communicate securely with Bob. The KDC generates a random session key K_{AB} [3], which is communicated to both parties. In particular, the KDC encrypts and sends the session key to Alice and Bob using the corresponding secret key shared with each of them. In this way, Alice and Bob obtain a session key, known only to them (and obviously to the KDC), that they use to encrypt the transmission of a further communication to be established. An additional result is that Alice and Bob are authenticated.

For years, researchers have been designing many different protocols to carry out these functionalities (i.e., use a KDC to provide two users with a random session key and allow them to authenticate each other). Among them, some pioneering works are [4–8]. It must be noted that the protocol to select for a particular application will basically depend on several factors, such as

- The underlying communications architecture;
- The goal to minimize either the size or the number of messages exchanged;
- Control as to whether all the parts intervening need to interact with one another during the distribution protocol.

Nevertheless, the use of a KDC also has some serious drawbacks, as pointed out in [9]:

- The KDC has enough information to impersonate a user. If an intruder obtains access to the KDC, all the encrypted documents circulating on the network will become vulnerable.
- The KDC represents a single point of failure. If it is rendered unusable, no one can establish secure communications within the network unless they use keys distributed before the failure (which is not desirable because, by definition, session keys should not be used for longer than the duration of a communication session between two users).
- The overall system performance may suffer when the KDC becomes a bottleneck, which is not an unlikely scenario, given that all the users need to communicate with it frequently in order to obtain session keys.

The most basic management system of this type uses a single KDC for the entire administrative domain. However, in this approach, the aforementioned drawbacks become even more obvious. One possible solution could be to have more distribution centers. In large networks, key management is normally organized in a hierarchical way in local, regional, and then in a single global KDC. In this way, a local KDC communicates with its local entities (i.e., with the set of users—the domain—registered in it). For the communication among users belonging to the same domain, key distribution procedures are the same as in the case of a single KDC. However, if users belonging to different domains wish to communicate, a common KDC of some higher level but within the predetermined structure is used. This hierarchical design can be extended to as many levels as are necessary [10].

A similar design is also applicable in the case of the certification authorities. As it is already well known, one of the main advantages of public key cryptography techniques is its asymmetry: while a particular user is the only one who, using his private key, can carry out a certain operation (either decrypt or sign a message), all other parties who know the associated public key can carry out the inverse operation (encrypt or verify the signature, respectively). Therefore, the distribution of keys is conceptually easier in public key cryptography because each user of the system is responsible for knowing his own private key, while at the same time public keys of all users can be retrieved from one place.

Nevertheless, this technology also has some disadvantages. For example, a user could gain access to the place where the public keys are stored and could illegitimately substitute his public key for that of another user (e.g., Alice). The rest of users of the system would believe they were obtaining access to Alice's public key when they would in fact be accessing that of the intruder. As a result, Alice would be being impersonated by that user, who would intercept the encrypted messages originally intended for Alice.

Therefore, Alice's public key, K_{Apu}, is useless to Bob unless he believes it is authentic (i.e., unless Bob can be persuaded that K_{Apu} really belongs to Alice and that only she knows the corresponding private key K_{Apr}). Hence, the main issue in public key cryptography is to provide mechanisms that allow users to obtain another user's public key together with proof of identity of that key. A typical solution to this type of problem is the use of a CA, which is in charge of issuing *identity certificates* that will also support authentication of users.

An identity certificate or a *public key certificate* is a digital signature on a message, issued by an authority, which proves that a certain public key belongs to a particular user (i.e., it creates a link between that individual and the corresponding public key). For this, the certificate contains information on the user identity as well as the public key and the authority who issued it. From this point on, we will use $Cert_X(Y)$ to refer to the public key certificate of user "Y" issued by a CA "X."

Therefore, all users of the system must know the public key of the CA, K_{CApu}, so that they can verify the signature of any certificate it issues. The signature performed by the CA provides the certificate with three important features:

- It guarantees the integrity of the certificate;
- Because only the CA knows its own private key, K_{CApr}, every user that verifies the signature on the certificate can be sure that only the CA could have issued it;
- For the same reason, the CA cannot deny having issued the certificate.

In this way, identity certificates constitute the basis of a systematic solution to public key distribution, and the use of the CAs is fundamental.

In a broad sense, and although not necessarily used in the same scenarios, from the key management point of view the use of CAs has certain advantages over the systems using KDCs—namely:

- A CA requires less trust from the user than a KDC because most of the information the CA possesses is totally public.
- The CA does not need to be online since the certificate is issued when the user is registered into the system (it is normally necessary for the individual to be physically present with some type of official document to identify herself).
- The network continues working even though the CA stops working (unlike the case of the KDC). The only operation that would be inoperative is the issuing of new certificates.
- Certificates are not security sensitive since the public keys they contain are not confidential and they are guaranteed by the CA signature. In this way, for example, if the certificates are stored in a directory service and an intruder has access to them, he will be able to delete them but neither modify the content nor create other new certificates because only the CA can produce the appropriate signature.
- If a CA is compromised or exposed, there will be no possibility of decrypting communications, whereas if a KDC controlling the communication between two users is exposed, that communication can still be decrypted.

9.2 Public Key Infrastructures

As stated previously, when a user wishes to validate the certificate issued by a CA, she must know and therefore have previously obtained the public key of this authority. If it were possible to establish a single CA that issued all Internet users with identity certificates, then the problem of public key distribution would be resolved as it would be easy to inform all users about the specific public key of the CA.

Unfortunately, this is not possible, since it is not practical for a single CA to have sufficient knowledge of all the users, and even less feasible for it to be able to establish a suitable relationship with all of them. It is therefore necessary to acknowledge two facts: (a) multiple domains, and consequently many authorities, are necessary; and (b) it cannot be assumed that a user will have previous knowledge of the public key certificates of all those users he might wish to communicate with.

Therefore, for example, Alice can trust on Bob's public key, K_{Bpu}, if this is given to her personally by Bob. Otherwise, Alice can trust the validity of that key if it is obtained from a certificate, $Cert_{CA3}(Bob)$, issued by the certification authority CA_3, which Alice knows and trusts. Alice has the public key of the CA, K_{CA3pu}, and she knows this is genuine, as it was either personally delivered by CA_3 or, more generally speaking, obtained following any equivalent secure way.

However, it may happen that Alice does not know CA_3. In this case, Alice does not trust it, though she may still be able to obtain its public key from a certificate, $Cert_{CA2}(CA_3)$, issued by authority CA_2, from which she obtained K_{CA2pu}. Similarly, this public key could have been delivered personally to Alice by CA_2, and, therefore, Alice will be convinced of its legitimacy.

Yet Alice may not know CA_2, and she might obtain its key from another certificate, $Cert_{CA1}(CA_2)$, issued this time by CA_1, which Alice definitely knows personally and for which she has learned its public key, K_{CA1pu}.

As we have seen, the certification scheme is used recursively to obtain the certificate of a remote user, using a certificate chain, also known as *trust chain* or *certification path*. The process begins by Alice having the public key of an authority in which there is direct trust. From there, the public key of the targeted user Bob is acquired through a chain of certificates that are progressively obtained and on which Alice trusts in an indirect way.

In summary, the certification path is determined by a sequence of CAs in such a way that Alice, who wishes to verify Bob's signature on a document, has a trusted copy of the public key of the first CA of the path (CA_1 in the example), while the last CA of the path (CA_3) is precisely the one who issued the certificate with the public key needed by Alice (i.e., K_{Bpu}, Bob's key). The result of this procedure is that users do not need to personally know the other users with whom they are interacting, they just need to validate the certificates of those yet unknown.

Consequently, it is necessary to design an infrastructure of CAs that allows users to establish those trust chains. Such an infrastructure is known as *public key infrastructure (PKI)*. Figure 9.1 shows a typical PKI. This is the underlying framework that allows public key technology to be widely used, since it provides the trust basis necessary for electronic correspondence among users who cannot interchange their keys manually. Through the use of a PKI and the administration of public key certificates, it is possible to establish and maintain a secure network environment, enabling the use of encrypted services and digital signatures in a wide range of applications.

However, services offered by the CAs of the infrastructure will only be of use if users consider them to be suitable, reliable, and practical. This opinion will partly depend on the operating procedures of the CAs being well documented, and such

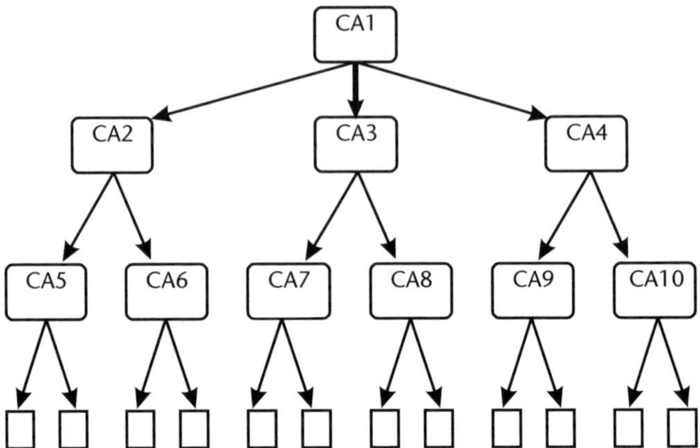

Figure 9.1 Example of PKI.

documentation must be available to any of the parties involved. To this end, the *certification policy statement (CPS)* has been developed. It is a declaration of the procedures a CA uses when issuing certificates and ensures that procedures are homogenous and compatible when numerous CAs coexist within the same system, as is the case of a PKI.

In general, the CPS must define or determine, among other things, the entity that is going to generate the pair <public key, private key>, the size of the prime numbers used in the key generation, the quality of the software used to generate keys as well as the parameters used, the identification and authentication requirements of the users, the issuing frequency of lists of certificates that are no longer valid as well as methods for their use, the validity period of the certificates, and so on.

The next sections introduce, first, the services that the PKI must provide and, second, the types of entities that operate in the infrastructure. This is presented from a global perspective, since both the structural model and the specific certification policies of every PKI determine, for every specific case, the services to be provided, the entities involved, and the requirements necessary.

9.2.1 PKI Services

The operation of a PKI is associated with a set of activities. These are determined by the specific necessities of the users, who require that the infrastructure provides them with a series of services that guarantee the security of the information systems and the communication that are based on public key cryptosystems. Description of the services follows.

9.2.1.1 Issuing and Distribution of Certificates

The CAs integrated in a PKI act as trust agents within that infrastructure, issuing the public key certificates of all related users. The process of issuing a digital certificate can be offline or online. In the first case, the physical presence before the CA of the user requesting the certificate is necessary. This process is similar to applying for a driver's license or a credit card, and it is necessary to present a proof of identity. However, in the second case, a web interface (or similar) is used to apply for the certificate. This interface must be able to guarantee to the CA the authenticity of the user, which in the absence of a physical presence is very difficult. Furthermore, this method is open to attacks.

A CA issues certificates by digitally signing a data structure, which includes information such as the user name, his public key, the validity period of the key, and the identity of the CA who issued the certificate. Figure 9.2 shows the format of the public key certificate. The format corresponds to the X.509 certificate standard. The latest version of this format, V3, evolved from V1 [11], which was subsequently revised in V2 [12] to address security problems [13] and finally extended to the actual version [14]. In fact, X.509 was initially conceived to support authentication in X.500 directory as a standard way to develop an electronic directory of people in organizations in order to build a global directory available to all Internet users. This influenced the X.509 certificate structure, even when it is now used with goals different than those in its conception. The fields of the certificate are as follows:

- *Version*: indicates whether the format corresponds to version 1, 2, or 3 of X.509;
- *Serial number*: unique identification number for the certificate, assigned by the CA;
- *Signature algorithm*: identifier of the algorithm used by the CA to digitally sign the certificate;
- *Issuer*: identity of the CA issuing the certificate;
- *Validity period*: validity period of the certificate (when the certificate becomes valid and when it expires);

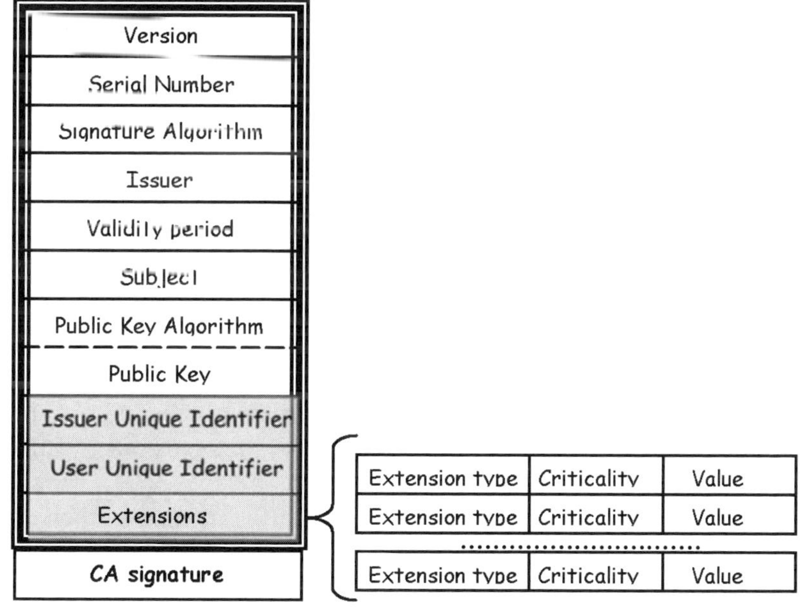

Figure 9.2 Structure of a X.509 v.3 public key certificate.

- *Subject*: identity of the user of the private key for which the corresponding public key is being certified;
- *Public key algorithm*: identifier of the algorithm with which this public key is to be used;
- *Key*: value of the user's public key;
- *Issuer unique identifier*: optional bit string used to make the issuing certification authority name unambiguous, in the case in which this can happen (not used in version 1);
- *User unique identifier*: optional bit string used to make the user name unambiguous, in the case in which this can happen (not used in version 1);
- *Extensions*: additional fields to the certificate, in which the criticality indicator is a flag that specifies if the extension is noncritical (it can be ignored if it does not recognize the extension type) or critical;
- *Signature*: digital signature of the CA to the hash value of all the previous fields.

Another issue to consider is how the certificates are distributed within the infrastructure. In other words, when Alice receives a message signed by Bob and wants to verify the digital signature of that message, she needs Bob's certificate, or as previously stated, the certificates of all intermediary CAs that form part of the certification path between the two of them.

The distribution of a certificate does not require confidentiality, given that it contains public information. Similarly, protection is not an issue, as it is signed by a reliable authority. Hence, CAs can send the certificates to directories (or repositories) that are universally available [11, 15] and that work as distributed databases designed to keep certain information of real-world objects. An alternative is that CAs do not store the certificates in repositories, but, rather, certificates are given to each user of the system (for instance, on registration) and the user stores it locally. In both cases, the certificates of the CAs and of the users are treated identically.

9.2.1.2 Obtaining Certificates

In order to verify the digital signature of a document, the public key of the signer is necessary. Therefore, and as already mentioned, the recipient must obtain the certificate of that key and almost certainly other additional certificates, like those of the CAs of the certification path between the verifier and the signer.

There are several possibilities for obtaining the certificates. For example, they can be obtained from the certificate repository. Although the database is distributed, the verifier communicates with the directory as if the information was centralized, meaning that the search for the information is carried out transparently. The verifier will have to acquire the certificates forming part of the certificate chain, one by one.

On the other hand, if users have their own certificates, then the issuer (signer) can send his certificate, together with the signed document, to the recipient (verifier); alternatively, the certificate and the document can be sent in separate messages. However, the signer might send not only his certificate but also all those that form part of the certification path. In these cases in which a repository is not used, the acquisition of the necessary certificates relies on a specific protocol integrated into the application that uses them.

9.2.1.3 Certificate Revocation and Suspension

Once a certificate has been issued, its lifetime is defined by the dates in the corresponding field of the certificate data structure. At the moment of expiration or, depending on the specific policy, sometime later, it will no longer be used because the link between the key and the subject disappears except in the case in which it is used in the future to validate a document signed while the certificate was still valid [16].

However, before the expiration date and under certain circumstances, it may be advisable to avoid the use of a certificate of a particular user. Some possible causes are as follows:

- The public key is no longer valid (e.g., when it is detected or even suspected that the corresponding private key has been exposed or compromised and it is known by someone other than its legitimate owner);
- The user identified in the certificate is no longer considered an authorized user of the corresponding private key (e.g., because he has moved to another company or remains within the same company but his privileges have changed).
- The information in the certificate changes (e.g., a change in the name, such as when a woman changes her last name to that of her husband after marriage), or when there is a change in some of the characteristics or attributes described in the certificate.

In these circumstances, the certificate must be invalidated. This is the *revocation* process and is the responsibility of the CA who issued the certificate. The revocation state of a certificate must be checked before it is used, and for this the PKI must provide a scalable revocation system. The CA must be able to publish information relating to the state of the certificate in the system. That is, it must be able to revoke the certificate, and this will allow users to verify whether or not the certificate is valid.

The Internet community and the CCITT have developed the concept of the *certificate revocation list* (CRL) as a revocation mechanism. A CRL is a list of revoked certificates, digitally signed by the authority who previously issued the certificates in question (Figure 9.3). The fields of the CRL are:

- *Version*: indicates whether the format corresponds to version 1 or 2 of X.509 CRL;
- *Signature algorithm*: identifier of the algorithm used by the CA to digitally sign the certificate;
- *Issuer*: identity of the CA that issued the list;
- *Actual update*: date and time of issue of the CRL;

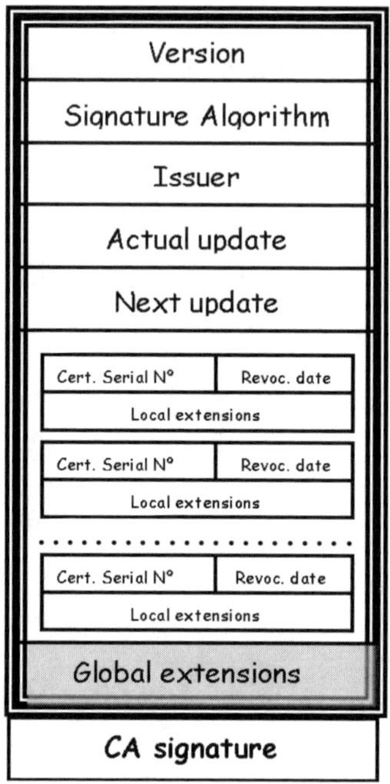

Figure 9.3 Structure of a CRL V.2.

- *Next update*: date and time when next CRL will be issued;
- *Certificate serial number*: serial number of the certificate being revoked;
- *Revocation date*: date of revocation of the particular certificate;
- *Local extensions*: additional fields that may be attached to the particular revoked certificate;
- *CRL extensions*: additional fields that may be attached to the CRL;
- *Signature*: digital signature of the CA to the value obtained by applying a hash function to all the previous fields.

These lists must be generated or updated periodically so that they are available to all the system users. This is important in the validation of a certificate or a certification chain since the verifier not only has to verify the certificate signature but also has to get the most recent CRL in order to check that the certificate in question has not been revoked.

Generally, users have two options to check whether or not a certificate has been revoked. First, when a CA generates a CRL it can be sent automatically to all the users; this is called the *push mode*. Alternatively, and a more common solution, is the *pull mode*, in which the CRLs are not distributed automatically. Instead, users access the list periodically issued by the CA, which is deposited in a directory server, for example. The list has the format determined by some of the versions of the X.509 standard for the revocation lists. In the pull mode, the PKI users are entirely responsible for obtaining the lists. In this way, when a user wishes to use a certificate, she must verify that the serial number of that certificate is not in the CRL. In any case, only the CA can revoke a certificate and can do so either on the request of the certificate owner or on the request of a third party with sufficient authority.

Sometimes a certificate is not revoked directly but is left in a state of suspension, indicating to users that the revocation is being considered because, for example, the revocation request has not yet been authenticated. The suspension has an expiration date, which is when it is deleted from the CRL. This time period must give the CA enough time to be able to check and confirm the request, after which it will have to decide either to revoke the request or grant the certificate.

It must be noted that the IETF community has developed an online alternative to CRLs. This is the *online certificate status protocol* (OCSP), used for determining the current status of a digital certificate without requiring CRLs [17, 18].

9.2.1.4 Cross Certification

Cross certification is a process used to extend trust relations between two different domains. A CA of domain *X*, for example, creates a certificate, called *cross certificate*, which will be used by its users and which contains the public key of the CA of a second domain (e.g., domain *Y*). Simultaneously, the CA of the domain *Y* creates another cross certificate, which will be used by its users and which contains the public key of the CA of domain *X*.

In this way, users of domain X can implicitly trust on the certificates of users of domain Y using the certificate issued by *X*'s CA regarding *Y*. Clearly, the opposing implicit trust process is also true.

Therefore, the cross certificate is the mechanism by which two CAs show the trust relation between them, offering users higher flexibility and creating shorter certification paths. This type of certification may break the model of the basic structure. However, many PKIs allow cross certificates to be created in justifiable circumstances, such as when there is a high degree of interaction between the users of two specific domains (e.g., in which the domains relate to two departments within a company and there is a steady flow of documents between them, or in which those domains relate to business partners who communicate frequently).

The use of cross certificates also allows a network of trust relations to be established, among not only CAs in the same infrastructure, but also those from different infrastructures. Nevertheless, it is not recommended that cross certification is used freely; it should be subject to some restrictions.

9.2.1.5 Key Generation

There are various alternatives for the generation of the <public key, private key> pair. The first alternative is when a pair is generated in the same system (possibly using the same hardware and software) in which the private key will be stored and used. It is as if the user generates her pair of keys. It should be observed that, in this case, the private key never leaves its native environment, which is the best way to guarantee that no other user can have access to it.

If the user generates her own pair of keys, the private key must be stored in a safe place in such a way that it will be neither exposed nor compromised. The user will have to present her public key to her CA so that this one can issue the corresponding certificate.

The CA can perform some security tests on the key to verify that it is a *strong key* and also to check that it has in fact been generated by the user wishing to register it. The quality of the key pair depends on the random features of the seed of the generation function [19]. Ideally, a truly random function should be used. However, this type of function is not always available, and it is therefore more common to use a pseudo random function on the secret seed, which is normally derived from the previous cycle of the same function. The security of the whole process depends on the difficulty of predicting the first seed. Once the CA is convinced of both the user identity and the validity of the key, it generates the certificate linking the user with the public key.

Another alternative in the generation process is in the case in which the user does not always have the software or the hardware available to generate her key pair. Sometimes it is simply not practical to do so, either because of the cost or because of the feasibility. In these cases, the PKI provides a central system to generate the keys in which the user can generate the key pair by herself. The system generates the keys and delivers them to the user. Later, the user presents the public key to the CA so that it can issue the corresponding certificate.

One last alternative is when the user is not authorized to generate the pair of keys, neither autonomously nor using the resources offered by the PKI. In this case, the user must use the key pair that the CA generates directly for her while the corresponding certificate is being issued. Obviously, with this method there is always a risk that some time in the future the user repudiates the signature of a document based on the fact that for certain period her private key was not under her direct control.

9.2.1.6 Key Update

All the <public key, private key> pairs have to be updated periodically, and, therefore, the PKI must provide an update service. The purpose of this service is to limit the possibility of a cryptanalytic attack and also to limit the use of keys to a predefined fixed period. Depending on the technology used and on the application environment, this update period may vary from weeks to a few years.

The validity period of a certificate indicates to other users that, during this period, they can assume that the public key is valid and that the link between the public key and the rest of the information within the certificate is also valid. As we have already seen, this validity period can be terminated by the revocation of the certificate.

However, the validity of the private key may differ from the validity period of the corresponding public key certificate. This is because when the key pair is updated, the private key will not be used further (and can be destroyed); however, the public key contained in the certificate will still be necessary in the future to verify the documents signed by her owner before updating or revoking the certificate. Therefore, the validity of the public keys (and hence, the certificates that contain them) must be extended further than the validity of the corresponding private keys, which means it is necessary to keep a historical record of the private keys of all users.

It is important to remember that the lifetime of a certificate is not only limited by its validity period, but also by the validity of the signature of the CA issuing it. Therefore, it is important that the public key of the CA is valid for longer time than the certificates the CA issues.

The updating process must be transparent to the user and must take place before the keys expire, so that the user never finds herself unable to operate (i.e., she should never be mistakenly denied a service). Once the update has taken place, the public key should automatically and transparently be stored in the records.

9.2.1.7 Key Protection and Recovery

So far we have discussed the role of the PKIs under the premise that user Alice needs to know the public key of user Bob in order to verify the signature of a document created by him. The certification chain containing not only Bob's certificate but also those of the CAs of the PKI between them allows Alice to learn Bob's public key and guarantees its authenticity.

The focus has been on the signature on documents. As already stated, however, the main feature of some of the public key cryptosystems is that they can be used not only to digitally sign documents, but also to encrypt them. In other words, Alice may need Bob's public key, not to verify his signature on a document, but in order to send him an encrypted message, which only Bob will be able to decrypt using his private key.

Nevertheless, looking more closely at the management issues, it can be seen that it is not practical to use a pair of keys twice, and, therefore, if a user wants to perform the two operations mentioned, then two different key pairs will be necessary—one for the encryption/decryption process and another one for the signing/verification process. More specifically, requirements vary from one case to another, and hence for the signature key pair, we realize that:

(a1) The private key of the pair must be stored in such a way that no other user can obtain access to it. It is advisable and indeed sometimes compulsory (e.g., in the standard ANSI X9.57 [20]) that the private key must never leave the device in which it is used. In this way the legitimate user of the key will never be able to deny having signed a document.

(a2) There should be no backup copy created or stored of the private key as a security measure (e.g., in case of loss), as this would be a violation of requirement (a1). If the user loses his private key, a new pair is simply generated and a new certificate issued.

(a3) The public key of the key pair does need to be filed in order to verify old signatures for a period of time after the corresponding private key has expired.

(a4) The private key of the key pair must be destroyed when its validity period has expired, as even much later it could be used fraudulently to falsify the signatures relating to old documents.

On the other hand, the management requirements for the encryption key pair are as follows:

(b1) It is necessary to file or to create a backup copy of the private key, since this may be the only way of recovering the encrypted information. It would not be acceptable that if the private key were lost, all the encrypted information would then be inaccessible. For example, an employee may forget his private key or could be fired and refuse to reveal it. In both cases, the company would be unable to recover the encrypted information unless there was a record of the user's key.

(b2) The public key of the pair need never be filed, since if it is lost a new pair of keys is generated.

(b3) The private key need not be destroyed when its lifecycle has finished. Furthermore, requirement (b1) implies that it never should be.

These two sets of requirements are in conflict with each other. If the goal is to use a single key pair for both the encryption and the signature, it is impossible to satisfy all the requirements. In addition, other arguments support the use of different key pairs for encryption and signing. The first one is that it is undesirable that the key pair changes too frequently, as this is detrimental to the PKI in relation to certificate storage and revocation. However, it is highly recommended that the key pair used for encryption is changed frequently. The more often it is used, the more it should be changed. The second argument is that not all the public key algorithms can be used both in encryption/decryption and signing/verification processes.

Consequently, it is necessary for the PKI to provide services for safeguarding and recovering keys. It is clear that this service only makes sense if the users of the PKI are going to carry out as many encryptions as signing operations. It is important not to confuse the safeguarding and recovery of keys with key escrow. The last one means that a third party can obtain the decryption keys necessary to access the encrypted information.

9.2.2 Types of PKI Entities and Their Functionalities

There are different types of entities in a PKI, each one having particular functions within the infrastructure. In the following, these types are described, specifying the tasks they carry out within the system. It is necessary to point out that some of these entities may not exist (not be necessary) within certain PKI implementations. Furthermore, the names with which they are referred in the following description are generic and may vary according to specific implementations.

9.2.2.1 Certification Authority

This entity is the typical kind of authority that we have described so far. Its scope is limited to the following tasks: issue final user certificates; generate CRLs; generate <public key, private key> pairs; confirm that any user requesting the issue of a certificate has the private key corresponding to the public key presented; perform backup copies and/or filing the encrypting keys; verify the uniqueness of any public key; and generate certificates that permit cross certification.

9.2.2.2 Local Registration Authority

Sometimes CAs require the physical presence of the final users. For instance, to verify the identity of a user who presents his public key when he requests the CA to issue a certificate. This makes it difficult for a CA to be able to provide a service to a wide community of users, especially if they are spread out geographically. One solution is to use intermediary authorities, subordinated to the CA, which fill in the gap existing among the final users and their corresponding CAs, especially when they are physically very far apart.

This type of entity is known as *local registration authority* (LRA). It cannot carry out typical CA services, such as issuing certificates or CRLs. It can, however, perform the following tasks: assist in registration tasks, users joining or leaving the system, and attribute changes of the final users; verify the identity of the final users; authorize and sign requests for the generation of key pairs, issuance of certificates, key recovery, and certificate revocation, which will later be transferred to the CA on which they depend (and which carries out these operations); and provide the user with the equipment necessary so that he can generate his own key pair.

9.2.2.3 Root Authority

Many PKIs incorporate an authority in charge of approving the global certification policy for the whole infrastructure (see higher node in Figure 9.1). This type of authority is called the *root authority* (RA). Where it exists, this authority is unique, and it establishes the guidelines that final users, groups of users, CAs, and other subordinate authorities must follow. This type of entity rarely certifies the users directly, which is a task normally left to the CAs. Generally, its public key must be recognized by all the other authorities and PKI users, as it is normally used in any certification path verification.

An RA normally performs the following tasks: establishes the policy and the general procedures for all the entities of the infrastructure; publishes its own public key appropriately; identifies and authenticates all the subordinated CAs and generates their corresponding certificates; identifies and authenticates any RA of another PKI with which it may establish cross certification and generates the corresponding certificate; and receives and authenticates revocation requests and generates the CRLs necessary for all the certificates that it has issued.

9.2.2.4 Policy Certification Authority

There are situations in which the RA approves specific certification policies for particular user subgroups, allowing these policies to be extensions but not limitations of the policy originally established. The expansion may be related with, for instance, the range of the size of the public key module, the way CRLs are managed, the validity period of the certificates, and so on.

In these cases, a new type of authority is established between the RA and the CAs. This authority is known as the *policy certification authority* (PCA), and there is one for every subgroup of users who work with a policy that is different from the initial one (see second level of authorities in Figure 9.1). It performs the following tasks: publishes the identification and location of all the CAs it certifies; publishes its security policy or, at least, those procedures that are different from the ones included in the original policy established by the RA; identifies and authenticates all the subordinated CAs and generates their corresponding certificates; and receives and authenticates revocation requests and generates the CRLs necessary for all the certificates it has issued.

In these cases in which PCAs exist within the PKI, the RA tasks described previously change as they are applicable to the PCAs and not to the CAs. Furthermore, the RA will have to receive and publish the policies related to those PCAs.

9.2.2.5 End Users

The end users or clients of the infrastructure will generally perform the following tasks: generate signatures (i.e., digitally sign documents); verify signatures performed by other entities in the PKI; interpret certificates; interpret CRLs; determine the certificate status; and obtain certificates of directory servers.

9.3 Privilege Management Infrastructures

As we have seen in the previous section, the authentication service helps the user to prove who he is, and public key certificates provide the best solution to integrate that basic security service into applications that make use of digital signatures. On the other hand, many applications additionally demand an *authorization service* (i.e., a service that describes what the user is allowed to do). In these cases, privileges to perform tasks should be considered. For instance, when a company needs to establish distinctions among its employees regarding privileges over resources (either hardware or software), the authorization service becomes important, and different sets of privileges are assigned to different categories of employees.

Traditional authorization solutions are not easy to use in application scenarios, where the use of public key certificates to attest the connection of public keys to identified subscribers is a must. One example of a traditional solution is the *discretionary access control* (DAC), which governs the access of users to information on the basis of the users' identities and authorizations. Authorizations specify, for each individual user and each object (resource) in the system, his access rights (i.e., the operations that the user is allowed to perform on the object). Each activity is checked against the access rights, which are held as access control lists within each target resource. If authorization stating the user can access the object in the specified mode exists, then access is granted; otherwise, it is denied.

Another scheme is *mandatory access control* (MAC), which governs access on the basis of the classification of resources and users according to security levels. Thus, access to a resource is granted if the security level of a particular user stands in accordance with the security level of that object. A classification list that is typically used in military applications is unmarked, unclassified, restricted, confidential, secret, and top secret [21]. It is reasonable to think that management of access rights under both DAC and MAC must be done by system administrators.

A *role-based access control* (RBAC) scheme is an alternative solution to discretionary and mandatory schemes [22]. A role policy regulates the access of users to information on the basis of the activities that the users perform in the system in pursuit of their goals. A role can be defined as a set of actions and responsibilities associated with a particular working activity. Instead of specifying all the actions that any individual user is allowed to execute, actions are specified according to roles [23].

Those schemes mainly focus on access control. However, in this section we focus on issues like group membership, role identification (collection of permissions or access rights and aliases for the user's identity), limits on the value of transactions, access time for operations, security clearances, and time limits that ITU-T generally treats as authorization. In order to provide support to applications dealing with those issues, *attribute certificates* come into place.

The U.S. American National Standards Institute (ANSI) X9 committee developed the concept of attribute certificate as a data structure that binds some attribute values with identification information about its holder. This type of certificate has been incorporated into both the ANSI X9.57 standard and the X.509-related standards and recommendations of ITU-T, ISO/IEC, and IETF. According to RFC 2828 [24], an attribute certificate is "a digital certificate that binds a set of descriptive data items, other than a public key, either directly to a subject name or to the identifier of another certificate that is a public-key certificate."

One of the advantages of an attribute certificate is that it can be used for various purposes. It may contain group membership, role, clearance, or any other form of authorization. Yet another essential feature is that the attribute certificate provides the means to transport authorization information to decentralized applications. This is especially relevant because through attribute certificates, authorization information becomes "mobile," which is highly convenient for many applications. Actually, the mobility feature of attributes has been used in applications since ITU-T 1997 Recommendation was published [25]. However, at that moment, the concept of attribute certificate was ill-defined, and because of this, as an alternative, one of the extension fields (*subjectDirectoryAttributes*) of the public key certificate itself has been initially used in a number of applications as support to store attributes of the user:

```
subjectDirectoryAttributes EXTENSION ::= {
SYNTAX AttributesSyntax
IDENTIFIED BY id-ce-subjectDirectoryAttributes }
AttributesSyntax ::= SEQUENCE SIZE (1..MAX) OF Attribute
```

Nevertheless, this solution does not make user attributes independent from user identity, which can cause problems. First, this is not convenient in situations in which the authority issuing the public key certificate is not the authority for assigning privileges, which occurs frequently. Second, life of public key certificates is relatively long when compared to frequency of change of user privileges. This means that every time privileges change, it would be necessary to revoke the public key certificate. Moreover, many applications deal with authorization issues like delegation (conveyance of privilege from one entity that holds a privilege to another entity) or substitution (one user is temporarily substituted with another user, and this one holds the privileges of the first one for a certain period of time). But public key certificates do not support delegation or substitution.

In a further improvement (i.e., in another recommendation [26]), the ITU-T has specified the format of attribute certificates as a separate data structure from the public key certificate of the user, but logically bounded as explained in the following paragraphs. To the eyes of the ITU-T, the use of a wide-ranging authentication service based on public key certificates is not practical unless it is complemented by a framework in which the PKI provides the efficient and trustworthy means to manage and distribute all certificates in the system. Similarly, the ITU has defined a framework for the use of attribute certificates on which a *privilege management infrastructure* (PMI) can be built. The same framework defines a new type of authority, the *attribute authority* (AA), for the assignment of privileges. A PMI is built to contain a multiplicity of AAs and end users, and the AA at the highest level is called *source of authority* (SOA). Revocation procedures are also considered by defining the concept of *attribute certificate revocation list* (ACRL), which is handled in the same way as CRLs published by CAs.

As mentioned, attribute certificates were designed to be used in conjunction with public key certificates. The way both types of certificates are bound is shown in Figure 9.4. The link between PKI and PMI is justified by the fact that authorization relies on authentication to prove who the user is. Although linked, both infrastructures can be autonomous and managed independently. Creation and maintenance of identities can be separated from PMI, as authorities that issue certificates in each of the infrastructures are not necessarily the same ones. In fact, the entire PKI may be existing and operational prior to the establishment of the PMI.

Figure 9.5 shows that the field *holder* in the attribute certificate contains the serial number of the public key certificate. However, there are other ways to describe the holder of the attribute certificate. In spite of using the serial number for the public key certificate, it is possible to bind the attribute certificate to any object by using the hash value of that object. For instance, the hash value of the public key, or the hash value of the identity certificate itself, can be used. All possibilities for binding can be concluded from the ASN.1 [27] specification for the field *holder* shown in Figure 9.5, where other related data structures are also specified.

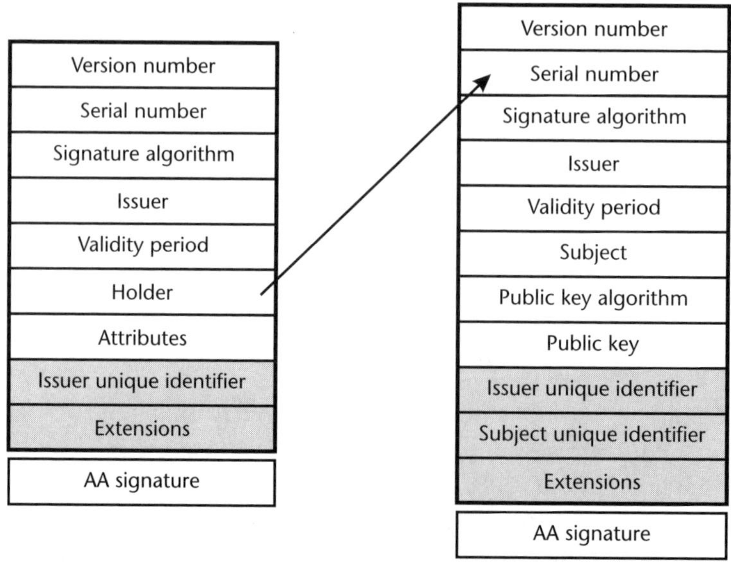

Figure 9.4 Attribute certificate structure and its link to public key certificate.

```
Holder ::=  SEQUENCE
{
            baseCertificateID       [0] IssuerSerial        OPTIONAL,
                -- the issuer and serial number of the holder's identity certificate
            entityName              [1] GeneralNames        OPTIONAL,
                -- the name of the entity or role
            objectDigestInfo        [2] ObjectDigestInfo    OPTIONAL
                -- used to directly authenticate the holder, e.g. an executable
                -- at least one of baseCertificateID, entityName or objectDigestInfo
                  present --
}

GeneralNames ::= SEQUENCE SIZE (1..MAX) OF GeneralName

GeneralName ::= CHOICE
{
   otherName                    [0]  INSTANCE OF OTHER-NAME,
   rfc822Name                   [1]  IA5String,
   dNSName                      [2]  IA5String,
   x400Address                  [3]  ORAddress,
   directoryName                [4]  Name,
   ediPartyName                 [5]  EDIPartyName,
   uniformResourceIdentifier    [6]  IA5String,
   iPAddress                    [7]  OCTET STRING,
   registe redID                [8]  OBJECT IDENTIFIER
}

ObjectDigestInfo   ::= SEQUENCE
{
   digestedObjectType  ENUMERATED {
        publicKey             (0),
        publicKeyCert         (1),
        otherObjectTypes      (2) },
   otherObjectTypeID          OBJECT IDENTIFIER  OPTIONAL,
   digestAlgorithm         AlgorithmIdentifier,
   objectDigest            BIT STRING
}
```

Figure 9.5 ASN.1 specification of the field *holder*.

The reason for the clear separation of functionalities between CA and AA, one dealing with user identities while the other deals with user attributes, is that an identity tends to have a global meaning; thus, public key certificates may be issued by a CA managed by an external organization (e.g., a national or regional government). On the contrary, an attribute tends to have a more local meaning because privileges are used in a more closed environment (i.e., inside an organization or among a group of them). In other words, authorization is merely valid for a specific application, scenario, or environment. Hence, it can rarely be considered to have a global meaning. In fact, it is reasonable to think that the same user will have several attribute certificates, for different applications, scenarios, or environments, while using only one identity certificate for all cases.

Consequently, there are numerous occasions in which an authority entitled to attest who someone is might not be the appropriate one to make statements about what that person is allowed to do. This argument is even more valid if user attributes are considered confidential information. The certificate may contain some sensitive information, and then attribute encryption may be needed, as proposed by PKIX [28]. That kind of confidential information should be solely managed by people belonging to the organization of the user.

In the extreme case, it may happen that some other authentication method different from that one based on public key certificates is used, while still using attribute certificates and the PMI for authorization purposes. In these cases, a PKI is not even used, and the name of the user could be a good option for the field "holder" of the attribute certificate.

It is important to note that the use of PMIs is very flexible because the ITU-T defines PMI models for different environments. Besides the general privilege management model, the ITU defines the following:

- *Control model*: This describes the techniques that enable the privilege verifier to control access to the object method by the privilege asserter, in accordance with the attribute certificate and the privilege policy.
- *Roles model*: Individuals are issued *role assignment certificates* that assign one or more roles to them through the role attribute contained in the certificate. Specific privileges are assigned to a role name through *role specification certificates*, rather than to individual privilege holders through attribute certificates.
- *Delegation model*: When delegation is used, a privilege verifier trusts the SOA to delegate a set of privileges to holders, some of which may further delegate some or all of those privileges to other holders.

In fact, the advantages of using attribute certificates to implement authorization have become clear to the security community because even the traditional access control solutions mentioned before have evolved in this direction. A clear example is the integration of role-based schemes with attribute certificates [29, 30]. Meanwhile, the ITU-T X.509 recommendation is still evolving and improving [31].

9.4 Conclusions

In this chapter we have focused on solutions for the provision of authentication and authorization services through the use of identity certificates and attribute certificates, as well as the frameworks provided for the extended use of those types of certificates—namely, PKI and PMI. Starting from the key management and users' authentication issues using the KDC as support, this chapter has elaborated on the evolution toward the use of public key–based cryptography to address those issues and, more precisely, the use of standard data structures: identity certificates and CAs in a first stage for authentication, and attribute certificates and AAs in a second stage for authorization. It is important to point out that the solutions explained in this chapter are those proposed in the recommendations of the standardization body ITU-T, which are the most widely used. However, other interesting solutions, like SPKI [32, 33], address the same problems from a more noncentric perspective.

References

[1] Fumy, W., and P. Landrock, "Principles of Key Management," *IEEE Journal on Selected Areas in Communications*, Vol. 11, No. 5, June 1993.

[2] Popek, G., and C. Kline, "Encryption and Secure Computer Networks," *ACM Computing Surveys*, Vol. 11, No. 4, December 1979, pp. 331–356.

[3] ANSI X9.17 (Revised), "American National Standard for Financial Institution Key Management (Wholesale)," American Banker Association, 1985.

[4] Needham, R., and M. Schroeder, "Using Encryption for Authentication in Large Networks of Computers," *Communications of the ACM*, Vol. 21, No.12, December 1978, pp. 993–999.

[5] Needham, R., and M. Schroeder, "Authentication Revisited," *Operating Systems Review*, Vol. 21, No. 1, 1987, p. 7.

[6] Otway, D., and O. Rees, "Efficient and Timely Mutual Authentication," *Operating Systems Review*, Vol. 21, No. 1, 1987, pp. 8–10.

[7] Neuman, B., and S. Stubblebine, "A Note on the Use of Timestamps As Nonces," *Operating Systems Review*, Vol. 27, No. 2, April 1993, pp. 10–14.

[8] Hardjono, T., and J. Seberry, "Authentication via Multiple-Service Tickets in the Kuperee Server," in *European Symposium on Research in Computer Security—ESORICS '94 Proceedings*, Springer-Verlag, 1994, pp. 144–160.

[9] Kaufman, C., R. Perlman, and M. Speciner, *Network Security: Private Communication in a Public World*, Upper Saddle River, NJ: Prentice-Hall, 1995.

[10] Study Group on Electronic Authentication, "Guidelines for Certification Authorities," Ministry of Post and Telecommunications of Japan, 1997.

[11] CCITT, "Data Communication Networks: Directory," Recommendation X.500–X.521, Blue Book, Volume VIII—Fascicle VIII.8, Consultation Committee International Telephone and Telegraph, International Telecommunications Union, 1989.

[12] ISO/IEC 9594 (multipart standard), "Information Technology—Open Systems Interconnection—The Directory," International Organization for Standardization, 1993.

[13] I'Anson, C., and C. Mitchell, "Security Defects in CCITT Recommendation X-509—The Directory Authentication Framework," *Computer Communications Review*, Vol. 20, No. 2, April 1990, pp. 30–34.

[14] ISO/IEC JTC1/SC21, "Draft Amendments DAM-4 to ISO/IEC 9594-2, DAM-2 to ISO/IEC 9594-6, DAM-1 to ISO/IEC 9594-7, and DAM-1 to ISO/IEC 9594-8 on Certificate Extensions," International Organization for Standardization, 1996.

[15] Sermersheim, J., "Lightweight Directory Access Protocol (LDAP): The Protocol," RFC 4511, June 2006.

[16] "Federal Public Key Infrastructure Technical Specification. Part E: X.509 Certificate and CRL Extensions Profile," Public Key Infrastructure Working Group, National Institute of Standards and Technology, March 1998.

[17] Myers, M., et al, "X.509 Internet Public Key Infrastructure Online Certificate Status Protocol—OCSP," RFC 2560, June 1999.

[18] Deacon, A., and R. Hurst, "The Lightweight Online Certificate Status Protocol (OCSP) Profile for High-Volume Environments," RFC 5019, September 2007.

[19] Federal Information Processing Standard (FIPS) Publication 186, "Digital Signature Standard (DSS)," U.S. Department of Commerce/National Institute of Standards and Technology (NIST), 1994.

[20] ANSI X9.57, "American National Standard, Public-Key Cryptography for the Financial Services Industry: Certificate Management," 1997.

[21] Chadwick, D., "An X.509 Role-Based Privilege Management Infrastructure," *Business Briefing: Global Infosecurity*, 2002.

[22] Ferraiolo, D., and R. Jun, "Role-Based Access Control," in *Proceedings, 15th NIST-NCSC National Computer Security Conference*, 1992, pp. 554–563.

[23] Sandhu, R. S., et al., "Role-Based Access Control Models," *IEEE Computer*, Vol. 29, No. 2, 1996, pp. 38–47.

[24] Shirey, R., "Internet Security Glossary," RFC2828, Network Working Group, Internet Engineering Task Force, May 2000.

[25] ITU-T Recommendation X.509, "Information Technology. Open Systems Interconnection. The Directory: Authentication Framework," June 1997.

[26] ITU-T Recommendation X.509, "Information Technology. Open Systems Interconnection. The Directory: Public-Key and Attribute Certificate Frameworks," March 2000.

[27] Kaliski, B., "A Layman's Guide to a Subset of ASN.1, BER, and DER," November 1993.

[28] Farrell, S., and R. Housley, "An Internet Attribute Certificate Profile for Authorization," RFC 3281, April 2002.

[29] Hwang, J., K. Wu, and D. Liu, "Access Control with Role Attribute Certificates," *Computer Standards and Interfaces*, Vol. 22, March 2000, pp. 43–53.

[30] Oppliger, R., G. Pernul, and C. Strauss, "Using Attribute Certificates to Implement Role-Based Authorization and Access Control," in *Proceedings of the 4. Fachtagung Sicherheit in Informationssystemen* (SIS 2000), October 2000, pp. 169–184.

[31] ITU-T Recommendation X.509, "Information Technology. Open Systems Interconnection. The Directory: Public-Key and Attribute Certificate Frameworks," August 2005.

[32] Ellison, C., "SPKI Requirements," RFC 2692, September 1999.

[33] Ellison, C., et al., "SPKI Certificate Theory," RFC 2693, September 1999.

CHAPTER 10

Smart Cards and Tokens

Manfred Aigner and Karl C. Posch

This chapter presents the main architectural and functional characteristics of smart cards and similar tokens. It focuses on the security features that they support. References to a number of application domains exploiting such technology are given. The chapter also highlights how smart-card technology is gradually moving to more advanced contactless technology like radio frequency identification (RFID) tags. The reader is guided through the evolution from the era of "people having tokens" to the new era of "things having tokens." An analysis of the security threats and privacy problems inherent to this new generation of tokens is provided, together with a description of possible solutions and current research results.

10.1 New Applications, New Threats

Smart cards are small, single-chip, stand-alone computing systems. They evolved from plastic cards with magnetic stripes for storing digital data. The problem with cards with magnetic stripes was that it was easily possible to read or modify the data stored on the magnetic stripe with standard equipment. In order to prevent this, the next step toward today's chip cards was the introduction of memory cards with digital semiconductor memory.

Soon the chip card was used as a personal token, suitable to carry sensitive information like credits for phone calls or even used as an authentication token for automatic teller machines (ATMs), where the card carried credentials (i.e., secret keys for cryptographic authentication). To enable protection of the card's memory, digital hardware implementing simple finite state machines was used to control access to the data. This approach was used for prepaid telephone cards, for example. As it turned out, soon this protection was not strong enough for preventing misuse, and forged cards circumventing the protection mechanisms became available on the gray market.

In order to enable secure applications of the card in offline payment terminals, or on SIM cards in GSM phones, better protection mechanisms became necessary. Cryptographic communication protocols between a card and the reader device were introduced to decrease the card's vulnerability. But these also came along with a rather high computational demand on the card itself. Many of these cryptographic protocols involve a secret key. This key is stored on the card in a way that it can be used to perform cryptographic operations directly on the card. Thus, extracting this key from the card is not necessary and typically should not even be possible. For these more sophisticated protocols, and in particular for the computation of cryptographic primitives, programmable microcontrollers were introduced on the cards.

A variety of such microcontroller cards exists for a broad area of applications. Cards used for pay-TV protection, banking applications, and digital signatures are currently those with the highest security level. Besides a microcontroller, those cards typically include coprocessors for cryptographic applications, sensors for active tampering detection, and countermeasures against known implementation attacks. In this section, we will talk only of smart cards with such high-security processor chips.

As shown in Figure 10.1, smart cards are available in different form factors. The International Organization for Standardization defines the various sizes [1]. Format ID-1, for instance, is the credit-card–sized plastic card with included chip. The SIM card used by GSM phones is defined as format ID-000. Other formats, like ID-00, exist but are less commonly used. The electrical contact interface of smart cards is specified in the standard ISO/IEC 7816-2. Eight contact pads are defined. Two of them are used to support the card with supply voltage (VCC and GND), and one to input the clock signal (CLK). One pad is intended to provide a programming voltage, which can be different from the supply voltage. Modern cards do not make use of this extra programming voltage input, because a very simple attack would be to disconnect this pin in order to prevent writing to the card's nonvolatile memory. One pin is used for serial communication in both directions. Another pin is dedicated for the reset signal (RST). Two pads are reserved for future use (RFU).

Cards with contacts are often used in which the use case requires an actively intended user interaction, like insertion of the card into a slot near a door to be opened. Unfortunately, the contact-based interface has several drawbacks. The electrical contacts of the cards and also those of the reader devices get worn out rather quickly by slipping the cards in and out. Readers in public places are an easy goal for vandalism, in which considerable damage can be produced by simply putting

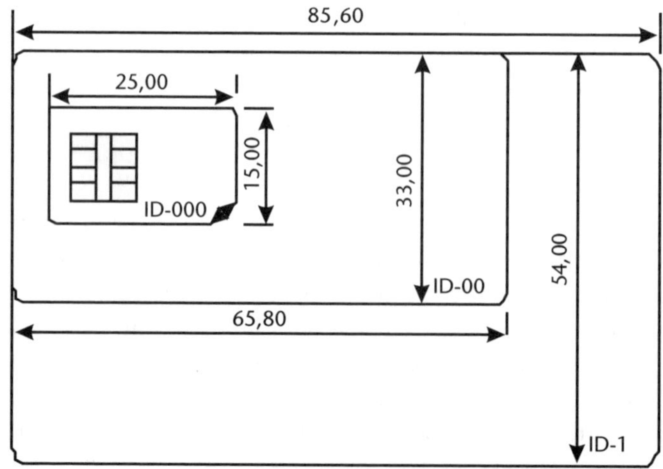

Figure 10.1 Form factors of smart cards.

other things (e.g., coins) into the slot. In addition, the time for interaction with the reader is quite often too long for some applications. Contactless smart cards that communicate with the reader via radio frequency (RF) can solve these problems. Such contactless cards also draw their supply energy from the electromagnetic field issued by the reader. A reading distance of some centimeters is achieved with these cards and readers. This is enough reading distance to leave the card in a wallet or keep it in a pocket while passing through a radio frequency gate. Functionality of such contactless cards is comparable to contact-based cards. ISO 14443 is the standard that defines the contactless interface for smart cards. The newest version of cards combines both interfaces, contact-based and contactless, on one card, and thus can be used for applications with different use cases.

Sometimes, the term *smart card* is also used for cards with any type of microchip (e.g., just a memory chip). In this chapter, we speak only of smart cards in their modern connotation: they have a full-fledged microprocessor with various types of memory and usually also a cryptographic coprocessor.

10.1.1 Typical Smart Card Application Domains

Secure smart cards are used as a root of trust in distributed applications. During the personalization phase, the card generates its private and public key pair in a protected environment, and then the public key is published, quite often after being signed by the card issuer. Secret keys for symmetric cryptographic schemes are generated and stored on the card. In another step of the personalization procedure, the card holder's identification data is written to the card; finally, the card is locked so that the secret keys cannot be changed or be read from outside the card.

The generation and verification of digital signatures constitute one class of applications. In Austria, for example, smart cards are issued to every citizen for easy access to health services. In addition, this health card offers also the functionality of generating digital signatures. As a result, new e-government services are offered increasingly granting authenticated web access to administrative services. Authentication is done by using digital signatures. The previous mandatory physical visit to an administration office is no longer necessary in many cases. The yearly tax declaration process, for instance, is one of such e-government services available via the Internet.

In online banking systems, transaction authentication numbers (TANs) are currently widely used to authenticate transactions on the web. The user gets a list of TANs through an authenticated secure channel (e.g., paper mail) from the bank. Each transaction is conducted with the one-time use of such a TAN. Such systems suffer from so-called *phishing* attacks, in which attackers try to cheat the user with faked web sites and spam mails to convince a potential victim to reveal his or her TANs. New generation online banking platforms already use smart cards and digital signatures instead of TANs and in this way prevent such attacks.

Another successful application domain for smart cards is secure authentication: the most successful example is the subscriber identity module card (SIM) of the GSM telephone system. Every subscriber holds a smart card issued by the phone operator; this SIM card is placed into the subscriber's mobile phone. The authentication of the mobile device to the network is performed with the keys stored on the

SIM card; this means that the subscriber is not authenticated by her mobile phone, but by her SIM card inside the phone.

Pay-TV systems also use smart cards for subscriber authentication. A smart card, which is placed into the receiver box, is issued to every subscriber. The card holds the necessary keys for the decryption of the pay-TV channel.

A widely used application of smart cards is payment. In public transportation systems, contactless smart cards with prepaid credits are widely used. Examples are the "ez-link card" in Singapore or the "oyster card" in London.

Modern ATM cards also use chips to prevent the production of illegal copies of the card. Old ATM systems used only the magnetic stripe and the personal identification number (PIN) to protect from fraud, but copies of cards could then be easily produced and a successful attack required only successful eavesdropping of the PIN. Newer ATMs perform authentication of the smart card, which makes successful attacks much more complicated. Additionally, the smart card for ATM access is often used in electronic purse applications as a replacement for cash.

Nowadays, smart cards serving as secure tokens are found everywhere. A typical European citizen usually has several smart cards with cryptographic functionality in her wallet. Credit cards, banking cards, health cards—all of them offer functionality for more than one application. Recently, applications using those cards for purposes other than issued for in the first place are seen more often. For instance, a youth protection law in Austria should prevent teenagers under 16 years of age from buying cigarettes. Thus, vendors of tobacco products are required to check the age of their customers. In order to prove permission to buy tobacco at a tobacco vending machine is given, the client has to introduce her banking card, which includes information about the age of the card holder.

10.1.2 The World of Tokens

Smart cards typically come in two varieties: with contacts and contactless. Contactless cards do not have to be inserted into a card reader's slot; rather, they need to be brought into the proximity of the reader (i.e., less than 10 cm). Many applications like public transport ticketing are benefiting from this.

Contactless cards get their energy through magnetic coupling with the reader's antenna. Reader and card communicate with each other by modulating this field. But from their functionality, the cards work quite similar to smart cards with contacts. They usually have enough computing power in order to be able to do cryptographic protocols for securing the communication between the reader and the card.

Contactless smart cards typically need electrical power on the order of 10^{-3}W. In a different development, very simple electronic chips have been used for preventing theft in retail stores for quite some time. These chips are attached to various sales items in shops and cause an alarm once they are taken through the magnetic field at the checkout counter without being deleted during the payment process.

In a more intelligent version, such chips not only have 1 bit of memory, but also are able to respond with an identification number on request. These so-called *transponder chips*, also called *tags* or *labels* (see Figure 10.2), are also typically consuming energy from the surrounding magnetic or electromagnetic field issued by the nearby device. This device is commonly referred to as the *reader*, since it reads

Figure 10.2 An RFID label.

out the tag's identification number. With decreasing size and price, and increasing energy efficiency, such transponder chips with more sophisticated communication and computation facilities will be attached to more and more items as a substitute to bar codes. Since these tags in the simplest case respond just with an identification number to a reader request, the term *radio frequency identification* (RFID) is commonly being used. One of the standards for such RFID tags is very similar to the one being used with the aforementioned contactless smart cards. They both use 13.56 MHz as carrier frequency. But tags have a much smaller chip size (typically less than 0.5 mm^2), are made for operating in the range of less than approximately 2m of the reader antenna, and, in order to reach that range, are built for a power budget of less than 20 μW. We thus see a development in which the technological gap between simple RFID tags used for identification and contactless smart card technology used for more fancy applications like authentication and access control is blurring.

The initial driving force for using such RFID tokens in connection with goods is the idea to augment and, in the long term, substitute bar codes, which are attached to virtually all items today. The usually manual work of scanning bar codes with a laser reader can then be avoided. Moreover, radio communication can be done without the need for a line of sight between the RFID token and the reader. This clear advantage for the logistics of handling goods creates a business model for the use of millions of tokens as soon as the price is affordable. At the time of writing this text, the clothing industry, for example, has started to attach such tokens to items.

As soon as you can think of creating such an "internet of things" in this way, a huge variety of new possible applications arises: think only of the automatic update of databases by using item-aware smart shelves in warehouses. Business management will be revolutionized once the gap between the real world of goods and the corresponding virtual database view of this world is closed [2].

10.1.3 New Threats for Security and Privacy

As soon as enough items are equipped with tags, we also will experience all well-known IT security problems with confidentiality (i.e., privacy of individuals carrying such goods), integrity (i.e., goods with counterfeited tags), and availability (denial-of-service attacks, for instance). Since electronic information then will be connected to the real world, a new dimension of security problems arises.

Currently, at the European level we see a discussion about the meaning of privacy in connection with items being attached to RFID tags. The question is whether a person's privacy is compromised when she carries tagged items and the identification number of these items is being communicated without the person's knowledge to a hidden nearby reader device. Since the tag's ID is not connected to the ID of the person, one could claim that privacy is not at stake. On the other hand, a linkage between these data can of course be done and probably will be done in order to create people's profiles. Think only of the moment, when tagged items are being paid for at the checkout counter of a retail store. There, payment is linked to a real person's bank account. This problem has created a lot of discussion in the concerned public and has also caused its "scandals," like the Metro Future Store incident [3] or the one with Benneton in 2004 [4].

RFID tags also have the potential to prevent counterfeiting of goods. According to an OECD study in 1998 [5], 5 to 7 percent of the world trade volume is due to faked goods. This rather large amount steers the interest of manufacturers for upcoming tag technology capable of authenticating itself. Good examples for the use of authenticated goods are pharmaceutical items. The benefit of avoiding incorrect medications is apparent. The percentage of phony drugs (i.e., drugs that are not the product they are believed to be) found on the market is shockingly high [6]. Currently, more people die due to incorrect medication than due to traffic accidents [7].

Counterfeiting needs unilateral authentication of the tag only. But once a tag is capable of doing cryptographic operations for authentication purposes, we can also introduce mutual authentication between tag and reader. Then, we can think of tags that talk to authorized readers only. From a person's point of view, the technical side of privacy could be managed in such a way. From an industry's point of view, the same problem is known as espionage and thus would also have a technical solution.

In this first section we now have had a look at the history and the development of smart cards, from magnetic stripe cards to high-security chip cards with either contact-based or contactless communication interfaces, and finally to RFID tokens or labels. Finally, we briefly looked at security and privacy issues. The rest of this chapter discusses the topic in more detail. In Section 10.2, the architecture of smart-card systems, their operating systems, and their communication protocols are studied in some detail. Section 10.3 discusses side-channel analysis, a class of attacks to smart cards that has steered a lot of interest in the industry as well as in the scientific community. The pervasive use of advanced contactless technology and its related security and privacy problems are the topic in Section 10.4. Finally, in Section 10.5, we draw conclusions.

10.2 Smart Cards

In this section we describe the typical differences between smart cards and ordinary microcontroller systems. We mainly point out the differences due to security requirements. After describing their hardware architecture, we have a look at card operating systems and communication protocols between card and reader.

10.2.1 Architecture

The size of a smart card chip is limited due to the necessary flexibility of the plastic card. When the card is bent, the chip must not break. With a maximum size of approximately 10 to 20 mm^2, the risk that the chip mechanically breaks during normal handling of the plastic card is low enough. To keep the chip size at a minimum, most smart cards build on 8-bit microcontroller architectures. Their functionality is reduced to the necessary minimum. Newer smart-card products use dedicated 16-bit or 32-bit controller cores. Cryptographic operations like symmetric and asymmetric encryption, random number generation, or cryptographic hash computations are mainly performed in dedicated coprocessors. On 8-bit platforms, it is necessary to include dedicated coprocessors for those operations in order to finish computation within reasonable time limits. For smart cards with contactless interfaces, energy efficiency is a mandatory requirement to include dedicated cryptographic coprocessors. Computations of symmetric crypto algorithms or cryptographic hash algorithms are less resource demanding and would allow for software implementations in order to fulfill the throughput and energy requirements; nevertheless, in many cases dedicated crypto hardware is being used. The reason for this is the high security demand of smart card applications. Appropriate protection of the crypto operations is crucial for the overall system's security, since the algorithm's operations involve the secret key, and leakage of intermediate data would facilitate attacks. Protection against side-channel attacks (see Section 10.3), especially, is difficult to achieve by software-only solutions. Another efficient attack against smart cards is probing of the bus signals between microcontroller and memory. The metal lines for bus signals are clearly visible in the chip layout when a smart card is dismantled. Due to the necessary strength of bus signals, they are easier to observe by an attacker. Therefore, special protection mechanisms need to be applied. Data bus and address bus of high-security smart cards are usually encrypted to impede probing attacks. All parts of the system are connected via so-called bus slaves, which perform encryption and decryption of the data transmitted over the bus. A bus master controls the communication on the bus and configures the bus slaves so that the encryption key for the bus encryption is correctly distributed throughout the system.

A smart card operates in a potentially hostile environment. For many applications, it has to be assumed that the owner of the card can also be a potential attacker. Thinking of an electronic cash application, any card owner could be interested in reloading his or her card with new electronic cash by hacking the card. Operating electronic circuits outside their specified operation conditions often helps to reveal secrets stored in the system. Wrong computations during a special operation can produce a failure in the output data, and thus might help the attacker to deduce the secret key. Sensors are necessary so that a card can react to such invasive or semi-invasive attacks. A security smart card typically features various sensors, like a supply-voltage sensor, glitch detectors for supply and clock signals, and a temperature sensor in order to detect such attacks. In case that a sensor detects suspicious operation conditions, the card refuses operation and resets itself.

The memory of a smart card is usually separated into three different types. The ROM includes the operating system and all additional programs. Volatile RAM is necessary for computations when the card is operating and nonvolatile memory is

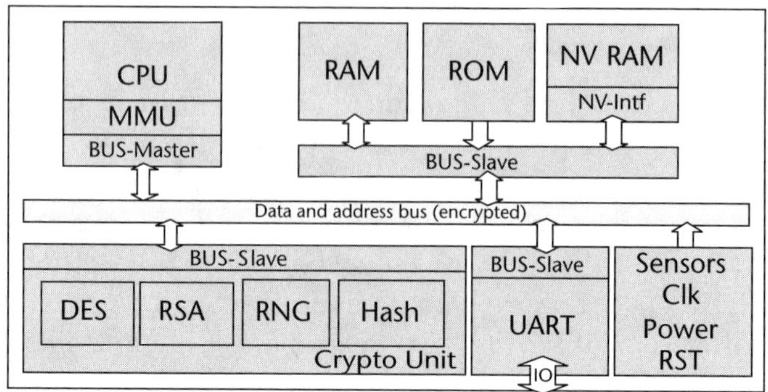

Figure 10.3 Typical hardware architecture of a modern smart card.

necessary to store data on the card. EEPROMs are typically used to store non-volatile data. The number of possible write operations to EEPROM cells is limited by some 10,000 write cycles, but this is high enough for the lifetime of a card. Access times to EEPROM, especially for write access, are very long compared to RAM or ROM access times. Special interfaces for accessing the EEPROM secure the microcontroller system against introduction of faults by intentional or unintentional shortage of the power supply during a write access. Due to their regular geometric structure, memory blocks are often targets of invasive or semi-invasive attacks and therefore need special protection mechanisms. Smart cards use encrypted memories with redundancy checks to detect possible modifications. A typical smart card architecture is depicted in Figure 10.3.

10.2.2 Smart Card Operating System

Operating systems for smart cards differ essentially from operating systems of other computers like personal computers. Nevertheless, an operating system is usually defined as "the set of computer programs that manage the hardware and software resources of a computer." A major difference between an operating system for smart card (card OS, or COS) and a PC is the size and consequently the overall functionality. While a standard operating system for a desktop PC like Windows XP or Linux easily uses hundreds of megabytes, a COS uses between 3 KB and 30 KB. A card OS provides no user interface and no network stack. The only communication interface is a simple serial communication for the access to and from the reader. Another main difference is the absence of peripherals, which makes a COS much simpler. On the other hand, security demands for the cards are typically much higher than for standard operating systems. Luckily, the smaller functionality makes security assessment easier also. Some smart card products allow programming of the card by the user (e.g., Java card). In this section, however, we refer to standard smart cards that are not programmable by the user. Thus, the COS and all executable programs are stored in the ROM. The main tasks of the operating system of such cards are described next.

- *Command processing*. Incoming commands from the reader need to be interpreted, and the corresponding responses need to be computed and issued. The communication is always triggered by the reader; thus, the card does not send results without any reader request.

- *Memory management.* During startup, the card performs a self-check of its ROM content to guarantee integrity of its content. Personalization data and configuration data of the card are often placed in a special section of the nonvolatile RAM that is implemented as a write-once-read-many (WORM) memory. Memory management of the RAM is less complex, since smart cards do not support multitasking. Only one program or application is typically running at any given time. Some controller architectures use internal and external RAM with different access commands. If a program uses more memory than available in the internal RAM, the COS can take over the management of the access to those external memory sections.
- *File management.* When accessing the nonvolatile RAM, the COS has to ensure that the application is allowed to read or modify certain data. Data stored in the nonvolatile RAM is organized as a simple hierarchical file structure. For security reasons, the files are separated into header and body, both stored in different locations. The header includes the access conditions, the file structure, and a pointer to the file's body. The body holds the stored data. Access to protected data is granted depending on successful authentication of the user with a PIN. Usually the directory structure and files are created during the personalization phase. The content of the files can be modified afterward, but it is not possible to delete files.

10.2.3 Communication Protocols

Communication between a smart card and an application running on a desktop PC involves more than one interface. The card itself is communicating with a terminal, called the reader, and the application requires access via the operating system of the PC to the card terminal peripheral. The name *reader* is misleading, since this terminal is actually reading data from and writing data to the smart card.

To allow application development on the desktop PC without depending on a specific device driver for the reader, a generic API for communication with the terminal and the card in the terminal is necessary. The PC/SC consortium developed a multipurpose interface specification for Windows platforms that allows convenient integration of smart cards into applications running on a PC. Meanwhile, a variety of readers are available that fulfill this specification. PC/SC allows the use of one reader and a smart card simultaneously for many applications, but it also supports several readers connected to a PC. APIs for various programming languages are available.

The communication between reader and card is a serial communication in half-duplex mode. The reader is the active communication part triggering the actions, and the card answers as the passive counterpart to requests. After insertion of a smart card into the reader, the card sends an *answer-to-reset* (ATR) string. The ATR is followed by a protocol negotiation procedure, where card and terminal agree on a specific transfer protocol. In most cases, this transfer protocol is either "T=0" or "T=1." "T=0" is a byte-oriented transfer protocol that has several weaknesses but is still in use. The protocol "T=1" is block oriented. A block, called transport protocol data unit (TPDU), consists of some addressing information, a control block, the actual data block, and a section for error detection. The data content of a block is transparent for the link layer; therefore, completely encrypted communication between application and smart card is supported. The data block of a TPDU is formatted as application protocol data unit (APDU), which is an international standard for data exchange on the application layer.

Secure communication between a card and an application is performed by a secure messaging mechanism. Man-in-the-middle attacks or eavesdropping of the serial communication are prevented by end-to-end data authentication and encryption between the application and the smart card. An attacker gaining control over the reader (e.g., by changing the driver code of the desktop's reader hardware driver) is therefore not able to interfere in the transactions between card and application if the card operates in the correct mode.

10.3 Side-Channel Analysis

In parallel to the evolution of magnetic stripe cards in the 1980s toward cards with microchip memory, and later to full-fledged systems on chip as we see them now, unwanted exploitations of weaknesses of these tokens have been developed. Since the cards have always served as a means to store or secure assets, they have been a target for criminal activities. Today we also see a race between more sophisticated implementations of security-related features on smart cards and the ever-increasing capabilities of those who want to bypass these. A comprehensive history of exploiting cards can be found in Ross Anderson's book [8].

One of the typical assets stored on a modern smart card is the secret key used in a cryptographic computation on the card. Whenever a crypto algorithm is being executed with such a key, the chip consumes electrical energy. The electrical current flowing into the chip depends on the microchip's activity and thus to a certain extent also on the data being processed in the chip. Measuring this electrical current during cryptographic operation using a secret key provides information about this key. This information channel is usually called the *side channel*. Thus, we speak of side-channel analysis, or side-channel attacks. This kind of noninvasive cryptanalytic attack using power consumption variations was first published by Kocher in 1998 [9]. The problem has already been briefly mentioned at the end of Chapter 7. Here, we are looking at it in more detail with regard to smart cards and other tokens.

Cryptologists historically concentrated only on mathematical properties of crypto algorithms, and "to break" an algorithm meant to find weaknesses in the algorithm itself. With the threat of side-channel analysis, we also need to check for implementation weaknesses in order to produce a secure system.

In addition to measuring the electrical current consumed by a smart card, called power analysis, the microchip also produces an electromagnetic emanation during computation. Again, if such a computation uses a secret key, the electromagnetic field in the vicinity of the microchip depends on the key data to some extent and can be exploited to reveal the key in the same fashion as when measuring energy consumption. Finally, the time a cryptographic computation consumes also might leak information about the key. Such attacks are called timing analysis.

During the past years, a lot of research on side-channel attacks was conducted. A comprehensive book on this topic was published in 2007 [10]. Apparently, smart

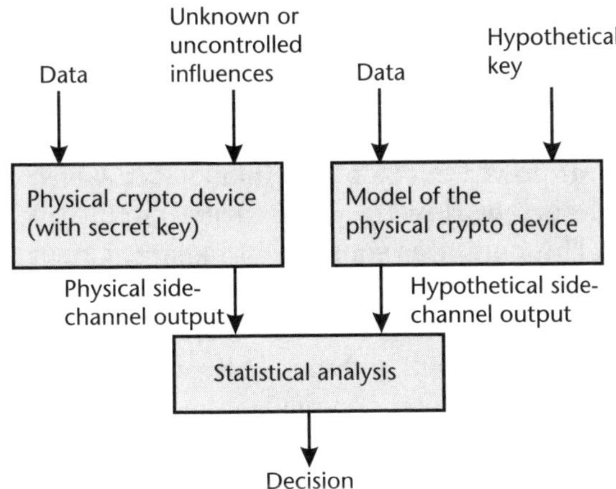

Figure 10.4 Overall setup of a differential power-analysis attack.

cards are the typical device for being attacked through side-channel analysis. But, also other systems like personal computers or RFID chips [11] may be attacked. Meanwhile, a whole variety of techniques to exploit side-channel information became known. In addition to timing analysis, we have simple power analysis, differential power analysis, and higher-order power analysis.

Figure 10.4 shows the overall setup of a differential power-analysis attack. A physical crypto device (e.g., a smart card) is being asked to perform a cryptographic computation by inputting "data." These data are usually a request within a communication protocol from the card reader to the card, like "here is a number, please encrypt the number and send back the result." Such a request is a typical part of an authentication protocol. In addition to these data, the physical system also produces all kinds of inherent "noise." In the Figure 10.4, this is indicated as "unknown and uncontrolled influences." While computing, the device leaks information through a physical side channel output. The attacker measures and records this output.

In addition, the attacker also has a model of the physical device. This model mimics the operation of the real device to a certain extent. In particular, it should somehow model the leakage of information through a side channel. The model is also fed with the same data as the physical device and with a hypothetical key (i.e., a guess for the secret key used in the physical device).

With statistical analysis, the two side-channel outputs—the physical one and the hypothetical ones with various key guesses—are compared until a strong correlation between them is detected. Then the guessed key corresponds to the secret key. Such analysis is typically done key bit by key bit or bytewise.

In parallel to the evolution of the knowledge of more sophisticated side-channel attacks, countermeasures have also been developed. In the remainder of this section, we will first discuss some technical details about power-analysis attacks and then talk about countermeasures.

10.3.1 Power-Analysis Attacks

From an attacker's point of view, the efficiency of a power-analysis attack lies in its simplicity, since almost no special laboratory equipment is needed. Power traces can be recorded with an appropriate probe connected to a digital sampling oscilloscope, and the traces can be analyzed with an ordinary computer. Since the attack is non-invasive, the card will not get destroyed during the attack either. The attack also does not involve any sophisticated physical tampering techniques.

Let us look in more detail at differential power analysis (DPA). Figure 10.5 shows a typical power trace of a CMOS circuit. Given a constant supply voltage, the electrical current flowing through the circuit over time constitutes such a power trace. This current is consumed to a large extent due to the switching activities of the transistors. In a synchronously clocked system, many transistors switch with this frequency. We thus see a continuously repeating waveform corresponding to the clock frequency. But, and this is also shown in Figure 10.5, the current trace also depends to a lesser extent on the Hamming weight of the data being processed. In this example, the data on an 8-bit data bus connecting the CPU and the memory is looked at. Each data wire of the data bus is rather long and thus constitutes a large electrical capacitance. Charging and discharging this capacitance consumes relatively much electrical current. The Hamming weight is the amount of 1 bit on this bus, and thus can be between 0 and 8.

The differences of the power trace due to the Hamming weight can be exploited with statistical analysis. As an alternative to looking at the Hamming weight, other properties of the power trace (e.g., the Hamming distance) could also be used successfully for an attack. In attacks using differential power analysis, the waveforms of the real device are merged with the side-channel waveforms of the model through a correlation method.

Power traces are usually rather noisy. This noise originates to some part from the statistical movement of physical particles due to temperature, so called *electronic noise*. But noise is also caused by the circuit's switching activity not connected to the secret key under attack. This noise is called *switching noise*. Some of

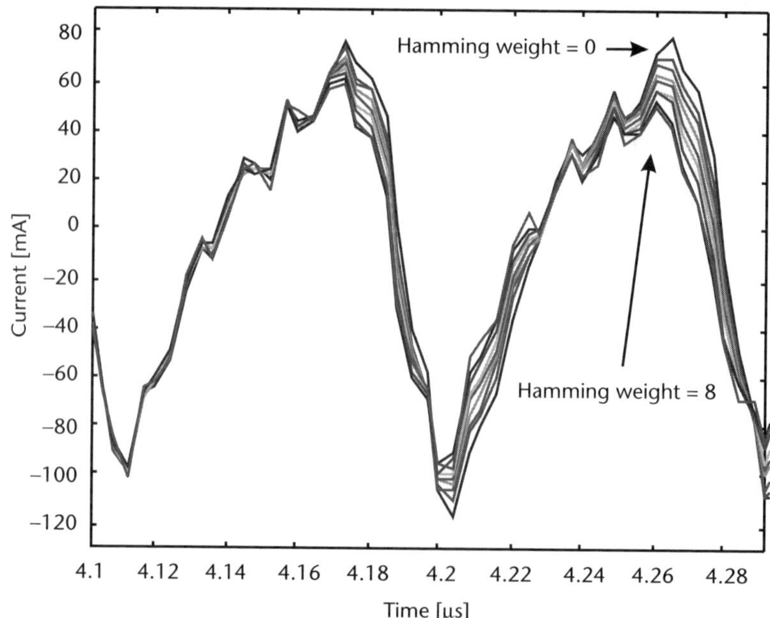

Figure 10.5 Electrical current consumption of a CMOS circuit depends on the Hamming weight of the data on an 8-bit bus.

this noise, in particular the noise due to temperature, can be removed by averaging many power traces. DPA attacks usually need hundreds, or even thousands, of traces with the same key data.

There exist a lot of tricks to reduce the complexity of the attack. In addition to removing electronic noise, power traces are also compressed by removing "unimportant" information. Important information with regard to the key can usually be found by looking at maximum current points or by integrating the electrical current over some time interval.

In a DPA attack, first an intermediate result of the crypto algorithm is chosen. This intermediate result needs to be a function of a part of the key under attack. After gathering power traces, hypothetical intermediate values are computed from the model. These are then mapped to hypothetical intermediate power consumption values. This process can be done by simulating an appropriate model.

The linear relation of the side-channel signal of the physical device and the hypothetical side-channel signal of the model is defined by the correlation coefficient. Attacks using this correlation coefficient are the most common ones. Correlation gets to a maximum once the guess matches the attacked part of the secret key.

10.3.2 Countermeasures Against DPA

Apparently, countermeasures against differential power analysis need to somehow hide or mask the data dependency. The power consumption should be equalized such that processing a 0 is identical to processing a 1 in the circuit. This goal cannot be achieved totally. It can only be approached, and for a realistic scenario it suffices to make the attack for an attacker just expensive enough so that it does not pay off.

Two different categories for countermeasures exist: hiding and masking. Quite often, a combination of the two is used. Hiding removes the data dependency of the power consumption from intermediate values, whereas masking randomized the intermediate values being processed in the crypto algorithm. Both hiding and masking can be done at various levels of abstraction: at the level of logic values, at the level hardware architecture, and also at the level of software.

Typical methods of hiding are the random insertion of dummy clock cycles or the random shuffling of clock cycles. Other countermeasures increase the noise of the side-channel information. In software, the random choice of machine instructions, changes in the program flow, randomizing address use, or parallel activity help to conceal the key. In hardware, a randomly changing clock frequency, an occasional random skipping of clock cycles, the existence of multiple clock domains, or noise engines might be used. Many circuits are built with special libraries of logic gates, like dual-rail logic.

Masking randomizes and thus conceals intermediate values. Bits can be masked by using an exclusive-or operation with a random value. In this way, a bit might or might not be inverted. But arithmetic masking may also be employed. For instance, in asymmetric cryptography, the method of *blinding* may be used. Masking only helps against so-called first-order DPA attacks. Attacks on masking are done with second-order DPA attacks. In hardware, masking can be done at the bit level by using masked logic styles, at the bus level by masking bus values, at the arithmetic level by masking multipliers, or at the algorithmic level by masking the whole algorithm (e.g., *masked AES*).

Maybe the most powerful countermeasure against side-channel attacks is the use of session keys that are changed often enough. By the time the attacker gets to know a secret key through a side-channel attack, this key is no longer valid.

The side-channel attack scenario shown in this section is only one among many possible attacks. By looking at DPA in some detail, we tried to show that for building secure systems, it is necessary not only to use proven algorithms, but also to have appropriate implementations of these, and—maybe most important—an appropriate use of them in proven protocols.

For a comprehensive study of countermeasures, we refer to [10]. A good source for information is also the Side Channel Cryptanalysis Lounge [12].

10.4 Toward the Internet of Things

Automated identification of goods at the item level is only the simplest application of tokens, and it clearly arises from the idea of substituting bar codes with more sophisticated technology. Instead of having the costly manual process of scanning bar codes, an automated readout of identification information can be done at any point in time and space. From a management point of view, automation—which allows a fine-grained knowledge about items, their location, their amounts, at almost any time, and an automated update of the corresponding database—is clearly a big advantage. Management can then control ordering, manufacturing, or shipping of goods in real time. Instead of the once-a-year inventory, tag-aware shelves provide live information about available items. As soon as these *smart products* are integrated into the virtual world, logistics becomes more efficient. In this way, the additional costs for the tags are covered.

Some industries have installed so-called closed-loop systems with tagged items and readers. In a closed loop (i.e., a proprietary solution within a single domain of responsibility), IT security is typically not considered to be a major problem. One example is the clothing industry. The benefits of tagging at the item level in this industry are higher than in, say, the food industry. This is due to the large amount of item variants, the high fluctuation due to seasonal and fashion reasons, and the rather high margins, which allow for the extra costs for the tags. In addition, prevention of theft seems also to play a substantial element [13].

But, also in the example of the clothing industry, we already see the benefits of substituting closed-loop systems with so-called open-loop systems. Tagged items should be useful over several domains of responsibility in the supply chain. In an open-loop system, smart services could be implemented. Such services could be the automated handover of items, automated payment, and later return of items. Even proper recycling of items can be thought of.

Immediately, we see the increased role of IT security. Just compare it with the classic open system, the Internet. On top of the basic service of identification we would like to have authentication, or prevention of espionage and denial-of-service attacks.

The substitution of bar codes with RFID is globally organized by EPCglobal [14]. By developing appropriate industry-driven standards, EPCglobal tries to provide a base on which open-loop systems can be built.

In the long run, tags not only will provide identification but will probably be equipped with sensors or even with displays. Sensors will sense temperature, pressure, light, and other physical quantities. Tags will not only play a passive role, but could also become active participants in the network. As technology advances, we soon will truly enter an era of ubiquitous computing, or pervasive computing in which things will be attached to the Internet.

10.4.1 Advanced Contactless Technology

Having understood the benefits of tagging items for the industry, we now have a look at some of the current issues in technology. As already explained in this chapter's introduction, we need to distinguish between contactless smart cards (which work like smart cards, with a relatively large silicon area, a reading distance less than 10 cm, and a relatively high energy consumption) and tags (small chips, reading distance in the range of several meters, available energy very low), although this distinction is not always made clear. From a technological point of view, current tags have severely limited computing power in comparison to contactless smart cards. For the sake of completeness, we should also mention the distinction between active and passive tags. Active tags contain batteries. Passive tags get their energy from the electromagnetic field issued by the reader. In the rest of this chapter, we will talk about passive tags only. Tags are also divided into classes with respect to the frequency from which they derive their energy and through which they communicate with the reader. High frequency (HF) tags operate on 13.56 MHz, while ultra-high-frequency (UHF) tags operate around 900 MHz and even higher frequencies. HF tags and UHF tags have different uses due to different ranges and due to different behavior in connection with liquids or metal. So far, we have only talked about silicon technology. A tiny microchip is attached to a much larger antenna. Several other technologies are currently being researched. Among these are polymer electronics (i.e., "plastic chips").

For the sake of clarity and brevity, the rest of the discussion in this section concentrates on 13.56-MHz tags manufactured with traditional silicon technology only. We thus talk about a situation in which we typically have less than 0.5-mm^2 chip size and 20-μW energy budget. The distance between the tag and the reader is typically less than 2m.

As soon as we discuss crypto technology for such tags, the constraints due to available energy, the computation time, and the limited chip size (necessary for a low price of the tag) become dominant. Options for building blocks are the use of passwords, cryptographic hash algorithms, and symmetric and asymmetric cryptography. In addition to "proven" algorithms, there is also ongoing research on new cryptographic algorithms that suit the severely restricted arena of tags.

In Table 10.1, implementations of various algorithms are compared with respect to chip area (measured in gate equivalents, GE), the mean value of the necessary electrical current (I_{mean} in μA), and the computation time measured in clock cycles [15].

Table 10.1 Comparison of Implementations of Various Algorithms [15]

Algorithm	*Chip Area (GE)*	I_{mean} *(μA@100kHz, 1.5V, 0.35 μm)*	*# Clock Cycles*
AES-128	3,400	3.0	1,032
SHA-256	10,868	5.83	1,128
SHA-1	8,120	3.93	1,274
ECC-192	23,600	13.3	5000,000
Trivium low-pow radix-1	3,090 [2,390]	0.68 [2.33]	(1,603)+176
Grain low-pow radix-1	3,360 [1,760]	0.80 [1.56]	(130)+104

The essence from this table is that cryptographic hash algorithms (like SHA-1 or SHA-256) are more expensive to implement than, say, AES. We also see that the only asymmetric cryptographic algorithm in this list, elliptic curve cryptography with 192 bits, is still out of reach at the moment. Newcomers like Trivium [16] and Grain [17] seem to have strong potential for being used in conjunction with tags. Figure 10.6 shows an implementation of AES-128 [18]. At the time of this writing, this chip is still a state-of-the-art implementation. It consumes 3 μA (at 100 kHz) with 1.5V on an area of 0.25 mm^2 in 0.3-micron technology.

For the following discussion, we will concentrate on using AES as the cryptographic building block on a tag.

10.4.2 Cloning and Authentication

Currently, most RFID chips provide identification only. On the request of a reader, they answer with an ID. This ID can trivially be eavesdropped by an attacker. If the attacker uses a programmable device capable of speaking the protocol, he can clone any RFID tag. In Figure 10.7, such a programmable RFID tag is shown. It consists of electronic standard parts, and the circuit diagram is readily available. Moreover, programmable tags can also be purchased easily. Thus, cloning identification-only tags is simple.

Currently, many deployed RFID systems, whether in libraries for simplifying borrowing books or parking-lot access, usually just do identification. As shown with the previous argument, RFID systems based on identification can only be used

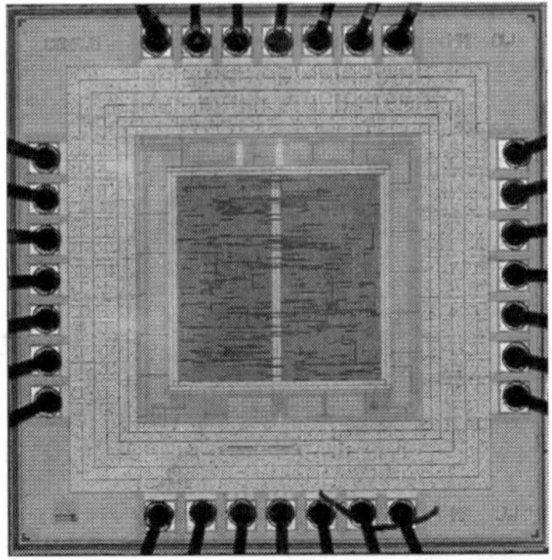

Figure 10.6 Chip photograph of TINA, a tiny implementation of the AES [16].

Figure 10.7 RFID demo tag [19].

for simplifying procedures like borrowing books, but do not at all prevent theft of books or illegal access to parking lots by themselves.

We need authentication in order to avoid the problems sketched earlier. Then cloning is no longer possible by trivial eavesdropping. In a standard unilateral authentication protocol using symmetric cryptography, the tag proves its identity by knowing a secret key. The reader device is connected with the network and thus does not need to know the key itself, but rather only needs access to the key in some database. The tag could even give an indication of where to find the key.

Tags that can authenticate themselves in this way are also useful for a manifold of other applications. High-price merchandise can be equipped with the proper manufacturer's tags and can thus prove its origin. Pharmaceutical items can be distinguished from phony lookalikes. And car parts will probably no longer be distributed through wrong distribution channels in high amounts like today.

10.4.3 Privacy and Espionage

Consumer concern has been rather loud in connection with the introduction of RFID technology in recent years. The possibility of tracking and tracing persons, creating profiles by linking items and persons, and many more arguments have lead to the term *spy chips*. The representatives of the member states of the European Union in the Data Protection Committee have been discussing this problem intensively [20]; the OECD has done a study on it [19]; and the RFID industry is currently learning about these concerns and trying to cope with them. Interestingly enough, the problem of privacy is rather similar to the problem of espionage, and the latter problem is of course of importance for many industrial users of RFID technology.

Many solutions have been proposed for coping with the privacy problem. Due to lack of space, we can only briefly mention them here. Of course, the problem of privacy also goes well beyond technology into regulations and law making. These topics we do not cover here. For a thorough discussion, we refer to [21].

In order to avoid tracing and tracking, a person could of course always rip off the tag. In order to do so, she must know of the tag's presence. Others proposed having half of the tag's ID on the item, with the second half on the package of the item. Opening the package destroys the tag. Some tags also come with a simple mechanism to detach the microchip from its antenna and make it dysfunctional in this way.

The EPC standard defines the kill command. After having purchased an item, a reader could issue this kill command to the tag, and the tag no longer identifies itself. Funnily enough, it is not forbidden to implement an "unkill" command, similar to the undo functions known from many computer programs.

Of course, a consumer can put items into a kind of "Faraday cage"–like shopping basket. This does not seem to be very useful. Others have proposed a "blocker tag" that consumers should carry with them. This blocker tag responds to readers by simulating the presence of zillions of tags and in this way mounts a denial-of-service attack against the reader. There is also the idea of programmable blocker tags that are not so severe in their hostile behavior against readers.

On the tag side, in the recent past a lot of proposals were made that use cryptographic hash functions. All these proposals were made under the assumption that hash functions are more efficiently implementable than strong symmetric algorithms like AES. With Feldhofer's work [15], this assumption no longer holds true. Let's have a look at the privacy problem by limiting our discussion to tags with AES encryption only. In addition, we assume that tags can also create real random numbers. With this, a tag could first authenticate the reader and only identify or authenticate itself to authorized readers. Of course, this alone provides a solution only at a technical level and would be useless without the necessary overall setup.

Authentication and privacy, or, in other words, cloning and espionage, are the two main security issues when introducing pervasive computing. For smart cards, either contact-based or contactless, many solutions exist and are also in place. We can expect a similar learning curve in the RFID arena.

10.5 Conclusions

In this chapter we had a look at the history of the development of smart cards, from magnetic stripe cards to high-security chip cards with either contact-based or contactless communication interfaces, and finally to RFID tokens or labels. These devices have in common that they are used to "identify" people and things (quite often carried by people). In some cases we want identification and authentication; in some cases we would like to avoid it and rather keep a person's privacy. For both problems there exists a broad variety of technical solutions, usually based on cryptographic algorithms, protocols, and mechanisms. Although some of these are computationally rather intensive, for most typical cases there exist appropriate solutions.

Unfortunately, whenever there is a technical system promising to provide security in some form, there also seems to be "the bad guys" trying to attack or exploit the system in a way unwanted by "the good guys." Therefore, the topic of smart cards or RFID tokens is "naturally" linked to the topic of preventing such misuse. As technology advances, the attack scenarios also change and get even more advanced. This seems to be a never-ending game.

References

[1] International Organization for Standardization, ISO/IEC 7810, 1995.

[2] Fleisch, E., "Die betriebswirtschaftliche Vision des Internets der Dinge," in *E. Fleisch, F. Mattern, Das Internet der Dinge*, Springer, 2005.

[3] FoeBuD e.V., http://www.foebud.org/rfid/metroskandal.

[4] Consumers Against Supermarket Privacy Invasion and Numbering, http://www.boycottbenetton.com/.

[5] Organization for Economic Co-operation and Development, *The Economic Impact of Counterfeiting*, 1998, http://www.oecd.org/dataoecd/11/11/2090589.pdf.

[6] Healthcare Distribution Management Association, *Pharmaceutical Product Tampering News Media Fact Sheet*, 2004, http://www.healthcaredistribution.org/resources/pdf_news/Product%20Tampering%20edit.pdf.

[7] IDTechEx, "Smart Healthcare USA 2004 at a Glance—RFID Smart Tagging and Smart Packaging," http://www.idtechex.com/smarthealthcareusa/index.asp.

[8] Anderson, R. J., *Security Engineering: A Guide to Building Dependable Distributed Systems*, Wiley Computer Publishing, 2001.

[9] Kocher, P. C., J. Jaffe, and B. Jun, "Differential Power Analysis," in *Advances in Cryptology—CRYPTO'99, Proc. 19th Annual International Cryptology Conference*, Vol. 1666 of Lecture Notes in Computer Science, M. Wiener (ed.), Santa Barbara, CA: Springer, 1999, pp. 388–397.

[10] Mangard, S., E. Oswald, and T. Popp, *Power Analysis Attacks*, Springer, 2007.

[11] Oren, Y., and A. Shamir, "Power Analysis of RFID Tags," http://www.wisdom.weizmann.ac.il/~yossio/rfid/.

[12] Side Channel Cryptanalysis Lounge, http://www.crypto.ruhr-uni-bochum.de/en_sclounge.html.

[13] Tellkamp, C., and U. Quiede, "Einsatz von RFID in der Bekleidungsindustrie—Ergebnisse eines Pilotprojekts von Kaufhof und Gerry Weber," in *Das Internet der Dinge: Ubiquitous Computing und RFID in der Praxis*, E. Fleisch, F. Mattern (eds.), Springer, 2005, pp 143–160.

[14] EPCglobal, http://www.epcglobalinc.org.

[15] Feldhofer, M., "Comparison of Low-Power Implementations of Trivium and Grain: The State-of-the-Art Stream Ciphers," Ruhr University Bochum, January 2007, http://www.ecrypt.eu.org/stream/papersdir/2007/027.pdf

[16] De Canniere, C., and B. Preneel, *Trivium Specifications*, Katholieke Universiteit Leuven, Dept. ESAT/SCD-COSIC, http://www.ecrypt.eu.org/stream/p3ciphers/trivium/trivium_p3.pdf.

[17] Hell, M., T. Johansson, and W. Meier, "Grain—A Stream Cipher for Constrained Environments," http://www.ecrypt.eu.org/stream/p3ciphers/grain/Grain_p3.pdf.

[18] Feldhofer, M., S. Dominikus, and J. Wolkerstorfer, "Strong Authentication for RFID Systems Using the AES Algorithm," in *Proc. Conference of Cryptographic Hardware and Embedded Systems, 2004*, Springer, 2004, pp. 357–370.

[19] Organization for Economic Co-operation and Development, "Radio-Frequency Identification (RFID): Drivers, Challenges and Public Policy Considerations," 2006, http://www.oecd.org/dataoecd/57/43/36323191.pdf.

[20] European Commission, "Towards an RFID Policy for Europe," http://www.rfidconsultation.eu/.

[21] Garfinkel, S., and B. Rosenberg, *RFID Applications, Security, and Privacy*, Reading, MA: Addison Wesley, 2006.

[22] Secure Information and Communication Technologies SIC, "Crypto Toolkit—RFID Demo Tag," http://jce.iaik.tugraz.at/sic/products/rfid_components/rfid_demo_tag_1.

CHAPTER 11

Privacy and Privacy-Enhancing Technologies

Simone Fischer-Hübner, Dogan Kesdogan, and Leonardo A. Martucci

Privacy as an expression of the rights of self-determination and human dignity is considered a core value in democratic societies and is recognized either explicitly or implicitly as a fundamental human right by most constitutions of democratic societies. In Europe, the foundations for the right to privacy of individuals were embedded in the European Convention on Human Rights and Fundamental Freedoms of 1950. In 1980, the importance of privacy protection was recognized by the OECD with the publication of the OECD Privacy Guidelines [1], which helped to raise the international awareness of the need for privacy protection and served as the foundation for many national privacy laws.

In our modern Information Society with mobile and pervasive computing applications, individuals are losing more and more control over who knows what about them and thus over their personal spheres, as masses of data can be easily collected about them and retained for ages without their knowledge. Both legal and technical means are needed to protect privacy and to (re)establish individuals' control and thus to protect privacy.

This chapter discusses privacy from the legal and technical perspectives. Section 11.1 will first give an introduction to the concept of privacy. In Section 11.2, we will then discuss privacy risks and challenges of location-based services (LBS) and radio frequency identification (RFID) as two examples of emerging technologies that are affecting privacy. After having discussed examples of privacy risks, the rest of this chapter will then discuss legal and technical means for protecting privacy and thus for addressing such risks. In Section 11.3 we will present legal privacy principles of the European Legislative Privacy Framework comprising the EU Data Protection Directive 95/46/EC [2] and the E-Communications Privacy EU Directive 2002/58/EC [3], as well as limitations to communication privacy encoded in the EU Data Retention Directive 2006/24/EC [4]. When discussing general privacy principles of the EU Data Protection Directive 95/46/EC, we will also refer to corresponding principles of the OECD Privacy Guidelines [1] as far as they exist. Finally, this section will provide an overview to privacy legislation in the United States.

Section 11.4 will provide a classification and overview of privacy enhancing technologies (PETs). Section 11.5 and 11.6 will then discuss a selection of PETs in more detail. Section 11.5 will present anonymous communication technologies, as they are the classical PETs that provide examples how privacy can be best protected by avoiding or minimizing the availability of personal data. Section 11.6 will give a short introduction to antispyware, as spyware nowadays is in practice a major privacy concern of many Internet users.

11.1 The Concept of Privacy

Privacy as a social and legal issue has for a long time been a concern of social scientists, philosophers, and lawyers. The first definition of privacy was given by the two American lawyers Samuel D. Warren and Louis D. Brandeis in their famous article "The Right to Privacy" published in the *Harvard Law Review* [5], in which they defined privacy as "the right to be let alone." The reason for this publication was the use of photography as a new technology by the yellow press, which was in the view of the authors an attack on personal privacy in the sense of the right to be let alone.

In the era of modern information technology, another definition of privacy was given by Alan Westin: "Privacy is the claim of individuals, groups and institutions to determine for themselves, when, how and to what extent information about them is communicated to others" [6]. According to Westin's definition, natural (individuals) as well as legal persons (groups and institutions) have a right to privacy. Today, in many legal systems, privacy is in fact defined as the right to informational self-determination (i.e., individuals must be able to determine for themselves when, how, to what extent, and for what purposes information about them is communicated to others).

In general the concept of privacy has several dimensions. Besides informational privacy, so-called spatial privacy can be defined as another dimension of the concept of privacy, which also covers the "right to be let alone," where spatial privacy is defined as the right of individuals to control what is presented to their senses [7, 8]. Further dimensions of privacy, which will not directly be subject of this chapter, are territorial privacy by protecting the close physical area surrounding a person and privacy of the person by protecting a person against undue interference, such as physical searches, drug testing, or information violating his moral sense ([9]).

Data protection is the protection of personal data in order to guarantee privacy and is only a part of the concept of privacy. Privacy, however, is not an unlimited or absolute right, as it can be in conflict with other rights or legal values, and because individuals cannot participate fully in society without revealing personal data. Privacy and data protection laws serve the purpose of helping to protect privacy rights, if personal data is collected, stored, or processed.

As the German constitutional court proclaimed in its census decision in 1983, privacy is not only a fundamental human right, but also an essential value for democracy, because an individual "who cannot with certainty overlook which information related to him is known to certain segments of his social environment, and who is not able to assess to a certain degree the knowledge of his potential communication partners, can be essentially hindered in his capability to plan and to decide. The right of informational self-determination stands against a societal order and its underlying legal order in which citizens could not know any longer who what and when in what situations knows about them."

11.2 Privacy Challenges of Emerging Technologies

In our network society, the use of various types of data processing equipment (including notebooks, cell phones, personal digital assistants, and the like) has become commonplace. All these systems are linked through the Internet and through telecommunication networks. We use these communication networks everywhere and every time. Since services (e.g., online banking, shopping) are mostly developed for distributed network-centric platforms, all confidential information may become visible to third parties by intercepting the messages in communication networks. Even if the services as such are secure, and if the data exchanged between the services are encrypted, the endpoints of the communication will still be observable. Having access to such information it is easy to determine (e.g., who is communicating with whom, or using which service, for how long, and from which location). Hence, if someone collected and accumulated this traffic information, our private and vocational activities would be visible. The protection of this precious right is a crucial task.

Modern and emerging technical developments such as mobile services and ambient intelligence also pose new privacy challenges. For exemplifying this, we will briefly discuss privacy aspects of two modern technologies, LBS and of RFID, in this section in more detail.

11.2.1 Location-Based Services

In LBS, information about the user's location (or more specifically, the user's device) is passed to the service provider. The location is either measured by the device itself using specialized hardware, such as GPS receivers, or by the mobile operator in the operator's domain. LBS applications can be divided into single-user applications and multiuser applications (also called peer-to-peer applications). In single-user applications, there is a direct relationship between the user and the LBS provider, and the user does not directly interact with other users. In contrast, in multiuser applications the users communicate directly with one another, while the role of the LBS provider is mainly to assist the users in establishing these relationships with one another. LBS applications can also be divided into applications in which the user either explicitly "pulls" the information from the LBS provider (by explicitly sending her location) and applications in which information is automatically "pushed" to the user by the LBS provider at regular intervals. Since in the second approach the LBS provider is automatically repositioning the user at regular intervals or on special occasions, the latter kind of application is also characterized as position-tracking or location-aware applications.

Some typical and useful LBS applications that are already offered today or are believed to be deployed in the near future are described in [10], including city guide, travel navigation, friend finder, mobile marketing, or disaster management (e-government applications that in the event of a disaster, a disaster manager could be used to locate mobile phones in certain sections of the surrounding areas of the disaster to help evacuation planning). They are classified in Table 11.1 according to single user/multiuser and pull/push LBS:

Table 11.1 A Classification of the LBS Applications [11]

	Pull LBS	*Push LBS*
Single-user LBS	City guide, travel navigation	Mobile marketing, disaster management
Multiuser LBS	—	Friend-finder, mobile dating

11.2.1.1 Exposed Personal Data

Well-known privacy problems of the traditional Internet, as caused by cookies, customer and communication profiling, or SPAM, are also an issue in the mobile Internet. In addition to that, the deployment of advanced LBSs for mobile Internet introduces new privacy issues, since new forms of sensitive data are collected and processed by service providers. One example of sensitive data needed for mobile Internet application is of course location data that can provide the precise geographic location of the mobile user's device.

Furthermore, information about the users' preferences and their devices' capabilities is needed by service providers for personalization to enhance usability and performance. In the mobile Internet, in which restricted devices with small screens are in use, personalization is a much bigger issue than in the traditional Internet, in which personalization of sites is a matter of convenience to the end user. In mobile Internet environments, capability and preference information (CPI) in so-called user agent profiles can be especially useful to allow the service provider to generate content tailored to the characteristics and user interface of the requesting device, and thus enhance the user's experience and minimize the use of bandwidth. Privacy problems of capability and preference information have been discussed in [12, 13]. The need for personalization also implies that profiling and tracking techniques such as cookies might be used more extensively.

11.2.1.2 Threats to Informational Privacy

If proper security and privacy measures are lacking, the LBS applications described here could be misused. Major threats caused by LBS to the user's right of informational self-determination are unsolicited profiling, location tracking, and the disclosure of the user's social network and current context.

Personal data such as location data, the user's preferences, business activities, and the kind of information that a user requested could be compiled and stored by service providers in detailed user profiles. Push LBSs often require to some extent user profiling in order to provide adequate information and are for this reason especially challenging for privacy. Examples of potential misuse scenarios of such profiles could be unwanted marketing, digging into the past, and blackmailing politicians.

Another problem is that location data also reveals information about the user's context, which can be sensitive (e.g., whether he/she is currently in a night club, church, or at a meeting of a political party).

Location data could also be misused for unsolicited location tracking by using the information about the movements of mobile users. If the location information is not properly protected, it could be misused to track persons for the purpose of robbery, kidnapping, or looting. If service providers cooperate with other service providers or network operators and they merge their data sources, the problems related to profiling and tracking may be further intensified.

Another problem is that exposed information about social networks that is often of a private nature can be revealed. These privacy problems described earlier are especially an issue for multiuser LBS applications, such as friend-finder in which the LBS provider receives detailed information about the users' friends and contacts to them. However, parties such as network operators or service providers that have access to location data of different mobile users could by comparing the location profiles of two mobile users derive information about the users' colocation (i.e., information about when and for what length of time two users have spent time or possibly have been traveling together). Hence location data for single-user LBS can also reveal information about social networks.

11.2.1.3 Threats to Spatial Privacy

Another problem is that the user's spatial privacy (i.e., the user's ability to control what is presented to his senses) could be affected by SPAM and a lack of efficient reachability management. Marketing information is of great value to spammers and, as in the traditional Internet, the obvious risk is that spammers will send unsolicited emails to persons with matching profiles.

Besides, for most of today's commercial multiuser LBS applications, such as friend finder, the user often lacks efficient control over her reachability, in which a user can turn off the possibility to be localized and thus become "invisible" if she does not wish her friends to know her location. It is, however, often neither possible with today's LBS applications for a user to select visibility for a subset of friends in her friend list, making her reachable by this subset of friends but unreachable by others, nor is it possible for a user to configure her reachability on subject matter.

11.2.2 Radio Frequency Identification

RFID tags were introduced in Chapter 10, where it was been briefly mentioned that consumers have voiced privacy concerns due to the tacking possibilities that RFID technology provides. In this section, the privacy issues of RFID details will be discussed further.

11.2.2.1 Exposed Personal Data

For discussing privacy implications, we first need to classify the cases in which the RFID tags can actually be associated with an individual to answer the question of whether personal data are actually processed (i.e., whether privacy is actually affected). First of all, we can differentiate the situations in which a RFID chip is implanted into a human body or affixed or incorporated to an object. In the first case, the data stored on the tag is definitively associated with the human and is thus personal data. If an RFID tag is affixed to or incorporated into an object, the data stored on it can be classified as personal data in the following three cases, as discussed in [14]:

1. Personal data is directly stored on the tag, such as biometric data stored on RFID chips in the new European passports.
2. Data (e.g., the electronic product code) stored on RFID tags can be linked to personal data, which are typically stored on backend databases (e.g., if

a customer purchases a product and pays with his credit card, the unique product code could be linked with the retailer customer database).

3. RFIDs are used for tracking without "traditional" identifiers being available (e.g., imagine the customer Mrs. Svensson has purchased a watch that she is wearing with an RFID tag with a unique code. She can then be tracked and her behavior can be profiled with the help of this unique code without her name or other identifying information about her being known).

11.2.2.2 Threats to Informational and Spatial Privacy

Privacy defined as the right of informational self-determination means that individuals should have control over the release and use of their personal data. However, by the very nature of RFID tags the reading of tags occurs from a distance, and may very well occur in a manner invisible to a person carrying or wearing RFID tagged items. Hence, if no proper protection is in place, information about a certain individual can be communicated and tracked without the individual's awareness or consent, which means that the individual has basically no control over the processing of his personal data associated with his tags.

Imagine for instance that Mrs. Svensson with her tagged watch enters a supermarket. Through the RFID tag in her watch, she could be identified as a specific returning customer (even though the supermarket might not know her by name). It could be monitored which RFID-tagged items she is touching or picking in order to profile her interests and customer behavior within the shop. These customer profiles could be used for the purpose of displaying targeted advertisements for her on video screens that she passes in the supermarket, which is affecting her spatial privacy. When she is using her credit card for payments, the unique identifier of her watch could be linked with her name or personal number. In this way, Mrs. Svensson's activities and whereabouts could be tracked.

If the secret service has access to the link between the personal number of persons and unique RFID codes of their belongings, they could identify participants at political meetings by secretly scanning their RFID tags (see [15]; a similar scenario is also described in [16]).

11.3 Legal Privacy Protection

After having discussed privacy risks, the remainder of this chapter will discuss means of privacy protection. The traditional approach for protecting privacy has been by legislation. This section discusses legal means for privacy protection in Europe and in the United States.

In Europe, important legal instruments for privacy protection are the general EU Data Protection Directive 95/46/EC [2] and the EU Directive 2002/58/EC [3] on privacy and electronic communications. The EU Directive 95/46/EC codifies general privacy principles, which partly corresponds to principles of the OECD Privacy Guidelines to which we will also refer. The so-called E-Communications Privacy Directive 2002/58/EC sets out more specific rules for privacy protection in the electronic and mobile communications sector. Important provisions of both directives are presented in Sections 11.3.1 and 11.3.2.

Another important directive in this context is the EU Data Retention Directive 2006/24/EC [4], as it is restricting electronic communications privacy. The scope of the Data Retention Directive will also be briefly outlined in Section 11.3.3.

Section 11.3.4 will then present and discuss the Federal U.S. Privacy Act and a selection of specific sectoral laws regulating privacy protection for parts of the private sector in the United States.

11.3.1 EU Data Protection Directive 95/46/EC

In 1995, the EU adopted the Data Protection Directive 95/46/EC, which is still the main foundation for data protection in Europe. Member states of the EU had to amend their respective national laws (where necessary) to conform to the directive within three years. The directive has the objective to provide for a high level of protection of the fundamental rights and freedoms of the individuals with regard to the processing of personal data, and the protection of privacy in particular. Besides, it requires a uniform minimum standard of privacy protection to prevent restrictions on free flow of personal data between EU member states for reasons of privacy protection.

The Data Protection Directive codifies basic privacy principles that need to be guaranteed when personal data are collected or processed, which include the ones listed here:

1. *Legitimacy*: Personal data processing has to be legitimate, which is according to Art. 7 usually the case if the data subject has given his unambiguous consent, if there is a legal obligation, or contractual agreement (cf. the Collection Limitation Principle of the OECD Guidelines).
2. *Purpose specification and purpose binding (also called purpose limitation)*: Personal data must be collected for specified, explicit, and legitimate purposes and may not be further processed in a way incompatible with these purposes (Art. 6 I b). The purpose limitation principle is of fundamental importance, as the sensitivity of personal data does not only depend on how "intimate" the details described by the personal data are, but is also mainly influenced by the purposes of data processing and context of use. In its census decision the German Constitutional Court proclaimed in 1983 that there were no "nonsensitive data," as dependent on the purpose and context of use all kinds of personal data can become sensitive. There are personal data that per se already contain sensitive information (e.g., medical data), but dependent on the purpose and context of use, such sensitive data can become even more sensitive and data that seem to be nonsensitive (e.g., addresses) can become highly sensitive as well. For example, audit data, which an operating system is recording to detect and to deter intruders, could be (mis)used for monitoring the performance of the users. Thus, if audit data were used not only for security purposes, but also for work place monitoring, the audit data would become more sensitive. For this reason, the data processing purposes need to be specified in advance by the lawmaker or by the data processor before obtaining the individual's consent or before contractual agreements are made, and collected data may not be (mis)used later for any other purposes (cf. Purpose Specification and Use Limitation Principles of the OECD Guidelines).

3. *Data minimization*: The processing of personal data must be limited to data that are adequate, relevant, and not excessive (Art. 6 I (c)). Besides, data should not be kept in a personally identifiable form any longer than necessary (Art. 6 I (e)). This data minimization principle derived from the directive also serves as a legal foundation for privacy-enhancing technologies that aim to provide anonymity, pseudonymity, unobservability, or unlinkability for users and/or other data subjects (see Section 11.4). Indeed, the privacy of individuals is best protected if no personal data about them are collected or processed at all (Cf. Data Quality Principle of the OECD Guidelines, which requires that data should be relevant to the purposes for which they are to be used).
4. *No processing of special categories of data*: According to Art. 8, the processing of so-called special categories of personal data revealing racial or ethnic origin, political opinions, religious or philosophical beliefs, trade union membership, or aspects of health or sex life (i.e., types of personal data that are per se perceived as very sensitive) are generally prohibited, subject to exemptions.
5. *Transparency and rights of the data subjects*: A society in which citizens could not know any longer who knows what about them, when, and in which situations, would be contradictory to the right of informational self-determination. Hence, the privacy principle of transparency of personal data processing should be guaranteed for the data subjects. For this reason, data subjects should have extensive information and access rights. Pursuant to Art. 10, individuals about whom personal data are obtained have the right to information about at least the identity of the controller, their data processing purposes, and any further information necessary for guaranteeing fair data processing. If the data are not obtained from the data subject, the data subjects have the right to be notified about these details pursuant to Art. 11. Further rights of the data subjects include the right of access to data (Art. 12 a), the right to object to the processing of personal data (Art. 14), and the right to correction, erasure, or blocking of incorrect or illegally stored data (Art. 12 (b)) (Cf. Openness and Individual Participation Principle of the OECD Guidelines). Today, the privacy principle of transparency is, however, at stake in ambient computing environments later traditional user interfaces are disappearing. As discussed for RFID applications, there is the risk that there is no longer transparency for individuals, as others could secretly read data on RFID tags that the individuals are carrying or wearing. For this reason, RFID readers and tags should at least be clearly labeled.
6. *Security*: The data controller needs to install appropriate technical and organizational security mechanisms to guarantee the confidentiality, integrity, and availability of personal data (Art. 17) (Cf. Security Safeguards Principle of the OECD Guidelines).
7. *Independent supervisory authorities and sanctions*: Independent supervisory authorities (so-called data protection commissioners) shall monitor compliance with the directive and act on complaints by data subjects (Art.

28). In the event of violation of the provisions of the directive, criminal or other penalties are envisaged.

8. *Limitations to the data transfer to third countries*: Art. 25 of Directive 95/46/EC allows the transfer of personal data to countries outside the European Union only if the third country in question ensures an adequate level of protection. Exceptions to this rule are possible according to Art. 26. This restriction should obviously prevent data controllers to circumvent the relatively strict European data protection legislation by outsourcing the personal data processing to countries with no or with inadequate levels of data protection.

11.3.2 EU E-Communications Directive 2002/58/EC

Since 1995, other directives regulating personal data processing have been adopted in specific areas. One important example is the Directive 2002/58/EC on privacy and electronic communications, which replaced the Data Protection Telecommunications Directive 97/66/EC and defines specific privacy rules for the electronic communications sector. The so-called E-Communications Privacy Directive deals with a number of issues, including the following ones:

1. *Confidentiality of communications*: Directive 2002/58/EC aims with its Art. 5 I to protect the confidentiality of communications by in general prohibiting the listening, tapping, storage, or other kinds of interception or surveillance of communications without the consent of the data subjects. Moreover, according to Art. 5 III, member states must ensure that the use of electronic communications networks to store information or to gain access to information stored in the terminal equipment of a subscriber or user is only allowed if the subscriber or user concerned is provided with clear and comprehensive information and is offered the right to refuse such processing by the data controller. This provision should protect users and subscribers against cookies, spyware, web bugs, and other hidden privacy-intrusive data collection techniques.
2. *Protection of traffic and location data*: Further important provisions for the electronic and particularly the mobile communications sector are rules concerning the protection of location data other than traffic data, allowing the exact positioning of a mobile user's device, and rules concerning the protection of traffic data, which could also include location data giving geographic information that is often less precise (e.g., the foreign network link to which a mobile node is currently attached). According to Art. 6 I, traffic data must in principle be erased or made anonymous when it is no longer needed for the purpose of transmission of a communication. This rule is subject to well-defined exceptions for traffic data processing for billing/payment purposes (Art. 6 II), value-added services (Art. 6 III, IV), and data retention by authorized authorities (Art. 15 I). If used for value-added services, location data other than traffic data have higher protection, as location data reveal sensitive information about the user's or

subscriber's movements and social networks (see Section 11.2.1). Whereas for traffic data, informed consent by the user or subscriber is required (Art. 6 III, IV), for location data other than traffic data, either anonymity or informed consent is required (Art. 9 I) with the possibility for users or subscribers that have given their consent to temporarily refuse the processing for each connection or transmission of a communication (Art. 9 II). In both cases the processing is only permissible to the extent and for the duration necessary for the provision of a value-added service. Anonymity techniques that can partly also be applied for making anonymous location data will be discussed in Section 11.5.

3. *Opt-in for SPAM*: Another important provision is Art. 13 regulating unsolicited communications by allowing communications including emails for the purposes of direct marketing only to subscribers who have given their prior consent or in customer relationships. It thus installs a restrictive "opt-in" regime for SPAM.

11.3.3 Data Retention Directive 2006/24/EC

A legal basis for data retention has been opened by Art. 15 I of the E-Communications Privacy Directive 2002/58/EC. It allows member states to adopt legislative measures to restrict among other things the provisions of confidentiality of communications (Art. 5) and protection of traffic (Art. 6) and location data (Art. 9), when such restriction constitutes a necessary, appropriate, and proportionate measure within a democratic society to safeguard national security and law enforcement activities.

In early 2006, the EU Data Retention Directive 2006/24/EC was adopted. It requires providers of publicly available electronic communications services or of a public communications network to retain traffic and location data for 6–24 months. The types of data to be retained include data necessary to trace and identify the source and destination of a communication, as well as data necessary to identify the date, time, and duration of a communication; the type of communication; the users' communication equipment; and the location of mobile equipment. Pursuant to Art. 5 II, data revealing the content of the communication should not be retained. However, it is not always possible to clearly distinguish between traffic and content data. An *http get request*, for instance, is considered traffic data according to the directive, but also usually contains information about the content of the request.

Law enforcement authorities claim that data retention is needed to effectively trace back and locate criminals and terrorists. Traffic and location data, however, contain not only digital "fingerprints" of suspected criminals or terrorists but also—and mainly—those of innocent users. Besides, with modern communication technology, people are much more connected and online, and traffic and location data reveal many more personal details than is the case with "classic" communications. Not all these data need to be retained for law enforcement purposes. It is doubtful whether data retention for an extended time period will be very effective, as criminals usually find ways around (e.g., by using stolen mobile phones or unprotected wireless networks), and whether it is an appropriate means in relation to the serious impacts it will have on online privacy.

11.3.4 Privacy Legislation in the United States

In the United States, the U.S. Privacy Act of 1974 [17], which covers only the federal public sector, lays down some general privacy principles that are also part of the OECD Privacy Guidelines. These principles in particular include the requirement of data minimization; purpose specification; the establishment of appropriate administrative, technical, and physical safeguards to ensure the security and confidentiality of personal data records transparency; and rights of data subjects to access their data. However, the act does not establish an independent supervisory authority to oversee privacy protection of data processing agencies and to act if there are complaints from data subjects about unfair or illegal use of their personal data. Consequently, the only way for data subjects to fight against data misuse is through the courts. The General Government Appropriations Act of 2005 [18] at least requires every federal agency to appoint its own privacy officer, and the Federal Information Security Management Act of 2002 requires the appointment of a senior information security officer by each agency.

In the private sector in the United States, there are no general data protection rules, only a patchwork of some specific laws regulating privacy protection for parts of the private sector, whereas other parts are not regulated by law. This patchwork of laws covers in particular the protection of financial records [19], health information and credit reports, video rental [20], children's online activities [21], educational records [22], motor vehicle registration [23], and telemarketing [24] (see also [25]). A selection of these laws is discussed in more detail here.

Legal privacy regulation for the medical sector was only introduced in 2001 with final rules governing the privacy of health records for the Health Insurance and Portability and Accountability Act (HIPAA) of 1996. This HIPAA privacy rule provides basic protection for personal health information and provides data subjects rights with respect to their data. It is permissive in nature, as it allows several types of data disclosures, which are only permitted to the data subject, his personal representative, and to the Secretary of Health and Human Services for the purpose of enforcement. State law provisions, dealing among others with access to medical information, mental records, and records related to conditions such as HIV and reproductive rights, remain in place as far as they provide greater protection [25].

In the commercial sector, the Fair Credit Reporting Act [26] of 1979 is a federal law that promotes accuracy, fairness, and the privacy of personal information assembled by credit reporting agencies (CRAs) by regulating the collection, dissemination, and use of consumer credit information. Under the Fair and Accurate Credit Transactions Act of 2003 [27], which amended the Fair Credit Reporting Act in 2003, new privacy rights were introduced, such as the right for individuals to obtain a free credit report from each of the CRAs once a year. Besides, credit reporting agencies are also required to disclose credit scores. The Fair and Accurate Credit Transactions Act also allows consumers to opt out of affiliate marketing—a company's use of an affiliate company's information about a consumer for marketing purposes—for a period of five years and also allows individuals to file fraud alert, which requires CRAs to inform others that fraud may be present [25].

State legislatures have recently addressed identity theft by introducing laws requiring notification of consumers after disclosure of financial and other personal data. First the state of California introduced a statute in 2003 that requires an entity storing personal data to inform California residents of a security breach of their unencrypted personal data [28]. Further state legislatures have passed or are considering passing similar statues [25].

The federal CAN-SPAM Act [29] from 2004 regulates unsolicited mails (spam) by requiring that spam mails must include an opt-out notice with a postal address of the sender. Unlike the EU's restrictive opt-in approach for regulating spam, which disallows to send unsolicited emails to users unless they have given their prior consents or are in a customer relationship with the senders, under the CAN-SPAM Act, no prior permission is required for sending commercial messages as long as the receiver has the chance to opt out of receiving future message from the same sender. It has been criticized that the U.S. legislation is not going far enough and could even make matters worse by approving spam that follows certain rules.

Since the United States has no comprehensive privacy protection law for the private sector and has not established independent supervisory authorities, it has been doubted that the United States can ensure an adequate level of protection in comparison to the data protection level provided by the EU Directive, as required according to Art. 25 EU Directive 95/46/EC for the flow of personal data to countries outside the EU. For avoiding trade barriers, the U.S. Department of Commerce has agreed with the EU Commission on "Safe Harbors" privacy principles, which are self-regulatory privacy guidelines that U.S. companies can voluntarily adhere to in order to continuously exchange data with European organizations. In an evaluation of the implementation of Directive 95/46/EC [30], it has however been criticized that the enforcement of Art. 25 and 26 is lax in some member states, with the effect that many unauthorized and possibly illegal transfers are being made to countries or recipients not guaranteeing adequate protection.

11.4 Classification of PETs

Privacy cannot be protected solely by legislation. In practice, often privacy laws are not properly enforced. Besides, as the Internet has no national boundaries, it has proven to be infeasible to harmonize privacy legislation on an effective level and on a wide international basis due to cultural differences. Privacy enhancing technologies (PETs), which will be discussed in the remainder of this chapter, can help to protect legal privacy principles by technology. The term *privacy-enhancing technologies* was first introduced within the report jointly published by the Dutch Registratiekamer and the Information and Privacy Commissioner in Ontario on "Privacy-Enhancing Technologies: The Path to Anonymity" [31]. In general, PETs can be grouped into the following three classes:

11.4.1 Class 1: PETs for Minimizing or Avoiding Personal Data

As also discussed previously, obviously, privacy is not or is the least affected if no personal data at all or as little data as possible are collected or processed. For this reason, Directive 95/46/EC includes the following principles from which the need for this class of PETs can be derived:

As discussed in Section 11.3.1, Art. 6 I embodies the principle of data minimization in its letter (c) by stating that the processing of personal data should be limited to data that are adequate, relevant, and not excessive. The letter (e) reinforces this idea by adding that data should only be kept in a form that permits identification of the data subject for no longer than is necessary for the purposes for which the data were collected or for which they are further processed;

PETs for minimizing or avoiding personal data include technologies that are providing anonymity, pseudonymity, unlinkability, or unobservability. According to [32], these protection goals can be defined as follows:

1. *Anonymity* is the state of being not identifiable within a set of subjects (e.g., set of senders or recipients), the so-called anonymity set. A special case is perfect sender (or receiver) anonymity: an attacker cannot distinguish the situations in which a potential sender (or receiver) actually sent (or received) a message or not.
2. *Unobservability* ensures that a user may use a resource or service without others being able to observe that the resource or service is being used.
3. *Pseudonymity* is the use of pseudonyms as identifiers. Pseudonymity is a useful concept for providing both privacy protection and accountability.
4. *Unlinkability* of two or more items (e.g., subjects, messages, events) can be achieved if within the system, from the attacker's perspective, these items are no more or less related after the attacker's observation than they were before. A special case is unlinkability of sender and recipient (so-called relationship anonymity), which is achieved if who is communicating with whom is untraceable.

PETs of this first class can be furthermore divided depending on whether they minimize or avoid personal data on the communication level or on the application level. Anonymous communication mechanisms are implemented by protocols such as mix nets [33], DC-networks [34], onion routing [35, 36], and crowds [37–39], which will be presented in more detail in the next section.

PETs for anonymous or pseudonymous applications can be implemented on top of anonymous communication protocols. Examples are anonymous e-cash protocols based on blind signatures [40], as well as anonymous credential systems (such as Idemix [41]), which can be used to implement anonymous or pseudonymous access control, e-health or e-government, and other anonymous or pseudonymous applications.

11.4.2 Class 2: PETs for the Safeguarding of Lawful Data Processing

It is not always possible to avoid the processing of personal data. Government authorities, health care providers, and employers are examples of organizations that still need to process personal data for various reasons. In some cases, it may also be in the interest of an individual to disclose personal details to others. If personal data are collected and processed, legal privacy requirements, such as the ones discussed in Section 11.3, need to be fulfilled. The second class of PETs comprises technologies that enforce legal privacy requirements in order to safeguard the lawful processing of personal data. The driving principles behind these types of PETs can also be found in Directive 95/46/EC: Art. 17, which requires that controllers implement security measures that are appropriate to the risks presented for personal data in storage or transmission, with a view to protecting personal data against accidental loss, alteration, unauthorized access (in particular in which the processing involves the transmission of data over a network), and against all other unlawful forms of processing (see Section 11.3.1; as discussed earlier, other privacy guidelines and privacy laws of other countries typically include similar provisions for implementing technical security mechanisms for privacy protection; cf. the Security Safeguards Principle of the OECD Guidelines).

Examples of class 2 PETs are technologies for stating or enforcing privacy policies: The platform for privacy preferences protocol (P3P) developed by W3C [42, 43] increases transparency for the end users by informing them about the privacy policies of web sites and helping users to understand them. P3P can hence be used to enforce the legal requirement to inform data subjects according to Art. 10 EU Directive 95/46/EC. Privacy policy models (such as the privacy model presented in [44]) can technically enforce privacy requirements such as purpose binding, and privacy policy languages, such as the enterprise privacy authorization language (EPAL) [45, 46], can be used to encode and to enforce more complex enterprise privacy policies within and across organizations. The concept of RFID proxies that are based on selective blocker tags has been developed for enhancing the control of individuals over their personal data associated with RFID tags. Installed on a personal device such as a smart mobile phone or PDA, such RFID proxies allow individuals to define and enforce policies about who may read tags that they are carrying or wearing, and for which purposes under which conditions [16].

Example of technologies belonging to this class 2 of PETs also include classic security technologies, such as encryption or access control, which protects the confidentiality and integrity of personal data. Security tools of practical importance to user privacy protection are antispyware tools, which will for this reason be discussed in more detail in Section 11.6.

11.4.3 Class 3: PETs Providing a Combination of Classes 1 & 2

The third class of PETs comprises technologies that combine PETs of class 1 and class 2. An example for PETs of class 3 are provided by privacy-enhancing identity management technologies, such as the ones that have been developed within the EU FP6 project Privacy and Identity Management for Europe (PRIME) [47]. Identity management (IDM) subsumes all functionalities that support the use of multiple identities by the identity owner (user-side IDM) and by those parties with whom the owner interacts (services-side IDM). The PRIME project addresses privacy-enhancing IDM to support strong privacy by particularly avoiding or reducing personal data and identification and by technically enforcing informational self-determination.

PRIME is based on the principle that design must start from maximum privacy. Therefore, with the help of anonymous communication technologies and anonymous credential protocols, a priori all interactions are anonymous, and individuals can choose suitable pseudonyms to link different interactions to one another or to make interactions unlinkable, can bind attributes and capabilities to pseudonyms, and can establish end-to-end secure channels between pseudonyms. Whether or not interactions are linked to one another or to a certain pseudonym is under the individual's control. For this, PRIME tools allow individuals to act under different pseudonyms with respect to communication partners, roles, or activities. Besides, policy management tools are helping them to define and negotiate privacy policies with services sides regulating who has the right to do what with one's personal data under what conditions and yielding to what kind of obligations. Those policies are enforced at the receiving end via privacy-enhanced access control, and there is enough evidence so that users can actually trust in this enforcement. Transparency end user tools allow users to be informed who has received what personal data related to them and to trace personal data being passed on, and they include online functions for exercising users' rights to access their data, object to data processing, or to rectify, block, or delete data (see [47] for more information).

11.5 Privacy Enhancing Technologies for Anonymous Communication

Anonymity as a protection goal has to comprise an attacker model. Such an attacker model defines how far a potential attacker can access or derive relevant information and what resources he can use. Therefore, anonymity of communication depends on how far an attacker can control or observe network stations, communication lines and communication partners. However, it is hard to estimate the resources of an unknown attacker. To be on the safe side, the resources of an attacker are always overestimated. Therefore, all strong anonymity techniques [33, 34, 48, 49] assume the so-called global attacker [51] model. The global attacker complies with the Dolev-Yao model [52] and assumes that the network is insecure and the attacker is only limited by the used encryption techniques and the trusted environments (i.e., the attacker is able to eavesdrop all network stations and all communication lines).

By choosing the powerful omnipresent attacker model, it follows that a single transmission by a single person can be neither anonymous nor unobservable. The global attacker can observe the sender of a message (the sending act) and follow the message to the receiver, thereby detecting the communication relation without needing to read the content of the message. Anonymity techniques providing security against the global attacker model date back to the 1970s and 1980s when David Chaum and others suggested some revolutionary techniques: broadcast and implicit addresses, mixes, and DC-networks. In [53] these works are first presented as basic techniques, and in [54] various enhancements and extensions in theory and technique were suggested. These seminal contributions have proven to be the basis for many of today's anonymity techniques. In more recent times, a new technique providing perfect protection was discovered by two different groups [48, 49]. Known as private information retrieval (PIR), this new technique has similarities to DC-networks.

However, overly strong attacker models often result in techniques with low performance. Thus, the local attacker model assumes that only parts of the communication links can be eavesdropped by the attacker. Since there is a restriction in the view of the attacker, the anonymity techniques make use of this lack of control. However, it is impossible to state precisely which lines are controlled by the attacker. Thus, it could be by chance that the anonymity system uses only these controlled lines, and thus the anonymity system provides no security. In the other extreme case, if the anonymity technique by chance uses only "uncontrolled" lines, the protection system provides full anonymity. Therefore, it makes sense here to randomize the strategy and provide a probabilistic security.

In this section we will first present basic techniques that provide anonymity against global attackers (i.e., broadcast and implicit addresses, DC-network, mixes, and PIR). Then we will present the most prominent ones that provide anonymity against local attacker model (i.e., onion routing, TOR, web mixes, crowds, and hordes [39, 55–58]. We will not present all of the details of the techniques since we are more interested in the basic features of the anonymity techniques.

11.5.1 Broadcast Networks and Implicit Addresses

Receiving a message can be made completely anonymous to observers on the network by delivering the same message (possibly end-to-end-encrypted) to all stations (broadcast). If the message has a specific intended recipient, an addressee, the message must contain some attribute by which the addressee alone can recognize the message as being addressed to him [59]. This message attribute is called an implicit address. It is only meaningful to a recipient who can determine whether he is the intended addressee. In contrast, an explicit address describes the specific place in the network to which the message should be delivered and, therefore, cannot provide anonymity. Implicit addresses can be further distinguished according to their visibility (i.e., whether or not they can be tested for equality. An implicit address is called invisible if it is only visible to its addressee and is called visible otherwise [53].

Invisible implicit addresses, which are unfortunately very costly in practice, can be realized using a public key cryptosystem: A message is encrypted with the public key of the recipient and then broadcasted to all network stations. Only the station that can successfully decrypt the message with its private key notices that it is the actual receiver

Visible implicit addresses can be realized more easily by having users select arbitrary names for themselves, which are then prefixed to messages. Another criterion of implicit addresses is their distribution. An implicit address is called public if it is known to every user (like telephone numbers today), and private if the sender received it secretly from the addressee. This private distribution can be accomplished in several ways, including outside the network, as a return address, or by a generating algorithm that the sender and the addressee agreed on [59, 60].

Public addresses should not be assigned using visible implicit addresses in order to avoid the linkability of the visible public address of a message and the addressee. Private addresses can be realized by visible addresses but then each of them should be used only once. Figure 11.1 summarizes this.

Example

If user A wants to keep the recipient B of a message secret, she chooses additional pseudo recipients (e.g., C and D). Together with the real recipient B, these additional recipients form the anonymity set. The message is then broadcasted to all members of the anonymity set (see Figure 11.2).

		Address distribution	
		Public address	Private address
Implicit address	Invisible	Very costly, but necessary to establish contact	Costly
	Visible	Not advisable	Frequent change after use

Figure 11.1 Combination of implicit addressing modes and address distribution [54].

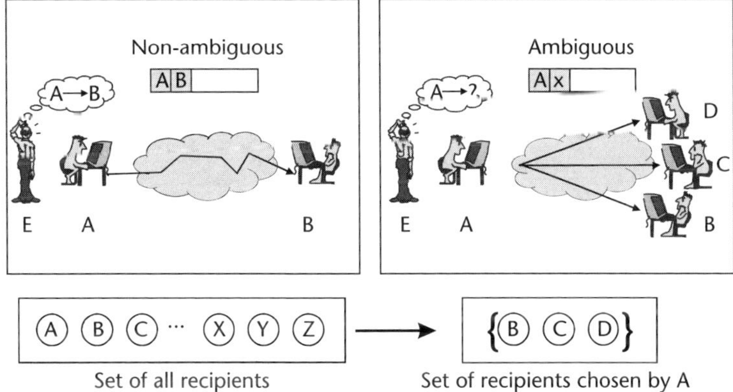

Figure 11.2 The idea of general recipient anonymity by broadcast and addressing attributes (E is the attacker) [61].

11.5.2 DC-Networks

While the previous method provides recipient anonymity, all observers know the identity of the sender. In [34] a powerful technique called DC networks is suggested for sender anonymity. The DC-network is time slotted, and in each time slot all participating senders send a message, although for successful transmission only one of them has to be the real message and the others have to be empty messages (consisting of only zeros). Now the task is to hide the real message in the cover of the empty messages. For this task n users exchange secret keys along a given key graph, each ($sender_i$) then locally exclusive-ors (XORs) all of the keys it has with the empty or real message *Mi* that it is to send, and finally all of the local results of the n users are XORed globally.

In [54] it is shown that this technique provides perfect protection if the computation is carried out over an Abelian (commutative) group. Here we formally present the technique for the simple case using binary numbers (bits):

1. *Initialization*: n users exchange secret keys (random bit strings as long as the messages) along a given key graph G (i.e., a graph with the users as nodes and the secret keys as edges). If a node X (i.e., user X) is connected to node Y, then they both share the same symmetric secret key, k_{xy}, such that all keys appear twice in the technique.

2. *Message transmission*: to send a message M including recipient address, a node X XORs M with all the keys k_{xj} that its shares with its *key graph neighbors*: $M \oplus \Sigma k_{xj}$.The result is sent as a communication packet.
3. *Cover traffic*: All other nodes that do not want to send a real message send an empty message (i.e., zeros) by calculating the XOR of all the keys they share with their key graph neighbors and sending the results as communication packets.

If all packets are XORed in this manner, and assuming that only one node sends a real message M, then only the message M will remain, since all keys occur exactly twice and will cancel each other. The result is broadcast to all participants. If more than one message was sent in the same slot, then each user will receive the XOR summation of these messages. Since no user will be able to decode it, the senders can recognize this *collision*. Detected collisions can be resolved by retransmitting the message after a random number of rounds (e.g., see the ALOHA protocol [63]).

DC-networks provide perfect sender anonymity since the fact that someone is sending a (nonempty) message is hidden by a one-time pad encryption (i.e., an encryption scheme that provides perfect secrecy in the information-theoretic sense). Perfect receiver anonymity can be provided by reliable broadcast, and message secrecy is enforced by encrypting messages.

Example

To send a message ("110101"), user A XORs the message with the previously exchanged secret key. Other users XOR an empty message in the same manner (Figure 11.3) with the keys that they share. All sums of all users are XORed succes-

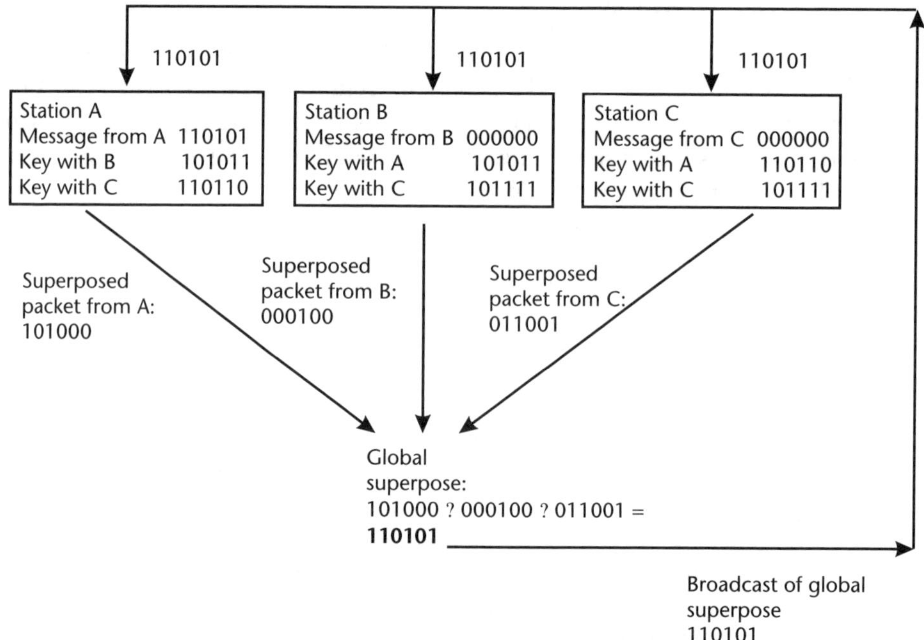

Figure 11.3 A DC-network.

sively. Because every secret key is added twice, the distributed message is the message of *A*.

11.5.3 Mix Nets

The anonymity-providing techniques presented so far suffer from the same major drawback: they require the preexistence of an organized group (i.e., forming the anonymity set) of participants (e.g., through the exchange of secret keys). The members of the anonymity set have to establish the anonymity set themselves and are then involved in every communication. This severely limits the flexibility of the involved users.

The mix method [33] avoids this drawback by shifting the task of generating anonymity sets from the user to special trusted third parties (TTPs) called mix nodes or mixes. Mix nets enable much larger networks of users because they eliminate the constraint that each user has to participate in every communication. By providing such flexible access to an anonymity service, the mix approach is the most interesting for open networks like the Internet. Here is a more formal description of the mix approach:

1. *Initialization*: A public key infrastructure (PKI) provides users with the public keys of the mix nodes, where PK_i is the public key of the ith mix node and SK_i the secret key.
2. *Message transmission*: Assume a data packet $M = [A_{recipient}, k(message)]$ containing recipient's address and an end-to-end encrypted message with a constant message length (messages are split or padded out to the specified length). The user performs the following recursive cryptographic operation (starting from the end of the message):

$$M_{N+1} = M$$
$$M_i = PK_i(A_{i+1}, r_{i+1}, M_{i+1}) \quad \text{for} \quad i = N, N-1, \ldots, 1 \tag{11.1}$$

   A_i is the address of mix_i, and r_i is a random number, which is included to cause any attempt at decryption to be nondeterministic. The user sends the message (e.g., $M_1 = PK_1(A_2,r_2,M_2)$), to the mix node with the address A_1. Since the packet contains all the hop-by-hop routing information (source routing) as readable only by the specified mix node, the message can be routed to the destination without revealing the complete address to any subgroup of the mix nodes.
3. *Cover traffic*: All other users send similarly prepared packets, which can be real, dummy, or a mix of real and dummy messages, all with equal length, to the mixes.
4. *Mix functionality*: Each mix node waits until *n* packets of equal length (preferably from *n* or at least a high number of distinct users) arrive and collects the packets in a so-called batch. All the packets are compared to the previously processed packets and any duplicates will be deleted (in order to thwart replay attacks). Then the mix node decrypts the packets with its private key, removes their random numbers, and outputs the

updated packets in a different order than that in which they arrived (e.g., lexicographically sorted).

If all packets flow through all mixes in a static order, then it is called mix cascade. In this case no address information for intermix routing is needed. Otherwise, if the route for each packet is chosen individually by the sender, then this mode is called mix network (or mix net) [33]. The general mix functionality hides the relationship between a sender and a recipient of a message from everyone but the mix and the sender of the message. This is because each mix in the chain (cascade or network) hides the relation between its incoming and outgoing packets by providing the following features:

1. The appearance (bit structure) of the packets is irreversibly changed by the mix through decryption (an attacker would have to guess the random number).
2. The packets cannot be distinguished by their length, since all packets have the same length (agreement).
3. Any time correlation (like FIFO) is avoided, as the mix reorders the packets that it collects in its batch before they are sent out.

A global attacker who can monitor all communication lines can only trace a message through the mix network if he has the cooperation of all mix nodes on the path or if he can break the cryptographic operations. Thus, in order to ensure unlinkability of sender and recipient, at least one mix in the chain has to be trustworthy.

A remaining problem is that an attacker can participate in this protocol too. If he is, for instance, able to contribute $(n-1)$ of the n packets, then the one remaining packet is observable. This so-called (n – 1) attack can be thwarted by a protection scheme presented in [50]. The first step is that every mix must know the sender anonymity set (i.e., the sending users). Thus, the first mix can apply direct identification techniques and has to ensure that it collects n packets from n distinct users before beginning operation. The following mixes have to ensure this security functionality without weakening the anonymity (i.e., untraceability) of the former mixes.

Example

Assume that A wants to send a message M to Z over the mix cascade (Figure 11.4). A must encrypt the message two times with the public keys, PK_i, of the respective mixes, and include the random numbers r_i in each encryption layer:

$$PK_1(r_1, PK_2(r_2, z, M)) \tag{11.2}$$

11.5.4 Private Information Retrieval

Private information retrieval (PIR) [48], also known as private message service [49], assures that an unbounded attacker is not able to discover the information that a user has requested. The goal is to request exactly one datum that is stored in a

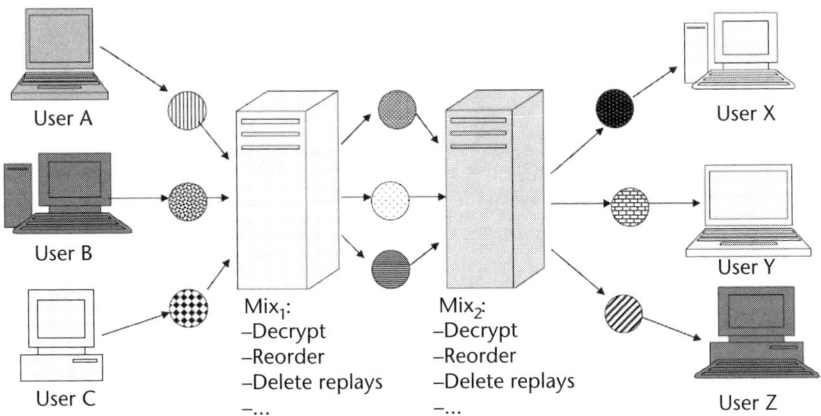

Figure 11.4 Cascade of two mixes.

remote memory cell of a server without revealing which datum is requested (protection of interest data). If the user requests all cells, then the interest data is easily protected using an efficient method suggested in [49]:

1. *Initialization*: N replicated servers, each with an identical copy of the database having n cells.
2. *Message request*: In order to read a cell at a position i, the user generates a vector V_i that has a "1" at the ith position and "0" otherwise.
3. *Cover traffic*: To protect his interest data, the user randomly generates a second vector V_{random} (i.e., for each ith position of the vector, a "1" or a "0" is chosen randomly) and XORs the two to form another vector:

$$V_1 = V_i \oplus V_{random} \quad (11.3)$$

   Sender then transmits V_1 via a secure channel (i.e., an end-to-end encrypted channel) to $server_1$.
4. *Additional requests*: Generate (N–2) further random vectors $V_2, \ldots, V_{N-1}$ and for calculating the Nth vector by the following formula:

$$V_N = \sum_{i=1}^{N-1} V_i \quad (11.4)$$

   Send each random vector over a secure channel to the corresponding server.
5. *Server algorithm*: If V_i has a "1" at position j, then the jth cell is read. After XORing all the read cells together to obtain one cell, the ith server responds with the result C_i using again a secure channel:

$$C_i = \sum_{\text{if } j^{th} \text{ position of } V_i = 1} cell_j \ (mod\ 2) \quad (11.5)$$

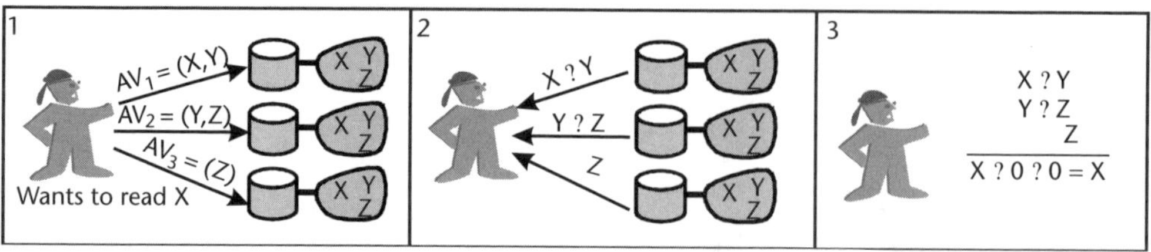

Figure 11.5 Basic concept of PIR and message service [61].

Example

Figure 11.5 shows an example of this method. Assume the cells are (X, Y, Z). If the user wants to read X then he creates the vector V_i=(1,0,0), a random vector V_{rand1}=(1,0,1), and calculates his first request vector $V_i \oplus V_{rand1} = V_1$=(0,0,1). Then he chooses another random vector, say, V_{rand2}=(0,1,1), and calculates a second vector as $V_2 = V_{rand1} \oplus V_{rand2}$=(1,0,1)⊕ (0,1,1) =(1,1,0). He sends the request vectors V_1, V_{rand2}, and V_2 end-to-end encrypted to the corresponding servers. Each server retrieves the cells corresponding to the 1s in its vector and XORs them to create one cell. The cells from all servers are returned (end-to-end encrypted) to the requestor who XORs all the cells and obtains the requested data cell X.

11.5.5 New Protocols Against Local Attacker Model: Onion Routing, Web Mixes, and P2P Mechanisms

This section presents some protocols designed to protect against the local attacker model, including onion routing (and its second generation TOR), web mixes, and P2P anonymous communication mechanisms. These protocols were designed for real-time communication and are commonly known as low latency protocols. They have two phases of communication: connection establishment and data sending phase. These low latency protocols do not apply the batching strategy (i.e., they directly decrypt the message and forward it to the next hop). Thus, an attacker observing such a node can principally relate the input to the output, but it is assumed here that the attacker cannot observe all used nodes.

11.5.5.1 Onion Routing, TOR, and Web Mixes

Onion routing [35, 36, 55] and TOR [56] (the second generation of the onion routing system) are distributed via overlay networks designed to add anonymity to real-time, bidirectional TCP-based applications. In onion routing, an anonymous bidirectional virtual path (or circuit) within link-encrypted connections already running between onion routers (ORs) is set up between communication partners (initiator and responder). Paths are constructed using "onions," which are objects with several public-key encrypted layers that are constructed via the recursive cryptographic operations as defined for mix nets and are encapsulating routes (i.e., the series of OR nodes that interconnect the initiator and the responder). The route is usually defined by the initiator's proxy, which constructs the onion and sends it with a "create" command to the first node on the path. ORs are currently (i.e., TOR) picked from a trusted directory server [56]. The onion is also distributing "forward" and "backward" symmetric encryption function/key pairs to the OR nodes on the path that, once that the virtual path has been set up, are applied by these nodes to encrypt data that will be sent forward and backward along the virtual path.

The onion structure is thus composed of layer on layer of encryption wrapped around a payload [35]. Each layer is encrypted with the receiving node public key and contains an expiration time for the onion (until that, an onion is stored by the respective node to detect replays), the next OR to which the payload is to be sent, the forward and a backward function/key pairs, and the payload. An onion received by an OR X has the following structure.

$$\{expiration\_time, next\_OR, F_f, K_f, F_b, K_b, payload\}\ PK_x \tag{11.6}$$

The structure of an onion is illustrated in Figure 11.6.

While the mix net concept based on layered public key encryption is used for setting up a virtual path, the more efficient symmetric encryption functions are applied for encrypting the data that are then communicated via the virtual path.

Since each OR knows only about the previous and next ORs along the path and nothing about the other nodes or their positions in the circuit, and since each node further encrypts multiplexed virtual circuits, traffic analysis is made difficult. However, if the first node behind the initiator's proxy and the last node on the virtual path cooperate, they can determine the source and recipient of the communication through the number of cells sent over the path or through the duration that the virtual path was used.

Onion routing currently has three generations. Onion routing generation 0 was proof of concept prototype consisting of a five-node system running on a single machine with proxies for web browsing [55]. The generation 0 of onion routing transmitted fixed-sized packets (cells) of 42 bytes of payload (for a perfect fit to ATM cells) between ORs. Onion routing generation 1 changes from generation 0 included: increased cell size to 128 bytes (for better performance); changed crypto engine; included proxies to different application protocols, such as SMTP, DNS, and Telnet; and reorganized the onion routing system internal architecture [35, 36, 55]. Generation 1 had a five-layer onion, and RSA was used for key exchange.

Onion routing system generation 2 is called TOR [56]. In TOR, the cell size increased even further, to 512 bytes; many TCP streams are allowed share a single

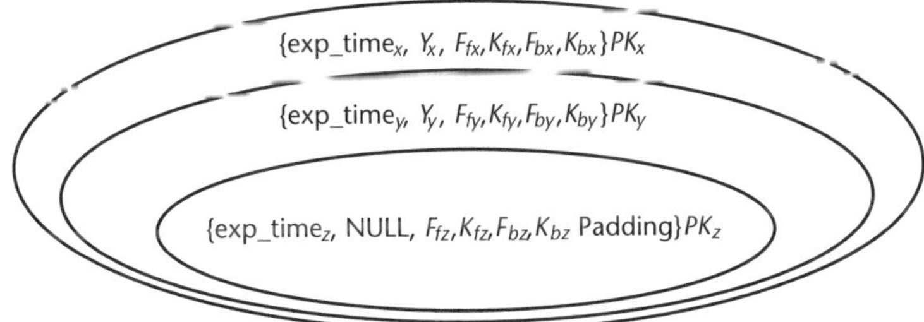

Figure 11.6 A forward onion (according to [35]). This onion was sent by an initiator to the responder Z, through two intermediary nodes X and Y. The outer layer of the onion is encrypted with the public key of OR X, PK_X, the middle layer is encrypted with Y's public key PK_Y and the inner layer with Z's public key PK_Z.

virtual circuit (which was not possible in the previous generations); trusted directory servers were added to the system to provide signed lists of ORs and their status; and TLS connections are maintained between ORs. The circuit build-up was modified to increase the privacy properties. Instead of using a single multiple data structure (as presented in Figure 11.6) to lay the circuit, TOR uses incremental (or telescopic) path establishment. In telescopic path building, the initiator negotiates session keys (using DH) with each successive hop in the circuit, increasing the circuit hop by hop. The use of telescopic encryption gives TOR a perfect forward secrecy property, since compromised nodes cannot decrypt old traffic once the session keys are deleted.

Location-hidden services (or responder anonymity) are used for offering TCP services, without revealing the address of the server. This mechanism also protects the server against network attacks, such as DDoS attacks. Access to hidden services is offered by rendezvous points. Hidden servers are advertised by several ORs functioning as introduction points. A client connects to one of these ORs and requests access to the hidden server in a rendezvous point (another OR) of the client's choice, and waits for the hidden server to connect to a rendezvous point.

The main difference between onion routing (and TOR) in contrast to web mixes is that web mixes provide a kind of guaranteed service, since their intermediated nodes are under the control of some professionals with a high-capacity access to the Internet. TOR nodes can be operated also from normal users (e.g., with a DSL connection). Thus, there are many TOR node operators scattered over the world, and only few web mix nodes. Consequently, the message delay by web mixes is nearly constant with a low rate message connection. In contrast to that, TOR offers usually a better connection performance that is highly dependent on the actual connection path.

11.5.5.1 Crowds

Reiter and Rubin developed *crowds*—a system based on the idea that users can make anonymous web transactions when they blend into a "crowd" [37–39]. Crowds was the first peer-to-peer (P2P) anonymous communication mechanism offered as an Internet service. Differently from mixes presented in Section 11.5.3, in P2P anonymous communication mechanisms there is no distinction between users and mixes (i.e., network nodes both generate their own traffic—the user role—and forward messages for other nodes—the mix role—simultaneously).

A crowd is a geographically diverse group that performs web transactions on behalf of its members. Each crowd member runs a process on his local machine called *jondo*. Once started, the jondo engages in a protocol to join the crowd, during which it is informed of the other current crowd members and the other crowd members are informed of the new jondo's membership. Besides, the user configures his browser to employ the local jondo as a proxy for all network services. Thus all his http-requests are directly sent to the jondo rather than to the end web server, and the jondo forwards the request to a randomly chosen crowd member. Whenever a crowd member receives a request from another jondo in the crowd, it makes a random choice to forward the request to another crowd member with a probability $p_f > 1/2$ or to submit the request to the end web server to which the request was destined with the probability $1 - p_f$. The server's reply is sent backward along the path, with each jondo sending it to its predecessor on the path until it reaches the originator of the request. All communication between jondos is encrypted with a symmetric key shared between the jondos involved [44].

Crowds provides sender anonymity against the end web server, which is "beyond suspicion" (i.e., web servers are equally likely to be the originator of a request), since the end server obtains no information regarding who initiated any given request. Second, since a jondo on the path cannot distinguish whether its predecessor on the path initiated the request or is forwarding it, no jondo on the path can learn the initiator of a request. Since all communication between jondos is encrypted, crowds also offers receiver anonymity against a local eavesdropper (e.g., local gateway administrator) that can observe the communication involving the user's machine unless the originator of the request ends up submitting the request itself. However, the probability that an originator submits its own request decreases as the crowd size increases.

Crowds enables very efficient implementations that typically outperform mixes using layered encryption techniques with costly public-key crypto operations. However, in contrast to mix nets, crowds cannot protect against global attackers. Besides, the set up is less practical, so that it has not been deployed much yet in practice.

11.5.5.2 Hordes and Other P2P Anonymous Communication Mechanisms

Hordes [58] is a crowds variant, and as crowds, it provides anonymity through plausible anonymity. Hordes uses the same forwarding mechanism of crowds, and multicast routing is used in the backward (reverse) path. Nodes have public-private key pair. Link encryption is used for message exchange between two consecutive nodes, and key establishment is done using Diffie-Hellman (DH) key exchange.

Using multicast routing in the backward path allows the sender to be indistinguishable within the multicast group, since multicast logical addresses do not refer to a specific device, but to a set of devices. The only requirement is that more than one device should be part of the multicast group. One of the design goals of hordes was to perform better than crowds and onion routing, and multicast routing proved to be computationally cheaper option for the backward path. Hordes offer the same anonymity properties of crowds.

Several other P2P anonymous communication systems were designed after crowds. But unlike crowds, which was deployed as an Internet service, most of the anonymous communication systems later proposed were theoretical designs with prototype implementations. A nonexhaustive list of those P2P anonymous communication mechanisms include the P2P anonymity system [64], GAP [65], MorphMix [66–68], Herbivore [69], and P5 [70].

11.6 Spyware and Spyware Countermeasures

Spyware in general refers to any privacy-invasive software [71], especially regarding informational and spatial privacy. Spyware initially referred to software that monitored the user behavior, in most cases without the user consent, and collected this data for the purpose of advertising. However, the concept of spyware became broader with the advent of new functionalities, such as the installation of other programs or the redirection of user activities, such as web browsing. The Anti-Spyware Coalition (ASC) is a group dedicated to the debate surrounding spyware and other potentially unwanted technologies. ASC defines spyware as "technologies deployed without appropriate user consent and/or implemented in ways that impair user control over (1) Material changes that affect their user experience, privacy, or system security; (2) Use of their system resources, including what programs are installed on their computers; and/or (3) Collection, use, and distribution of their personal or other sensitive information" [72]. An important aspect regarding spyware is the user informed consent. A piece of software that is installed with user consent is not considered a spyware, because it is running with the user's knowledge and permission. This aspect is usually abused by spyware producers to trick a user to consent to an end user license agreement that hides text about the spyware functions among other long legal phrases in order to obtain a legal umbrella to protect them against law suits.

There are basically three known ways to distribute spyware nowadays [71]. The first is to bundle spyware with another piece of software that users are willing to install. The second is to exploit security vulnerabilities in target systems. The third method is to deceive users to click on links on websites [73]. The challenge of antispyware programs is to distinguish spyware from other legitimate software in the system. The ASC risk model description [72] provides guidelines to identify suspicious software behavior that may impact users, including software installation and distribution methods, program identification and controls, usage of network resources, collection of personal data, computer security implications, and impact of the user experience. These guidelines are meant to be used by antispyware companies to define their policies for spyware identification and removal.

Basically, three different ways for identification of spywares are in use nowadays [74]: through filenames, through MD5 hashes of the spyware program (as a signature), or by deep scanning the system. The first method's advantages are the easy set up of a large database of spyware programs and the possibility of fast system scans, and the disadvantage is that this method is easy to bypass (by changing file names) and risks the possibility of false positives. The second method's advantages are the fast system scans and better reliability than filename comparison, but this detection method fails if a program changes a single bit of its code. The last method is in general similar to antivirus detection (but not equal, since spyware usually does not share some basic virus and worm distribution methods, such as code replication), and it takes a longer time to be completed than the first two methods, but the results are usually more reliable than the two previous methods [74]. These identification methods can be used for both real-time protection of the system (i.e., preventing spyware from being installed in the system) and also for detection and removal of spyware already installed into the target system. It is worth noting that removing every software considered spyware from a target system may impact legitimate programs, and safeguarding all programs considered legitimate may cause some spyware to be left untouched [71]. Therefore, it is left to antispyware software companies to decide which software is considered spyware and which software is legitimate. The arbitrariness of this decision resulted in legal disputes between some software houses, whose programs were labeled as spyware, and antispyware software producers [71].

11.7 Conclusions

In this chapter, we have defined the concept of privacy and discussed privacy risks that individuals are facing in our networked society, taking the privacy risks associated with LBS and RFID as emerging technology examples. Then, we proceeded with discussing legal privacy principles as existent privacy laws means for privacy protection. Privacy, however, cannot be solely protected by legislation. This chapter is therefore also classifying and presenting privacy-enhancing technologies, which provide important means for technically enforcing legal privacy requirements. Privacy can be best protected if personal data are avoided or minimized. Anonymous communication technologies, which were discussed in more detail in this chapter, are therefore essential PETs for protecting the privacy of communication partners (allowing, for instance, LBS users to send requests to LBS servers anonymously) and are providing important building blocks for privacy-enhancing applications. Still, further PET research and development is needed for adequately addressing privacy threats of emerging ambient technologies.

References

[1] Organisation for Economic Co-operation and Development, "Guidelines on the Protection of Privacy and Transborder Flows of Personal Data," *OECD Guidelines*, September 1980.

[2] European Union, "Directive 95/46/EC of the European Parliament and of the Council of 24 October 1995 on the Protection of Individuals with Regard to the Processing of Personal Data and on the Free Movement of Such Data," *Official Journal L*, No. 281, November 1995.

[3] European Union, "Directive 2002/58/EC of the European Parliament and of the Council Concerning the Processing of Personal Data and the Protection of Privacy in the Electronic Communications Sector," *Official Journal L*, No. 201, July 2002.

[4] European Union, "Directive 2006/24/EC of the European Parliament and of the Council of 15 March 2006, on the Retention of Data Generated or Processed in Connection with the Provision of Publicly Available Electronic Communications Services or of Public Communications Networks and Amending Directive 2002/58/EC," *Official Journal L*, No. 105, April 2006.

[5] Warren, S. D., and L.D. Brandeis, "The Right to Privacy," *Harvard Law Review*, No. 5, 1890–1891, pp.193–220.

[6] Westin, A. F., *Privacy and Freedom*, New York: Atheneum, 1967.

[7] Hogben, G., "Annex A," PRIME project deliverable D14.0a.

[8] Good, N., et al., "Users Choices and Regret: Understanding Users' Decision Process About Consentually Acquired Spyware," *I/S: A Journal of Law and Policy in the Information Society*, Vol. 2, No. 2, 2006.

[9] Rosenberg, R., *The Social Impact of Computers*, Boston: Academic Press, 1992.

[10] Kramer, G., et al., "Section 4.2: Location Based Services Application Scenarios," PRIME project deliverable D14.0a.

[11] Fischer-Hübner, S., and C. Andersson, "Privacy Risks and Challenges for the Mobile Internet," *Proc. 2nd IEE Summit on Law and Computing*, London, England, November 2004.

[12] Fischer-Hübner, S., M. Nilsson, and H. Lindskog, "Self-Determination in Mobile Internet," *Proc. 17th IFIP TC11 International Conference on Information Security (SEC 2002)*, Cairo, Egypt, May 2002.

[13] Nilsson, M., H. Lindskog, and S. Fischer-Hübner, "Privacy Enhancements in the Mobile Internet," *Proc. IFIP WG 9.6/11.7 Working Conference on Security and Control of IT in Society*, Bratislava, Slovakia, June 2001.

[14] European Commission, "Working Document on Data Protection Issues Related to RFID Technology," Art. 29 Data Protection Working Party, January 19, 2005.

[15] Fischer-Hübner, S., T. Holleboom, and A. Zuccato, "RFID Tags—More Than a Replacement of the Bar Code," *RFID Nordic Newsletter*, No. 04/2005, November 2005.

[16] Juels, A., "RFID Security and Privacy: A Research Survey," *IEEE Journal on Selected Areas in Communication*, Vol. 24, No. 2, February 2006, pp. 381–394.

[17] US Privacy Act, Pub. L. No. 93-579 (1974), codified at 5 USC § 552a, http://www.epic.org/privacy/laws/privacy_act.html.

[18] Transportation, Treasury, Independent Agencies, and General Government Appropriations Act, § 522, http://frwebgate.access.gpo.gov/cgi-bin/getdoc.cgi?dbname=108_cong_ bills&docid=f:h4818enr.pdf.

[19] Right to Financial Privacy Act, Pub. L. No. 95-630 (1978).

[20] Video Privacy Protection Act, Pub. L. No. 100-618 (1988), http://www.epic.org/privacy/vppa/.

[21] Children's Online Privacy Protection Act, Pub. L. No. 105-277 (1998), http://www4.law.cornell.edu/uscode/html/uscode15/usc_sec_15_00006501——000-.html.

[22] Family Educational Rights and Privacy Act, Pub. L. No. 93-380 (1974), http://www.epic.org/privacy/education/ferpa.html.

[23] Drivers Privacy Protection Act, Pub. L. No. 103-322 (1994), http://www.epic.org/privacy/laws/drivers_privacy_bill.html.

[24] Telephone Consumer Protection Act, Pub. L. No. 102-243 (1991), http://www.epic.org/privacy/telemarketing/.

[25] EPIC and Privacy International, Privacy & Human Rights 2006—An International Survey of Privacy Laws and Surveys, 2006.

[26] Fair Credit Reporting Act, Pub. L. No. 91-508 (1970), amended by Pub. L. No. 104-208 (1996), http://www.ftc.gov/os/statutes/fcra.htm.

[27] Fair Accurate Credit Transaction Act, Pub. L. No. 108-159 (2003).

[28] California Civil Code, §§ 1798.29 and 1798.82, http://www.privacy.ca.gov/code/cc1798.291798.82.html.

[29] CAN-SPAM ACT, Pub. L. No. 108-187 (2003), http://www.spamlaws.com/pdf/pl108-187.pdf.

[30] European Commission, "European Commission First Report on the Implementation of the Data Protection Directive 95/46/EC," May 2003.

[31] Registratiekamer, Information and Privacy Commissioner/Ontario, "Privacy-Enhancing Technologies: The Path to Anonymity," *Achtergrondstudies en Verkenningen 5B*, Vols. I & II, Rijswijk, August 1995.

[32] Pfitzmann, A., and M. Hansen, "Anonymity, Unlinkability, Unobservability, Pseudonymity, and Identity Management—A Consolidated Proposal for Terminology," Version 0.28, May 2006, http://dud.inf.tu-dresden.de/Anon_Terminology.shtml.

[33] Chaum, D. L., "Untraceable Electronic Mail, Return Addresses, and Digital Pseudonyms," *Communications of the ACM*, Vol. 24, No. 2, February 1981, pp. 84–88.

[34] Chaum, D. L., "The Dining Cryptographers Problem: Unconditional Sender and Recipient Untraceability," *Journal of Cryptology*, Vol. 1, No. 1, January 1988, pp. 65–75.

[35] Goldschlag, D. M., M. G. Reed, and P. F. Syverson, "Hiding Routing Information," *Proc. of the 1st International Workshop on Information Hiding (IH 1996)*, Springer-Verlag, LNCS 1174, Cambridge, UK, May 1996, pp. 137–150.

[36] Reed, M. G., P. F. Syverson, and D. M. Goldschlag, "Anonymous Connections and Onion Routing," *IEEE Journal on Selected Areas of Communications*, Vol. 16, No. 4, May 1998, pp. 482–494.

[37] Reiter, M., and A. Rubin, "Anonymous Web Transactions with Crowds," *Communications of the ACM*, Vol. 42, No. 2, February 1999, pp. 32–48.

[38] Reiter, M., and A. Rubin, "Crowds: Anonymity for Web Transactions," *DIMACS Technical Report 97-15*, 1997, pp. 97–115.

[39] Reiter, M., and A. Rubin, A"Crowds: Anonymity for Web Transactions," *ACM Transactions on Information and System Security*, Vol. 1, No. 1, November 1998, pp. 66–92.

[40] Chaum, D. L., "Security Without Identification: Card Computers to Make Big Brother Obsolete," *Informatik-Spektrum*, Vol. 10, 1987, pp. 262–277.

[41] Camenisch, J., and E. van Herreweghen, "Design and Implementation of the Idemix Anonymous Credential System," *Proc. 9th ACM Conference on Computer and Communications Security (CCS 2002)*, Washington, DC, November 2002.

[42] Cranor, L., *Web Privacy with P3P*, Cambridge, MA: O'Reilly, September 2002.

[43] W3C—World Wide Web Consortium, "The Platform for Privacy Preferences 1.0 (P3P1.0) Specification," *W3C Recommendation*, April 2002.

[44] Fischer-Hübner, S., *IT-Security and Privacy: Design and Use of Privacy Enhancing Security Mechanisms*, Berlin: Springer-Verlag, LNCS 1958, 2001.

[45] Karjoth, G., M. Schunter, and M. Waidner, "Platform for Enterprise Privacy Practices: Privacy-Enabled Management of Customer Data," *Proc. 2nd Workshop on Privacy Enhancing Technologies (PET 2002)*, San Francisco, CA, April 2002, pp. 69–84.

[46] Powers, C., and M. Schunter (eds.), "Enterprise Privacy Authorization Language (EPAL 1.2)," *W3C Member Submission*, November 2003.

[47] PRIME Project, Privacy and Identity Management for Europe, https://www.prime-project .eu/.

[48] Chor, B., et al., "Private Information Retrieval," *Proc. 36th Annual Symposium on Foundations of Computer Science (FOCS'95)*, Milwaukee, WI, October 1995, pp. 41–51.

[49] Cooper, D. A., and K. P. Birman, "Preserving Privacy in a Network of Mobile Computers," *Proc. IEEE Symposium on Security and Privacy*, Oakland, CA, May 1995, pp. 26–83.

[50] Kesdogan, D., J. Egner, and R. Büschkes, "Stop-and-Go-Mixes Providing Probabilistic Anonymity in an Open System," *Proc. Information Hiding 1998 (IH98)*, Springer-Verlag, LNCS 1525, 1998, pp. 83–98.

[51] Danezis, G., *Better Anonymous Communications*, PhD Thesis, University of Cambridge, Cambridge, UK, July 2004.

[52] Dolev, D., and A.C. Yao, "On the Security of Public Key Protocols," *IEEE Transactions on Information Theory*, Vol. 29, 1983, pp. 198–208.

[53] Pfitzmann, A., and M. Waidner, "Networks Without User Observability—Design Options," *Proc. Workshop on the Theory and Application of Cryptographic Techniques (EUROCRYPT '85)*, LNCS 219/1986, Linz, Austria, April 1985, pp. 245–254.

[54] Pfitzmann, A., *Dienstintegrierende Kommunikationsnetze mit teilnehmerüberprüfbarem Datenschutz*, Heidelberg: Springer-Verlag, IFB 234, 1990 (in German).

[55] US Naval Research Laboratory, "Onion Routing," http://www.onion-router.net/.

[56] Dingledine, R., N. Mathewson, and P. Syverson, "Tor: The Second-Generation Onion Router," *Proc. of the 13th USENIX Security Symposium*, San Diego, CA, August 2004, pp. 303–320.

[57] Berthold, O., H. Federrath, and S. Köpsell, "Web Mixes: A System for Anonymous and Unobservable Internet Access," *Proc. International Workshop on Design Issues in Anonymity and Unobservability*, Springer-Verlag, LNCS 2009, 2001. pp. 115–129.

[58] Levine, B., and C. Shields "Hordes: A Multicast Based Protocol for Anonymity," *ACM Journal of Computer Security*, Vol. 10, No. 3, 2002. pp. 213–240.

[59] Farber, D. J., and K. C. Larson, "Network Security Via Dynamic Process Renaming," *Proc. 4th Data Communication Symposium*, Quebec City, Canada, October 1975, pp. 8.13–8.18.

[60] Karger, P. A., "Non-Discretionary Access Control for Decentralized Computing Systems," Master Thesis, Report MIT/LCS/TR-179, Massachusetts Institute of Technology, Laboratory for Computer Science, Cambridge, MA, May 1977.

[61] Kesdogan, D., and C. Palmer, "Technical Challenges of Network Anonymity," *Computer Communications*, Vol. 29, No. 3, February 2006, pp. 306–324.

[62] Denning, D. E., *Cryptography and Data Security*, Reading, MA: Addison-Wesley, 1982.

[63] Tanenbaum, A. S., *Computer Networks*, Englewood Cliffs, NJ: Prentice-Hall, 1996.

[64] Freedman, M., and R. Morris "Tarzan: A Peer-to-Peer Anonymizing Network Layer," *Proc. 9th ACM Conference on Computer and Communications Security (CCS 2002)*, Washington, DC, November 2002.

[65] Bennett, K., and C. Grothoff, "GAP—Practical Anonymous Networking," *Proc. Int. Workshop on Privacy Enhancing Technologies (PET 2003)*, Dresden, Germany, March 2003, pp. 141–160.

[66] Rennhard, M., and B. Platter, "Practical Anonymity for the Masses with MorphMix," *Proc. 8th Int. Conference of Financial Cryptography (FC04)*, Key West, FL, February 2004.

[67] Rennhard, M., and B. Platter, "Introducing MorphMix: Peer-to-Peer Based Anonymous Internet Usage with Collusion Detection," *Proc. Workshop on Privacy in Electronic Society (WPES)*, Washington, DC, November 2002.

[68] Rennhard, M., "MorphMix—A Peer-to-Peer-Based System for Anonymous Internet Access," PhD Thesis, ETH Dissertation No. 15420, ETH Zurich, Switzerland, TIK-Schriftenreihe No. 61, 2004.

[69] Goel, S., et al., "Herbivore: A Scalable and Efficient Protocol for Anonymous Communication," *Technical Report 2003-1890*, Cornell University, Ithaca, NY, February 2003.

[70] Sherwood, R., B. Bhattacharjee, and A. Srinivasan, "P5: A Protocol for Scalable Anonymous Communication," *Proc. IEEE Symposium on Security and Privacy*, Oakland, CA, May 2002, pp. 55–70.

[71] Boldt, M., *Privacy-Invasive Software—Exploring Effects and Countermeasures*, Licentiate Thesis Series No. 2007:01, School of Engineering, Blekinge Institute of Technology, Sweden, 2007.

[72] Anti-Spyware Coalition, "ASC Risk Model Description," Working Report, June 2006.

[73] Moshchuk, A., et al., "A Crawler-Based Study of Spyware on the Web," *Proc. 13th Annual Network and Distributed System Security Symposium (NDSS 2006)*, San Diego, CA, 2006.

[74] McAfee, Inc., "Anti-Spyware Testing Methodology: Methodology for Comparing Anti-Spyware Products," McAfee System Protection Solution Series, October 2005.

CHAPTER 12

Content Filtering Technologies and the Law

Stefanos Gritzalis and Lilian Mitrou

As the Internet and the World Wide Web have grown from an academic experiment to a real mass media, they are often criticized for providing an integrated information infrastructure that can be misused for illegal and harmful content distribution (e.g., child-pornographic images and video, propagandistic material, instructions to make drugs). Several technical approaches have been introduced to protect users. In this chapter we address technical proposals, recent research trends, as well as legal and ethical issues raised.

12.1 Filtering: A Technical Solution as a Legal Solution or Imperative?

The Internet is exerting a significant impact on global patterns of access to information. The digital revolution makes possible widespread cultural participation and interaction that previously could not have existed, at least on this scale. However, at the same time, it creates new opportunities for limiting and controlling the flow of information and/or access to it, resulting in controlling and limiting of these new forms of cultural participation and interaction [1]. The introduction of new technologies usually tends to be accompanied by a desire on the part of public authorities to control and regulate their use and access to them.

The number of states that routinely block websites and limit access to Internet content has risen rapidly in recent years [2]. National states deploy national-level Internet filtering systems to limit access to websites containing material deemed to be unacceptable to the state [3]. Governments offer a range of justifications for web content filtering, which are often powerful and compelling rationales: "protecting national security," "preserving cultural norms and religious values," "securing intellectual property rights," and "shielding children from pornography and exploitation" rationales.

Techniques for blocking undesirable content are growing ever more sophisticated. Their use has increased as the "mandatory response to the current plagues of society, namely pornography, violence, hate and in general anything seen to be unpleasant or threatening" [4]. Technical measures are considered necessary in order to promote the safer use of online technologies, to protect the end user from unwanted content, and in the end to encourage the exploitation of the opportunities offered by the Internet [5]. Controlling access to the Internet by means of filtering has become a growth industry. Internet filtering products are increasingly being marketed as a way of monitoring and safeguarding children's access to the Internet. Filters can serve to supervise not only children but also any other person whose Internet access is controlled. Organizations such as government departments and businesses deploy organizational-level Internet filtering systems to stop civil servants and employees from accessing material that would violate acceptable use policies [3].

12.1.1 Filtering Categories

A filter is a "device or material for suppressing or minimizing waves or oscillations of certain frequencies" [6] and a content filter is defined as "one or more pieces of software that work together to prevent users from viewing material found on the Internet." Filtering software purports to identify the content of online material and accordingly separate the "good" from the "bad" [7]. There are different strategies for limiting access to websites and newsgroups. Internet filtration can occur at any or all of the following four nodes in network: (a) Internet backbone, where national content filtering schemes and blocking technologies may affect Internet access throughout an entire country, (b) Internet service providers (ISPs), who often implement either government-mandated or self-regulated filtering, (c) organizations, using technical blocking and/or induced self-censorship, and (d) the home or individual computer level with the installation of filtering software.

At the heart of the software usually lies lists of URLs and IP addresses or an algorithm that reflects the choices and decisions of the code's designers [4, 7]. Filtering software is intended to (a) restrict access to Internet sites listed in an internal database of the product, (b) restrict access to Internet sites listed in a database maintained external to the product itself, (c) restrict access to Internet sites that carry certain ratings assigned to those sites by a third party or that are unrated under such a system, (d) scan the contents of Internet sites that a user seeks to view, and (e) restrict access based on the occurrence of certain words or phrases on those sites [4, 6].

The term *censorship* describes the suppression of ideas, documents, letters, pictures, movies, or any other type of information. The word comes from ancient Rome, where two magistrates, called *censors*, compiled a census of the citizens and supervised public morals. The technical solutions that have been proposed for censorship/content filtering are separated in two main categories: *content blocking* and *content rating and self-determination*:

- The content blocking approach makes the ISPs responsible for deciding which content is allowed to be forwarded to their subscribers. In this case, the ISPs have to design and implement technological solutions so as to make unreachable those websites that provide or render available material with ambiguous content to their subscribers.

- The content rating and self-determination approach makes the subscribers themselves responsible for the selection of the content they do not want to access. In this case, the content providers have to evaluate and rate the material, and the end users protect themselves by setting up their navigation programs in such a way as to deny the receipt of the inappropriate material. While labeling provides the user with enough information to decide whether to open it, rating involves assigning a value to the data file based on defined assumptions and criteria [8].

12.1.2 A Legal Issue

From the time of the initial moral panic concerning the Internet, it has been suggested that the control of access to inappropriate content could be resolved, mainly—or even only—through the implementation of filtering software [8]. *In this perspective, filtering seems to be a solution to support the law in face of new technologies and the problems they pose.* Does filtering constitute a protective shield against the "unlawful" and "bad" [7], a technical instrument that enables users to make their own decisions on how to deal with unwanted and harmful content [5]?

The control of Internet content and access presents very particular difficulties. A central question is to determine what effects filtering measures are having on freedom of expression and freedom of access to information on the Internet. The question of illegality or harmfulness raises the constitutional issue of determining which content no one should have access to and that deserves no protection and which content enjoys constitutional protection, as it is harmful to some users but not to others [7]. Another crucial point is who is legitimated to determine blocking decisions and take the respective technical measures.

Defining illegal or offensive, harmful content inevitably encounters problems due to differing legislative regimes, which—usually—reflect differing cultural values and norms. However, content transcends national boundaries and legislative jurisdictions. The Internet as a global operating network exceeds national territories: every act in the cyberspace has virtual effects everywhere in the world [9–11]. The conflict between (free) access to information and possible limits to this (free) access raises significant questions about the legitimate character of legal and technical restrictions on access as well as about jurisdiction and enforcement of the differing legal frameworks.

Are these technical solutions, originally hailed as offering "the solution" to the problem of Internet control, the remedy of choice of keeping Internet a free and safe space [5]? Are the complex sociological and legal issues to be solved through the enactment of filtering measures? Are filtering systems effective tools, achieving the aims they promise, or have they evidenced serious limitations and shortcomings [8]? Is the use of filtering systems a "social need," an "obvious solution" [4], the "ultimate answer to information related problems" like (harmful) information flood [9] or does it represent a modern form of censorship, in which governments and ISPs, as the modern-day censors, compile and online census of the citizens to supervise public morals?

12.2 Content Filtering Technologies

12.2.1 Blocking at the Content Distribution Mechanism

Illegal or offensive content can be blocked either in the packet level or in the application level [12]:

- Content blocking at the packet level requires the existence of screening routers, which examine the IP address of the incoming packet, searching against a "black list" or a "white list," and either forwards the packet or discards it.
- Content blocking at the application level requires the existence of application gateways and proxy servers that examine the resources, or the information about the resources, in order to decide if the request of the corresponding application protocol is permitted. For example, a most common approach is the determination of the URLs that should not be accessible and their placement in a "black list" that is installed on proxy servers.

12.2.1.1 Packet-Level Blocking

The most appropriate mechanism that will support us in an effort to distinguish the packets that will be forwarded from the ones that will be discarded is an access control list (ACL). This separation is based on the information in the headers of the IP packets, such as the source and target IP addresses. Packet-level blocking can be performed from any ISP, which compares the IP source address of the packet against a "black list" or a "white list" of IP addresses. This method can easily be implemented with the use of appropriate ACL characteristics of the routers that are used from the Internet backbone service providers.

In general, ACLs may adopt one of the following approaches:

- Permission is granted to all IP addresses, except the ones that belong to the list (default permit, black list).
- Permission is granted only to addresses referred in the list, and all the others are discarded (default deny, white list).

In any case, the effectiveness of packet-level blocking is an ambiguous matter. There are some technical issues, which one must keep in mind [13]:

- Packet-level blocking devices can easily be deceived or bypassed. IP tunneling could easily evade ACL control. Moreover, it is not hard for a web page to change its IP addresses in order to bypass the ACL control.
- The packet-level blocking process may also affect other TCP/IP services beside HTTP (e.g., FTP, SMTP, NNTP), because packet-level blocking decisions are based on IP addresses.
- Packet-level blocking does not discriminate; thus, when an enterprise web page is blocked, it will then be invisible and unreachable for all Internet users.
- Packet-level blocking requires extra capabilities from the screening routers, which means that older routers need to be either upgraded or replaced.

12.2.1.2 Application-Level Blocking

Application-level blocking requires the existence of proxy servers and application gateways that check the resources (or information for those resources) in order to decide if the specific request has to be forwarded or not. Therefore, ISPs protect their customers by forcing them to access the Internet through a proxy server that handles blocking and can store material often requested. The proxy server compares the customer request with a "black list" containing web pages in case of the HTTP protocol or newsgroups in case of the NNTP protocol.

Application-level blocking is widely used in corporate intranets to control access to specific web pages.

At the same time, some countries try to enforce application-level blocking technologies in order to censor information citizens would like to access. At the end of 2001, the non-profit organization Journalists Without Limits, residing in Paris, reported examples of Internet censorship in many cases: One country filtered access to media, news, and web pages of the United States, European countries, as well as other countries and human rights organizations. Moreover, the same country didn't allow the use of the Google search engine, forbade the establishment of new Internet cafés, and recorded those already established for further control. Another country allowed free Internet access only to selected government employees and businessmen. A third country filtered the traffic through a central node, prohibiting chat services, while later prohibiting access to web pages that were considered to contain subversive content for the country and the religion.

On the other hand, from the end of 2000, Australia, in order to deploy an effective access content control, developed a complete legislative framework; the aim was the prohibition of Internet access in material that is considered pornographic, violent, and related to narcotics and criminal actions.

As far as the efficiency is concerned, similar to packet-level blocking, the effectiveness of the application-level blocking is ambiguous. In the next paragraphs some issues are described:

- Application-level blocking can be deceived or bypassed in many ways, as described in [13].
- There is the possibility to entirely bypass application-level blocking, in case the content is delivered to users which haven't requested it explicitly.
- Another problem is the possibility of rejecting access to material that is not offensive. For example, in 1996, NYNEX discovered that all the web pages that concerned its ISDN services had been blocked by programs of censorship. The pages had been developed using names of type ISDN/xxx1.html and ISDN/xxx2.html, while the blocking software had been configured to block and bypass the web pages that include "xxx" in their address.
- The blocking software development companies can also censor pages for reasons different from those that it officially declared. For example, there are recorded cases in which companies that develop blocking software have excluded ISPs because the ISPs had host pages that were critical to the particular blocking software. In other cases, various organizations, like the National Organization of Women in the United States, had been excluded by

blocking software that was supposed to block pages with sexual content. Some companies that sell blocking software consider that they have the exclusive privilege to syntax and exploit their own lists with excluded pages, and for this reason the customers can't have any information about the pages which belong to the "black list."

- The policy whereby all users who want to access the Internet must access a proxy server decreases dramatically the reliability of the connection, because this policy introduces a single point of failure.

As was the case with packet-level blocking, application-level blocking increases the complexity of related firewalls, as well as the total operating cost. Moreover it is possible to be technically bypassed, following various ways of attacks.

12.2.2 Blocking at the End-User Side

Another solution to deploy censorship, content filtering is the idea of combining *content rating* and *content filtering*:

- *Content rating* concerns the classification of content according to a specific filtering point of view. The concept of content rating is not new. For some media, such as cinema movies, we have been accustomed to content rating for many years. Moreover, in many countries, television shows have ratings. Web content rating may follow the *self-rating* approach, which means that rating is carried out by the website owners. Alternatively, a *third-party rating* approach is also useful; rating may be carried out by third-party external organizations. As far as time is concerned, content rating is not performed at the time an access request is submitted, as this could result in a considerable delay to the whole process.

  1. According to the *traditional rating strategy* [14], websites are rated either manually or automatically and classified into a set of categories. Subscribers can then select which site categories they do not want to access. Filtering services usually provide customized access to the web, according to which some website categories are considered inappropriate by default for certain user categories. Following this approach, some search engines return only websites belonging to appropriate categories.
  2. Another rating strategy is to attach a *label* to websites consisting of metadata that describe their content. The most famous approach is the Platform for Internet Content Selection (PICS) [15], a World Wide Web Consortium (W3C) standard [16], which will be described in detail in the following subsection.

- *Content filtering* is the process that includes the use of filtering systems, mechanisms, and techniques in order to manage access requests to websites on the basis of a given set of policies. Filtering services are responsible for allowing or denying access to online documents, denoting which users can or cannot access which online content, and the ratings associated with the requested web resource. Filtering systems can be classified [17] into *indirect filtering* and *direct filtering*.

1. According to indirect filtering, the filtering process is performed by evaluating website ratings, based on "white lists" and "black lists." Some services, known as *walled gardens*, allow users to navigate through a set of accepted websites. Rating is conducted following the third-party rating approach.
2. Direct filtering process is executed by evaluating web pages with regard to their actual content or the associated metadata. These direct filtering systems use two technologies. *Keyword blocking* prevents sites that contain any of a list of selected words from being accessed by users. *PICS-based filtering* verifies whether an access to a web page can be granted by evaluating not only the content description provided in the PICS label, but also the filtering policies specified by the end user. PICS-based filtering services adopt a self-rating approach by providing an online form that allows website owners to automatically generate a PICS label.

12.2.2.1 Content Rating Label Strategy

As introduced earlier, PICS is a general-purpose system for label placement in content of documents that is presented in the World Wide Web. PICS *labels* contain one or more ratings that are published by a *rating service*. PICS constitute a platform in which services of content rating can be built.

The software that materializes the PICS has important technological advantages compared to simple products of blocking software. Some of these are the following:

- It allows blocking per document.
- It makes it possible to get blocking ratings from more than one source.
- Since PICS constitutes a general framework for ratings information about the web, different users can place different rules for access control.

PICS can be used for assigning many kinds of labels. PICS labels [18]:

- Can specify the type or amount of sex or profane language in the document;
- Can rate whether a photograph is overexposed;
- Can specify whether a document includes hate speech;
- Can specify the political leanings of a document;
- Can indicate if a chat room is moderated;
- Can indicate the year that a document was created; thus they can denote copyright status.

The PICS anticensorship arguments include the following [18]:

- It can disallow access to the Internet entirely.
- It can disallow access to any site suspected of having objectionable material.
- Digital signatures allow labels created by one rating service to be cached or even distributed by the rated website, while minimizing the possibility that labels will be modified by those distributing them.

A lot of suppliers of online services have announced their support in the PICS and the disposal of software compatible with the PICS in their subscribers. An important rating scheme and service is RSACi [19]. The RSACi was developed by Recreational Software Advisory Council (RSAC), a nonprofit independent organization. The RSACi system provides to the consumers information regarding the levels of violence, which is rated between 0 and 4 (for violence, nudity, sex, and offensive language) in games and web pages. These levels are summarized in Table 12.1 [19].

RSACi is the predecessor of ICRA [20] and is no longer widely available.

Generally, PICS can be used in order to offer support not only in the characterization from the autonomous content owner or online editor, but in the characterization from a third person, as a specialized office of labels:

- A content owner or an online editor who wishes to characterize his own content should first choose the vocabulary of rating that it will use.
- Apart from the service of self-characterization, as an agency of independent ratings, it is not essential to collaborate with each content owner or editor that characterizes the content. Whoever makes independent characterizations, instead of attaching these labels in the documents, distributes the labels via one separate server, which is called office of labels (*label bureau*).

Rating Services

The PICS rating service specifications have been designed to ease many different kinds of ratings in the web. Rating services could be any person, organization, or any other entity that publishes ratings. Ratings could be disposed either directly with the document that is rated, or from a web page of a third institution, or with a CD-ROM, or with another electronic media.

The PICS specification determines a syntax for text files, which describes the different types of ratings a rating service can publish. This allows programs to analyze the syntax and deduce the types of ratings that a rating service provides.

Table 12.1 The RSACi Levels for Violence, Nudity, Sex, and Language

Level	*Violence Rating Descriptor*	*Nudity Rating Descriptor*	*Sex Rating Descriptor*	*Language Rating Descriptor*
4	Rape or wanton, gratuitous violence	Frontal nudity (qualified provocative display)	Explicit sexual acts or sex crimes	Crude, vulgar language, extreme hate speech
3	Aggressive violence or deaths to humans	Frontal nudity	Nonexplicit sexual acts	Strong language or hate speech
2	Destruction of realistic objects	Partial nudity	Clothed sexual touching	Moderate expletives or profanity Mild expletives
1	Injury to human being	Revealing attire	Passionate kissing	Mild expletives
0	None of the above or sports related	None of the above	None of the above or innocent kissing; romance	None of the above

In the papers that introduce PICS, P. Resnick and J. Miller [21, 22] create a sample rating service based on the plan of film rating from the MPAA.

```
((PICS-version 1.0)
(rating-system http://moviescale.org/Ratings/Description/)
(rating service "http://moviescale.org/v1.0")
(icon "icons/moviescale.gif")
(name "The Movies Rating Service")
(description "A rating service based on the MPAA's movie rating scale")
(category
(transmit-as "r")
(name "Rating")
(label (name "G") (value 0) (icon "icons/G.gif"))
(label (name "PG") (value 1) (icon "icons/PG.gif"))
(label (name "PG-13") (value 2) (icon "icons/PG-13.gif"))
(label (name "R") (value 3) (icon "icons/R.gif"))
(label (name "NC-17") (value 4) (icon "icons/NC-17.gif"))))
```

The description of the rating indicates the web place where information can be found about the system and the rating service, it gives it a name, and it creates a unique category of rating named "Rating." The objects that are rated can have one of five characterizations—ratings: G, PG, PG-13, R, or NC-17. The model gives each one from these ratings a price and a related description.

The description of rating service PICS is fixed to have a file type of MIME application/pics-service.

The PICS makes extensive use of pairs' name/price. These have the form (name price). They are interpreted as "name has the price."

PICS Labels

The PICS labels specification defines the syntax for the labels of document. Labels can be obtained either from the web using a search engine using an extension of HTTP that is described in the PICS standards, or the labels can be incorporated automatically as part of document heading.

For example [23], here is a PICS label that classifies a URL using the service mentioned before:

```
(PICS-1.1 "http://moviescale.org/v1.0"
Labels
on "2002.6.01T00:01-0500"
until "2003.12.31T23:59-0500"
for "http://www.missionimpossible.com/"
by "J.B." ratings (r 0))
```

This label describes the web page of the movie film *Mission Impossible* using the virtual service of characterization that was described previously. The label was created in June 1996 and was valid up to the end of 1996. The label corresponds to the information stored in the URL http://missionimpossible.com/, it has been written by J.B., and gives the rating "(r, 0)." Even if the film *Mission Impossible* was rated "R," the web page was rated "G." Value "G" is transmitted with 0 using the rating service http://moviewscale.org/v1.0.

Ratings may include more than one transmitted value. For example, if a rating service specifies two scales, a label of rating can be as follows: "(r 3 n 4)".

Labels can be compressed by removing all information apart from the characterization. For example, the label mentioned before could be transmitted as follows:

(PICS-1.1 "http://moviescale.org/v1.0" r 0)

Moreover, the labels can optionally contain a hash of the message using a hash function. This allows the software to determine the integrity of the document after the creation of the label. Digital signatures can also be attached, fulfilling integrity and authenticity requirements. That fact allows a web page to distribute labels for its content, which have been created by a third service, and to assure the users that the labels have not been altered.

PICS Implementations

Today there are rating services, compatible with the PICS, allowing the suppliers of content or the online editor to self-characterize their content (RSACi). Some companies, such as EvaluWeb, Net Shepherd, and NetView provide labels that have been given by third institutions using their own systems of rating that are compatible with the PICS. The list of PICS-compatible services and products are available at [24].

Nevertheless, the use and growth of PICS and PICS-compatible rating services still have not been developed widely, despite the fact that Microsoft Internet Explorer provides support for both PICS and RSACi.

PICS Advantages and Disadvantages

Compared to other models, PICS-based rating systems are semantically rich. For example, the most commonly used PICS-compliant rating system, which was developed by Internet Content Rating Association (ICRA) [25], provided 45 ratings grouped into macro categories, such as chat, language, nudity and sexual material, violence, and other topics including drugs use, weapons use, alcohol, and tobacco. On the other side, the category model used by RuleSpace and adopted by Yahoo! for parent control services makes use of 31 website categories [26]. In general, the PICS standard may be considered the best that can address the filtering issues.

Nevertheless, the available PICS-based rating services have many drawbacks [14]. The content description does not use ontologies, which could develop a more accurate approach. Furthermore, it is limited to filtering content domains according to the Western system of values and liability. The PICS-based approach is less diffused; the reason is that it requires websites to be associated with content labels, but up to now only a very small part of web pages is rated.

Resource Description Framework Labels

Resource Description Framework (RDF) [27, 28] is a well-known set of W3C specifications, which were designed at first as a metadata model. Eventually, RFD has come to be used as a general method of modeling information through a variety of syntax formats. The RDF metadata model is based on the term of "triples": the subject, the predicate, and the object. If we would like to make statements about resources, we must follow this set. For example, the notion "The sky in the Aegean Archipelagos has the color deep blue," in RDF is a triple of the following strings: a subject denoting "the sky in the Aegean Archipelagos," a predicate denoting "has the color," and an object denoting "deep blue."

The RDF labels may express everything that the abovementioned PICS labels can express, but will also permit string values, structured values, and other features. In 2000, the W3C proposed an RDF implementation of PICS, which provides a more expressive description of websites content, enabling more efficient filtering [29].

12.2.3 Recent Research Trends: The Multistrategy Web Filtering Approach

Recently, interesting research work in the web filtering area has been proposed. In the framework of the EUFORBIA project, funded by the EU Safer Internet Action Plan [30], a new filtering approach has been introduced [14, 20, 31].

A general-purpose rating system was developed, which allows all potential users to accurately describe website structure and content. The objective was to design a general filtering framework addressing both the flexibility and protection issues, which can possibly be customized according to users' needs by using a subset of its features.

This approach aims to improve and extend the available techniques by enforcing two main principles:

- Support should be provided to the different rating and filtering techniques, in order to be able to use them individually or in combination according to users' needs.
- Users' characteristics must be described accurately in order to provide more effective and flexible filtering.

EUFORBIA-proposed labels may be compared with the metadata scheme of a digital item in MPEG-21 [32, 33], although MPEG-21 focuses only on the problem of describing the structure and the access rights of a digital item.

Moreover, the possibility for end users to freely decide what objects they wish and what they do not wish to access could not only solve the ethical disputes regarding web filtering but also to reduce the end user's possible resistance in adopting them [24].

This multistrategy filtering model is considered the most promising attempt to describe a framework that is compliant with the efforts of the W3C in defining standard architectures for web services.

12.3 Content-Filtering Tools

A lot of content-filtering products are available, covering parents' needs, or school needs, or providing enterprises integrated solutions. Nonprofit organizations, such as the Electronic Frontiers Australia [34], have dealt with this issue. Moreover, in

[35], a thorough review of many web-filtering products is described, taking into account the plethora of criteria that include, among others, ease of use, ease of setup, filtering algorithms, filtering capabilities, reporting capabilities, management capabilities, target group, technical support options, supported browsers and configurations, foreign language filtering, port filtering and blocking, and cost.

12.4 Under- and Overblocking: Is Filtering Effective?

The use of filters has inspired much criticism. A first shortcoming relates to "imperfect technology" [36]. Filtering technologies are inclined to two simple inherent flaws: underblocking and overblocking. They filter much more and, at the same time, much less.

Underblocking refers to the failure of filtering to block access to all the content targeted for censorship [2]. The current level of technology does not filter out 100 percent of illegal or harmful content. A number of recent empirical and independent studies have found significant errors in underblocking and overblocking performance of Internet filtering software [3, 37]. According to some surveys, current software fails in filtering up to 20 percent of the pornographic sites [7]. Inevitably, web proxy programs have been designed to bypass web-filtering controls. Noteworthy is that web proxy programs, such as the Circumventor, have been specifically designed to "get around web-blocking software." Every few days the Circumventor operators send out a message referring to new Circumventor sites, "staying ahead of the Internet filtering software companies until they update their blacklists" [3].

At the same time, filtering technologies often block content they do not intend to block or they should not block (overblocking). Numerous examples illustrate shortcomings and restrictions. Pertaining to blocking of pornographic websites, there is long list of websites that were erroneously blocked, ranging from sites on breast cancer, sexual education, gays and lesbians, and planned parenthood, to sites of political organizations and candidates, a site that watches and criticizes the filtering software industry, and even some sites containing legal documents and cases [7]. Searches of British regions and universities such as Essex and Sussex have been blocked because of the inclusion of "sex" in the search keywords.

A system of keywords is fundamentally limited and it can be operated in an irrational way: As noted in the Public Policy Report of FEEP [23], blocked sites were—among others—part of the City of Hiroshima site, Vincent Van Gogh sites, the Declaration of Independence, Shakespeare's complete plays, the University of Kansas Archie R. Dykes Medical Library. Detecting the phrase "at least 21" has resulted to the blocking of news items on various sites, reporting war events, and referring to the number of killed persons [23]. Filtering has been proven unreliable also in cases of hate speech "since they ban speech on the basis of words that may be present in anti-hate propaganda as well" [36]. Many blacklists are generated through a combination of manually designated websites as well as automated searches and, thus, often contain websites that have been incorrectly classified. Filtering methods such as IP blocking can filter out harmless and legitimate sites, on the simple grounds that they are hosted on the same IP address as a site with restricted content.

Filters and ratings ignore context. Their use is often based on the assumption that websites can be reduced to a single letter or symbol [4]. Overblocking is not a technological problem, nor is it a temporary feature of filtering. Filtering is inherently not perfect: a simple technological selection system can never give back the variety of human thinking and value judgments [38]. Overblocking will be intensified due to features of the growing filtering industry. The fierce competition among the producers of filtering software will probably drive them to block more rather than less [7].

12.5 Filtering: Protection and/or Censorship?

From the standpoint of democracy, filtering undoubtedly constitutes a major issue. Does a filtering system realizes important values such as the end-user autonomy and the protection of certain categories from harmful content, or does it incorporate a new censoring instrument? Is it an important and needed response to widely shared concerns about minors' access to illegal or controversial web content, or a threat to freedom of expression and access to information? The outcome of such questions and the respective legal analyses hinge on who carries the filtering act and the criteria applied to filtering operations.

12.5.1 The U.S. Approach

The American discourse relating to the freedom of speech stems from the First Amendment. The language seems to cover the rights of the "speakers." As to the "listeners'" rights, theory and jurisprudence have recognized them in an indirect manner and mostly as an important interest in receiving information [7]. In recent years the U.S. Supreme Court has extended First Amendment protection to the Internet.

American legislators sought initially to protect minors from harmful material on the Internet through the provisions of Communications Decency Act of 1996 (CDA). The CDA prohibited "knowing" that the sending or displaying to a person under 18 any message "that, in context, depicts or describes, in terms patently offensive as measured by contemporary community standards, sexual or excretory activities or organs" and criminalized the "knowing" transmission of "obscene or indecent" messages to any recipient under 18 years of age. The Supreme Court eventually overturned the law in Reno v. American Civil Liberties Union. The Court found that the CDA is a content-based regulation of speech, which raises special First Amendment concerns because of its obvious chilling effect on free speech. According to the Court, "the interest in encouraging freedom of expression in a democratic society outweighs any theoretical but unproven benefit of censorship" (Supreme Court, *Reno v. ACLU*). Many of the organizations proposed the developing technology of filtering programs or filters as a means of empowering parents and local officials with the technical ability to limit access to unaccepted material [4].

Following the defeat of the CDA, Congress crafted the Children's Online Protection Act (COPA). COPA, which like the CDA used a broadcast regulation model by addressing the source [39], was also challenged in court as affecting the First

Amendment Rights. In *Ashcroft v. ACLU* (2004) the Supreme Court eventually held that filtering software could be a less speech-restrictive alternative. "They (filters) impose selective restrictions on speech at the receiving end, not universal restrictions at the source . . . Above all, promoting the use of filters does not condemn as criminal any category of speech, and so the potential chilling effect is eliminated, or at least much diminished." The Court came to the conclusion that filters may well be more effective than the law (COPA), as they can prevent minors from seeing all pornography, not just pornography posted to the web from America.

The most recent legislative attempt to curtail online indecency, the Children's Internet Protection Act (CIPA), requires schools and libraries with Internet access to certify to the FCC that they are enforcing a policy of Internet safety. These Internet safety policies consist mainly in the use of filters to protect against access to visual depictions that are obscene or harmful to minors. In order to ensure compliance, receipt of Internet funding depends on the fulfillment of filtering obligations imposed by the law. However neither the FCC nor CIPA mandates that public libraries use a particular Internet filter or that the chosen filter be completely effective. The compliance of these provisions with First Amendment rights has also been challenged. In the case *American Libraries Associations (ALA) v. U.S.*, the Supreme Court agreed that "the government has broad discretion to make content-based judgments in deciding which private speech to make available to the public" and characterized filtering in ALA as "reasonably effective" [39]. The Court held that libraries were not "public fora" and analogized content filtering to the professional judgments librarians make, when they choose not to purchase certain books for their library [6]. CIPA did not violate the freedom of speech. The Supreme Court recognized and, at the same time, dodged the constitutional problem of inaccurate filtering or "overblocking" by permitting the libraries to disable filters to "enable access for bona fide research or other lawful purposes."

12.5.2 The European Approach

By the mid-1990s the European Union expressed its interest in the control of harmful and illegal Internet content in order to protect the interest of minors. The European Commission initiated (1996) the public discussion with its Communication Paper on Illegal and Harmful Content on the Internet, which proposed that the solution to controlling access to illegal and harmful content lies in a combination of self-control of the service providers, new technical solutions such as rating systems and filtering software, awareness actions, information on risks, and possibilities to limit these risks [40]. Following this initiative the European Parliament called on the Commission to propose "a common framework for self-regulation," which was to include objectives in terms of the protection of minors and human dignity as well as "measures to encourage private enterprise to develop message protection and filtering software" [8, 36].

The starting point for the European Union's approach is that measures are needed to promote their safer use and protect the end user from unwanted content, in order to encourage the exploitation of the opportunities offered by the Internet and the new online technologies [5]. The new multiannual program on promoting safer use of the Internet stresses the need for technical tools, which could enable users to make their own decisions on how to deal with unwanted and harmful content (user empowerment). EU will therefore provide funding for technological measures that meet the needs of users and enable them to limit the amount of unwanted and harmful content, including (a) assessing the effectiveness of available filtering technology and providing this information to the public, (b) increasing adoption of content rating and quality site labels by content providers and adapting content rating and labels to take account of the availability of the same content through different delivery mechanisms (convergence), (c) contributing to the accessibility of filter technology, notably in languages not adequately covered by the market to enable users to select the content they wish to receive and provide European parents and educators with the necessary information to make decisions in accordance to their cultural and linguistic values [5].

As far as it concerns the conflict between the fight against illegal and harmful content and the freedom of expression, the European Union is bound by the provisions of the European Convention on Human Rights (ECHR). Art. 10 (1) instructs that everyone has the right to freedom of expression. This right shall include free dom to hold opinions and to receive and impart information and ideas without interference by public authority and regardless of frontiers. Balancing of competing rights and interests is a concept embedded in Art. 10 of the ECHR itself [7]. According to Art. 10 (2), the exercise of these freedoms may be subject to formalities, conditions and restrictions or penalties as are prescribed by law and are necessary in a democratic society. The European Court of Human Rights interpreted the term "information" to also include information and ideas that "offend, shock or disturb" (ECHR, *Handyside v. UK/Aksoy v. Turkey/Da Silva v. Portugal*). The Court regards freedom of expression as one of the essential foundations of a democratic society, which is based on "pluralism, tolerance and the spirit of openness" (ECHR, *Aksoy v. Turkey*). Subject to the restrictions of par. 2 of Art. 10, "it (freedom of expression) applies not only to "information" and "ideas" that are viewed favorably or regarded as inoffensive or immaterial, but also to those that are conflicting, shocking and disturbing." The right to receive information, as long as it is not illegal, is thus considered to be an inseparable part of freedom of speech [38].

However, the Council of Europe's position with respect to hate speech is very strict, as it considers "racism not as an opinion but as a crime" and adopted the approach that "also the dissemination of hate speech against certain nationalities, religions and social groups must be opposed" [41]. The issue is addressed by the First Additional Protocol to the Convention of Cybercrime, which imposes obligations on state parties to criminalize—among others—the dissemination of racist and xenophobic material. The Council of Europe itself recognized that the Additional Protocol actually "will have no effect unless every state hosting racist sites or messages is a party to it" [36, 41].

12.5.3 Filtering As Privatization of Censorship?

Filtering constitutes a significant parameter of regulating expression and communication in the society. As analyzed, the filtering software has demonstrated considerable technical limitations, and its potential impact on freedom of expression and freedom of access to information has been questioned. Individual filtering is mostly recommended for parents who want to prevent their children from viewing sites that contain pornography, violent, hateful, or other problematic material. However, even by filtering operated on a voluntary basis, the crucial problem remains of the criteria applied and the lack of transparency in restrictions to information access.

Filtering constitutes in the final analysis a regulatory decision. Moreover, filtering and rating systems "constitute fundamental architectural changes that could facilitate the suppression of speech far more effectively than national laws alone ever could" [42]. The use of filtering tools, offered by providers or integrated in Internet browsers, represents the move toward the privatizing of censorship. This raises the questions "of the point of the flow or information at which censorship takes place and who should have the responsibility for filtering decisions: the government, ISP's or the end-user" [8].

Filtering tools are not neutral about the values incorporated. Even tools such as the PICS technology, which allows multiple independent rating systems to be standardized and read by different screening software packages in accordance with personal preferences, have been strongly criticized, as they enable both user control and upstream control [4, 36]. Experts argue that the use of ICRA implies that the standard or rating scheme is not defined by the user, but basically by private organizations [43, 44]. Inevitably, a filtering system that describes, classifies, or blocks content is dependent on the subjective judgment of the filterer. Sites belonging to organizations that promote strong free speech viewpoints, such as ACLU, have been blocked by more conservative organizations. Even objecting to the patterns of exclusion adopted by certain filtering companies may result in the critical source being added to a blacklist [4]. It is highly questionable if it is possible for ISPs to assess the legality and the harmlessness of content. A major problem relates to the legitimization of private entities to evaluate and rate content, thus exercising regulatory and executive power and restricting fundamental rights. Filtering could result in concentration of power, which would have further implications for the society and the nature of the Net. Where rating systems are integrated in Internet browsers, the risk exists, furthermore, that all sites "have to" be screened by the criteria used by these systems to avoid being blocked [9]. The W3C has directed attention to the dangers of anyone labeling standard becoming too powerful, by noting that "if a lot of people use a particular organization's labels for filtering, that organization will indeed wield a lot of power" and adding that "Such an organization could for example, arbitrarily assign negative labels to materials from its commercial or political competitors" [8].

Private "regulation" of speech on the Internet has grown pervasive. The consequences of filtering must be taken into account, as the vast majority of speech in our days takes place on the Internet and within private places and spaces—places and spaces that are owned or administrated by private entities such as online service providers (like Google, Yahoo!, AOL) or private "pipeline providers" (like Comcast or Verizon) [45]. From this perspective, filtering is strictly relating to the conditions and the quality of public discourse in "public sphere." A democratic society is democratic in the sense that everyone has the freedom and the chance to participate in the development of ideas [46]. Many authors argue that filtering allows people to "customize their own communication universe" [47] and avoid "confronting the unfiltered," ignoring controversial ideas [48]. Internet filtering has long-term implications, as the wide-scale deployment of filtering schemes bears the risk that the Internet becomes "bland and homogenized" [49].

12.5.4 ISPs' Role and Liability

In the very center of the discussion relating to filtering stand issues pertaining to the role of ISPs. Imposing a general obligation of prior monitoring on the ISPs is difficult from a purely technical point of view [9]. The difficulty to identify and block the undesirable content is becoming even more acute, when the legality refers not only to the content but also to the user. Current technology does not permit an ISP to easily identify minors. In case of doubt the ISP would apply the restrictions on content and "click on the delete key" [7].

Furthermore, such a duty would raise a host of complex questions, as to the effect on the rights of the ISPs themselves (i.e., their right to property and contract) and effects on the costs, which are associated with imposing liability. Such a general obligation, leading to generalized preventive censorship in order to avoid liability, would have a "chilling effect" to the ISPs and consequently to end users [7, 9]. The introduction of a general obligation would imply a liability system based on risks with far-reaching consequences, especially for smaller ISPs.

In the United States, there are two separate legal regimes that govern liability of ISPs: one in regard to copyright law and one for all other kinds of content. The European Union opted for a rather general structure of liability. The European e-commerce directive (Directive 2000/31/EC) determines a unified system: the European legislators have accepted that they should not impose a general obligation to ISPs either to monitor the information they transmit or store, or to seek facts and circumstances indicating illegal activity. There is, however, a differentiation on the kinds of service provided: the directive assures immunity to ISPs that are mere conduits if they are not involved in initiating or editing the message transmitted and do not determine the parties of the transmission. Further, the directive provides ISPs with immunity for caching and for content stored on their servers by users. This immunity is subject to lack of knowledge on behalf of the ISP and to the condition that once knowledge about illegal or harmful content is acquired the content will be immediately removed.

12.6 Filtering As Cross-National Issue

Conflict of values has long been identified as a major issue of information policy. Filtering is a sound parameter of "regulating" the flow of information across the Internet. The filtering must confront the "clash of local standards and norms" [2]. Globalization and the Internet, a medium whose design resists boundaries and blocks, enhance the complexity and controversies by creating a "new type of information policy conflict: cross-national conflict" [50]. Since all content is potentially accessible "one mouse click away" on the Net, should we accept, as a technically determined blessing or fatality, the consequences of the globalization of information?

According to findings of recently conducted surveys, filtering is currently increasing worldwide [2]. Although much has been written about restrictive measures being adopted by more overly authoritarian regimes, such as China [50], with the aim to regulate and control illegal or dubious content, numerous countries, including countries with a strong commitment to democracy, filter or block access to Internet content in some way.

Traditionally, states claim sovereignty over their territory and the population. Does the Internet vitiate the responsibility and the power of a state to control such activities, undermining the territoriality and sovereignty principles? On the other side, when a country filters information packets within its geographic boundaries, it affects the communications of the citizens not only in its own territory but potentially worldwide [2]. Taking into account the characteristics of the Internet, does a country, aiming to restrict access to illegal content, need to block access to the network completely or to impose its views on the Internet as a whole?

12.6.1 Differing Constitutional Values: The Case of Yahoo!

The case of Yahoo! (i.e., the judgment of French Judge Gomez of Tribunal de Grande Instance de Paris) has been a landmark [17], a "mundane exercise in the analysis of territorial sovereignty and jurisdiction" [51]. The case raised many issues, including opposing conceptions in relation to freedom of expression, differing legal regime, multijurisdictional compliance, technical specificity, and architecture of the Net.

Under a special procedure that allows a judge to obtain preventive measures, the plaintiffs, the Union of Jewish French Students and the International League against Racism and Anti-Semitism, demanded the removal of all links to "Holocaust negationist sites" and the elimination of two sites, including one in French, that offered the text of Hitler's "Mein Kampf" on Yahoo! Inc.'s geocities.com subsidiary. On May 22, 2000, Judge Gomez ruled that the sales were an "offense to the collective memory of a nation profoundly wounded by the atrocities committed in the name of the Nazi criminal enterprise" and ordered Yahoo! "to take all measures of a nature to dissuade and to render impossible all consultation on Yahoo.com of the online sale of Nazi objects and of any other site or service that constitutes an apology of Nazism or a contestation of Nazi crimes" (Tribunal de Grand Instance de Paris, UEJF et LICRA c. Yahoo!) [17, 51].

Following the reactions of Chief Yahoo! Jerry Yang, who characterized the French court "very naïve," Judge Gomez decided in August 2000 to designate three experts to prepare a report on the technical feasibility of blocking content. In Court they agreed that no technical measures could ensure that Yahoo! would succeed in keeping all French surfers away from such sites [17]. Ultimately, the conclusion was that 60 percent of the targeted information could be blocked on the basis of geographical location and the nationality of the surfers. Through a combination of identification of the user on the basis of IP address, the nationality declaration and the use of filter systems based on keywords, 90 percent of blocked content could be reached [9]. On November 20, 2000, Judge Gomez ruled that Yahoo! should satisfy the terms of the previous order within three months.

Yahoo!, denying the jurisdictional authority of France, pursued the case to the U.S. courts. The U.S. Federal District Court reflected on the global nature of the Internet, accepted France's authority to prescribe objectionable content within its territory, but refused to allow France to enforce its law in the United States through any judgment against Yahoo!'s American assets, as it would not be compliant with the free speech embedded in the First Amendment [2]. This judgment resides on the classic U.S. free speech doctrine, which holds that "offensive speech is the price a society must pay for freedom of expression" [17].

The order of Judge Gomez reflected in general the European approach: shortly after the French judgment the Bundesgerichtshof, the German Supreme Civil Court decided that German laws prohibiting racial hatred clearly apply to Internet material created outside of Germany . . . but . . . are accessible to German Internet users (i.e., they are also clearly applicable on websites that are located abroad and thus not in Germany). However when German authorities required German ISPs to block a magazine published in a website hosted in The Netherlands, which allegedly included "terrorist violence," the Dutch host service provider put forward the argument that the action of the German authorities constitutes "an interference with the free movement of services within the EU" [36].

12.6.2 Territoriality, Sovereignty, and Jurisdiction in the Internet Era

In the final analysis, the main issue posed in the case of Yahoo! was the "idea that the Internet would also help "to break free from the body of rules that govern life in society" [52]. Since the Internet is a global network, the problem of sovereignty and jurisdiction arises in a different way: Internet "separatists," consider Internet as a network designed not to bear authoritarian control, but as a separate jurisdiction that transcends national borders and consequently the control of nation states [53].

Yahoo!'s position was not a separatist one: Yahoo! relied on the argument that local sites should be governed by local law. The French court never denied that, under U.S. constitutional law, Yahoo! has a legal right to defend the free speech approach. Yahoo! argued that the U.S. Constitution is applicable to its activities worldwide [17, 51], a disputable argument taking into account the policy of Yahoo! in China [50]. In any case, this constitutional regime right remains a territorially restricted regime. Internet content, and in this case Internet hate speech, is inevitably a universal problem. Every state has the right to defend its own constitutional values and protect its citizens, at least if the questionable web content is designed to reach a global audience [2, 51]. Undoubtedly conflicts between different national laws, which are an emanation of the states' authority based on the territoriality principle, are likely to arise.

A critical point arising from the Yahoo! case refers to the relation of public values and Internet architecture. Yahoo!'s argument, that it could not technically filter out French web users because of the geographic indeterminacy of data transmissions on the Internet, results in abandoning important information policy rules to technological choices. In this perspective, the Yahoo! conflict revitalized the discussion about the "democratization of Internet rules and design features" and also about the context of filtering [51]. The argument of the "technical feasibility" of imposing national constitutional values on the Net should not be underestimated but—technically—is not insurmountable. However, the solution envisaged by the French Court (i.e., the geographic determinacy and the localization of users in order to protect them from illegal content) raises significant concerns in relation to the protection of their anonymity and privacy [9, 54].

12.7 Conclusions

In this chapter, useful technological solutions, which are capable of filtering content on the Internet and the web, have been presented. More specifically, packet-level blocking, as well as application-level content rating, self-delimitation, and filtering, have been described. Packet-level blocking renders the ISPs responsible to decide for the content that will be provided for their subscribers, while on the other hand, the idea of application-level content rating, self-determination, and filtering renders the subscribers responsible for the content that they may access. Furthermore, modern research work, such as the results of the EUFORBIA project, has been described, presenting new trends in this area.

In the near future, further development of the semantic web, as well as the adaptation of standards such as MPEG-7 and MPEG-21 [7], may overcome existing problems. Since most online available information is either semistructured or often unstructured at all, emergent research work needs to be developed in order to deploy accurate filtering of information and especially multimedia content.

Taking into account the difficulties related to technical efficient monitoring the question could be raised whether it is possible for states to continue to think nationally considering the worldwide availability and dissemination of information [9]. Whether filtering and blocking measures are employed by ISPs or by governments, they are likely to fail unless there is international conformity in legal regulation of online acceptable speech [36]. A global solution seems to be the only solution, which would make sense, but this is not really feasible for every possible issue due to different legal regimes, cultures and perceptions. Conflicts could be partly solved by harmonization and approximation of national laws through conventions, which could clarify and reduce cross border legal conflicts.

A "good filtering system" should realize several important values: end-user autonomy, respect for freedom of expression, ideological diversity, transparency, respect for privacy, interoperability, and compatibility [55]. Free speech values must be protected through technological design and through administrative and legislative regulation of technology [46]. In any case both technological architectures and legal regimes must be structured in a way to make possible full and robust participation by individuals.

References

[1] Balkin, J. M., Digital Speech and Democratic Culture: A Theory of Freedom of Expression for the Information Society, 79 NYUL Rev 1 (2004), pp. 1–55.

[2] OpenNet Initiative, A Starting Point: Legal Implications of Internet Filtering, *A Publication of the OpenNet Initiative*, September 2004, http://www.opennetinitiative.org.

[3] Carey-Smith, M., and L. May, "The Impact of Information Security Technology upon Society," Paper presented to the Social Change in the 21st Century Conference, Center for Social Change Research, Queensland University of Technology, October 27, 2006.

[4] Rosenberg, R. S., "Controlling Access to the Internet: The Role of Filtering," *Ethics and Information Technology*, Vol. 3, 2001, pp. 35–54.

[5] European Union, Decision No 854/2005/EC of the European Parliament and the Council of 11 May 2005, Establishing a Multi-Annual Community Programme on Promoting Safer Use of the Internet and the New Online Technologies.

[6] Miltner K. T., "Discriminatory Filtering: CIPA's Effect on our Nation's Youth and Why the Supreme Court Erred in Upholding the Constitutionality of the Children's Internet Protection Act," *Federal Communications Law Journal*, Vol. 57, 2005, pp. 555–578.

[7] Birnhack, M. D., and J. H Rowbottom, *Shielding Children: The European Way*, Chicago: Kent Law Review, Vol. 79, 2004, pp. 175–227.

[8] Cooke, L., "Controlling the Net: European Approaches to Content and Access Regulation," *Journal of Information Science*, Vol. 33, No. 3, 2007, pp. 360–376.

[9] Bodard, K., "Free Access to Information Challenged by Filtering Techniques," *Information and Communication Technology Law*, Vol. 12, No. 3, October 2003, pp. 263–279.

[10] Foerstel, H. (ed.), *Banned in the Media: A Reference Guide to Censorship in the Press, Motion Pictures, Broadcasting and the Internet*, Westport: CT, Greenwood Publishing Group, 1998.

[11] Price, M. (ed.), *The V-Chip Debate: Content Filtering from Television to the Internet*, Mahwah, NJ: Lawrence Erlbaum Assoc., 1998.

[12] McRea, P., B. Smart, and M. Andrews, "Blocking Content on the Internet: A Technical Perspective," *Technical Report*, Australian National Office for the Information Economy, 1998.

[13] Oppliger, R., *Security Technologies for the World Wide Web*, Norwood, MA: Artech House, 2003.

[14] Ferrari, E., and B. Thuraisingham, *Web and Information Security*, London: IRM Press, Idea Group. Publ., 2006.

[15] www.w3.org/PICS (November 2007).

[16] www.w3.org (November 2007).

[17] Reidenberg, J. R., "The Yahoo! Case and the International Democratization of the Internet," *Fordham University School of Law—Research Paper*, Vol. 11, April 2001.

[18] Garfinkel, S., and G. Spafford, *Web Security and Commerce*, Cambridge, MA: O'Reilly & Associates, 1998

[19] http://www.rsac.org (November 2007).

[20] http://www.saferinternet.org/filtering/euforbia.asp (November 2007).

[21] Resnick, P., and J. Miller, "PICS: Internet Access Controls Without Censorship," *Communications of the ACM*, Vol. 39, No. 10, 1996, pp. 87–93.

[22] Resnick, P., "Filtering Information on the Internet," *Scientific American*, March 1997, pp. 106–108.

[23] Bayer, J., *The Legal Regulation of Illegal and Harmful Content on the Internet*, Central European University, Center for Policy Studies, Open Society Institute, Florence, 2002–2003.

[24] Ying Ho, S., and S. Man Kui, "Exploring the Factors Affecting Internet Content Filters Acceptance," *ACM SIG e-Com Exchange*, Vol. 5, No. 1, 2003, pp. 29–36.

[25] www.icra.org (November 2007).

[26] www.rulespace.com (November 2007).

[27] http://en.wikipedia.org/wiki/RDF (November 2007).

[28] www.w3.org/RDF (November 2007).

[29] www.w3.org/TR/rdf-pics (November 2007).

[30] www.saferinternet.org (November 2007).

[31] Bertino, E., E. Ferrari, and A. Perego, "Content-Based Filtering of Web Documents: The MaX System and the EUFORBIA Project," *International Journal of Information Security*, Vol. 2, No. 1, Springer, 2003, pp. 45–58.

[32] International Organization for Standardization ISO. Information Technology—Multimedia Framework (MPEG 21) Part 2: Digital Item Declaration, 2003.

[33] http://www.efa.org.au/Issues/Censor/cens2.html (November 2007).

[34] http://internet-filter-review.toptenreviews.com/ (November 2007).

[35] Timofeeva, Y. A., "Hate Speech Online: Restricted or Protected? Comparison of Regulations in the US and Germany," *J. Transnational Law and Policy*, Vol. 12, No. 2, 2003, pp. 252–286.

[36] Hunter, C. D., "Internet Filter Effectiveness: Testing Over and Underinclusive Blocking Decisions of Four Filters," *Proc. 10th Conference on Computers, Freedom and Privacy: Challenging the Assumptions*, Toronto, Ontario, 2000, pp. 287–294.

[37] Heins, M., "On Protecting Children from Censorship: A Reply to Amitai Etzioni," *Kent Law Review*, Vol. 79, 2004, pp. 229–255.

[38] Trevor-Hall, R., and E. Carter, "Examining the Constitutionality of Internet Filtering in Public Schools: A US Perspective, *Education and the Law*, Vol. 18, No. 4, 2006, pp. 227–245.

[39] European Commission, Communication to the European Parliament, the Council, the Economic and Social Committee and the Committee of Regions, *Illegal and Harmful Content on the Internet*, COM (96), p. 487.

[40] Council of Europe, *Racism and Xenophobia in Cyberspace*, Recommendation, 2001.

[41] Electronic Privacy Information Center (EPIC), Filters and Freedom, "Free Speech Perspectives on Internet Content Controls," EPIC, Washington, D.C., 2001.

[42] Lessig, L., "Tyranny in the Infrastructure: The CDA Was Bad—But PICS May Be Worse," www.wired.com/wired/5.07/cyber_rights.html.

[43] Lessig, L., "Law Regulating Code Regulating Law," Vol. 35 *Loyola University Chicago Law Journal*, 2003, pp. 1–14.

[44] Nunziato, D. C., "The Death of the Public Forum in Cyberspace," *Berkeley Technology Law Journal*, Vol. 20, 2005, pp. 1–78

[45] Hoanca, B., "Freedom of Silence vs. Freedom of Speech: Technology, Law and Information Security," *Technology and Society*, 2005.

[46] Sunstein, S. R., "Democracy and Filtering," *Communications of the ACM*, Vol. 47, No. 12, December 2004, pp. 57–59.

[47] Shapiro, A. L., "The Control Revolution: How the Internet Is Putting Individuals in Charge and Changing the World We Know," Century Foundation/Public Affairs, New York, 1999.

[48] American Civil Liberties Union (ACLU), "Fahrenheit 451.2: Is Cyberspace Burning? How Rating and Blocking Proposals May Torch Free Speech on the Internet," 1997, http://archive.aclu.org/issues/cyber/burning.html.

[49] Zheng, L., "Cross-National Information Policy Conflict Regarding Access to Information: Building a Conceptual Framework," ACM International Conference Proceeding Series, Vol. 228, *Proceedings of the 8th Annual International Conference on Digital Government Research: Bridging Disciplines & Domains*, Philadelphia, PA, 2007, pp. 201–211.

[50] Menestrel, M. L., M. Hunter, and H.-C. de Bettignies, "Internet E-Ethics in Confrontation with an Activists' Agenda: Yahoo! on Trial," *Journal of Business Ethics*, Vol. 39, 2002, pp. 135–144.

[51] Gomez, J. J., Statement, 23rd International Conference of Data Protection Commissioners, Paris, September 24–26 2001.

[52] Johnson, D. V., and D. Post, "Law and Borders—The Rise of Law in Cyberspace," *Stanford Law Review*, Vol. 1367, 1996.

[53] Data Protection Working Party, "Opinion 2/2002 on the Use of Unique Identifiers in Telecommunication Terminal Equipments: The Example of IPv6," WP 58, Brussels, 2002.

[54] Bertelsmann Foundation, "Self-Regulation of Internet Content," September 1999, http://www.stiftung.bertelsmann.de/internetcontent/.

CHAPTER 13

Model for Cybercrime Investigations

Ahmed Patel and Norleyza Jailani

This chapter examines and explains the issues relating to devising a comprehensive model to investigate cybercrimes in a consistent and iterative manner. The actual model of investigation is explained in some detail. This model of investigation is used to demonstrate the fundamental aspects of investigation and evidence gathering for a variety of purposes. The benefits of the model are explained through some examples of potential evidence in digital media.

13.1 Definitions

Cybercrime investigations, with its elements of forensics and digital evidence, is presently an inexact science with little rigorous formalism. It has emerged in recent years as the field dealing with the acquisition of evidence from information technology systems for use in investigations of abuse of those systems. Users include police investigators, auditors, and system managers. This evidence must be acquired in accordance with strict rules, meeting requirements that depend on the type of investigation. Forensic computing provides a second line of defense after information technology security. When security fails to prevent an abuse, investigation of what has happened can lead to sanctions against those responsible, which acts as a deterrent to others. Furthermore, certain types of abuse, especially by authorized insiders, are difficult to prevent by security techniques, while others, such as the distribution of illegal content, cannot be dealt with by any security system. In dealing with the problem of computer crime and abuse, then, forensic computing is an important field. In this chapter we make no further distinctions between the various terms, synonyms, definitions, and contexts. It all boils down to virtually the same exercise that can be extended as explained by Judd Robbins [1]:

> "Computer forensics is simply the application of computer investigation and analysis techniques in the interests of determining potential legal evidence. Evidence might be sought in a wide range of computer crime or misuse, including but not limited to theft of trade secrets, theft of or destruction of intellectual property, and fraud. Computer specialists can draw on an array of methods for discovering data that resides in a computer system, or recovering deleted, encrypted, or damaged file information. Any or all of this information may help during discovery, depositions, or actual litigation."

Computer forensics is the application of the scientific method to digital media in order to establish factual information for judicial review. This process often involves investigating computer systems to determine whether they are or have been used for illegal or unauthorized activities. Mostly, computer forensics experts investigate data storage devices, either fixed like hard disks or removable like compact discs and solid-state devices.

Computer forensics is done in a fashion that adheres to the standards of evidence that are admissible in a court of law. Evidence gathered from computers is increasingly important in criminal investigations, and forensic examination of computer and other digital data has become an indispensable tool for law enforcement, corporate security, and intelligence gathering.

It is also a well-known fact that information and networking security focuses on defending systems against various types of attacks, vulnerabilities, and abuse before they happen. Technically, intrusion detection systems (IDS) should recognize and take action against attacks, but when they fail to do so and allow cybercrime to take place, then post incident (after-the-fact) investigation and evidence-gathering methods and techniques are the only available sources for catching and prosecuting the criminal within the framework of law enforcement. However, it is becoming quite common to use after-the-fact information to devise proactive (before-the fact) policies, methods, techniques, and mechanisms to protect all forms of digital assets. The method of investigation is typically a *forensic* exercise that has many synonyms, definitions, and contexts that examine digitally based information for producing digital evidence of crime.

In general, cybercrime evidence gathering and presentation is a mandatory legal concern and requirement that is carried out by expert investigators and practitioners. These experts can do many things and carry out many functions, such as the following:

- For a vital link between legal, auditing, and ICT fields;
- Secure computer and other electronically resident data;
- Interpret the data resident on electronic devices;
- Rapidly search and analyze vast volumes of data;
- Recover deleted material and overcome encrypted data.

An important feature of cybercrime investigations is to develop a life cycle for using the results as input to the development of security and management technologies, beyond post investigations. The forensic elements resulting from cybercrime investigation must be proactive, rather than being only a post mortem activity as at present, so that it can help prevent crimes and protect assets by narrowing the crime invocation versus crime prevention gap. This is one of the major challenges facing both subject domains: security and cybercrime investigations.

In fact, cybercrime investigation is deployed across many areas either individually or in combination with activities such as hacking, cracking, invading, virus spreading, malicious code, computer misuse, fraud, perjury, forgery, copyright offences, organized crime, murder, divorce, defamation, immigration fraud, narcotics trafficking, human trafficking, snooping, bogus e-trading, privacy invasions and violations, electoral law violations, obscene and illegal publications, pedophile

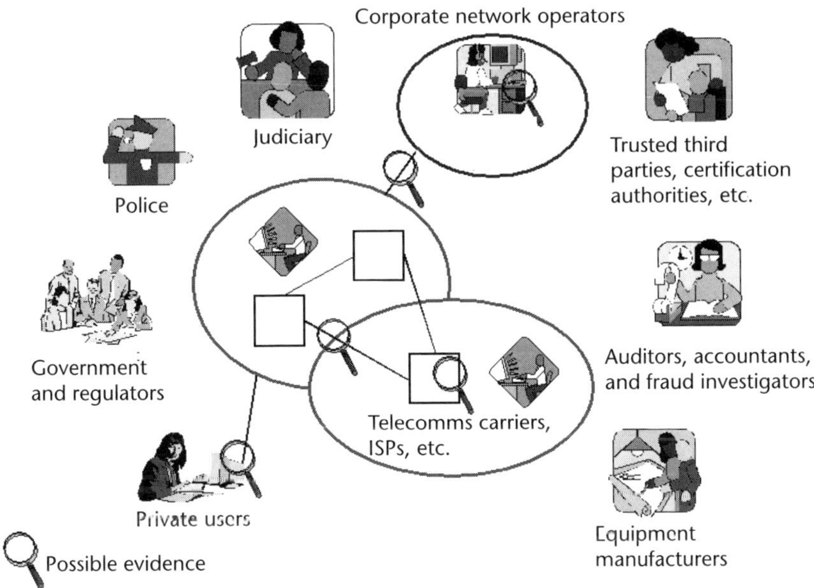

Figure 13.1 Users and providers of cybercrime investigations.

rings, sexual harassment, illegal private work, illicit software piracy, industrial espionage, data theft, credit and debit card cloning, global terrorism, and sabotage.

There are many users and providers of cybercrime investigations, as shown in Figure 13.1.

Some of the providers, like the trusted third party CA and ISPs, can provide vital information in an investigation. Much of the information that is collected must be secured and protected as part of the evidence-gathering process. Typically one would use one or more security methods to protect this information, as explained in Section 13.3.

13.2 Comprehensive Model of Cybercrime Investigation

A comprehensive model of cybercrime investigations is important for standardizing terminology, defining requirements, and supporting the development of new techniques and tools for investigators. In this section, we present and explain a model of investigations that combines the existing models, generalizes them, and extends them by explicitly addressing certain activities not included in them. Unlike previous models, the extended model explicitly represents the information flows in an investigation.

A fully defined model of cybercrime investigations is important because it provides an abstract reference framework, independent of any particular technology or organizational environment, for the purpose of discussion of techniques and technology and for ultimately supporting the work of investigators. It can provide a basis for common terminology to support discussion and sharing of expertise. The model can be used to help develop and apply methodologies to new technologies as they emerge and become the subject of investigations. Furthermore, the model can be used in a proactive way to identify opportunities for the development and deployment of technology to support the work of investigators, and to provide a framework for the capture and analysis of requirements for investigative tools, particularly for advanced automated analytical tools. At present, there is a lack of general models specifically directed at cybercrime investigations. The available models concentrate on part of the investigative process (dealing with gathering, analyzing, and presenting evidence), but a fully general model must incorporate other aspects if it is to be comprehensive.

We note also that such a model is useful not just for law enforcement—it can also benefit IT managers, security practitioners, and auditors, who are increasingly in the position of having to carry out investigations because of the escalating incidence not only of cybercrime but also of breaches of company policies and guidelines (e.g., the abuse of Internet connections in the workplace).

13.2.1 Existing Models

There are several models for cybercrime investigation in the literature. We give a brief description of the most important ones next.

13.2.1.1 Interpol

The "Interpol Computer Crime Manual" [2] discusses the structure of investigations in the context of developing standard methods and procedures that can ensure the acceptability of evidence to a court and facilitate the exchange of evidence internationally. The model is intended to describe the simplest type of investigation in which computer systems located in a single place are being searched and perhaps seized for later examination. The steps identified in the model are as follows:

1. Preinvestigation;
2. Search and seizure;
3. Investigation of seized material.

The preinvestigation stage deals with the collection of background information and the preparations for the search operation, such as briefing staff, arranging for necessary tools and expertise, and obtaining legal authorizations. The search and seizure stage deals with the collection of evidence at a site, either by seizing the computer systems or by making image copies of data stored in the systems. The investigation stage is concerned with the detailed forensic examination of the results of the search and seizure stage.

13.2.1.2 Casey

Casey [3] presents a model for processing and examining digital evidence. This has the following key steps:

1. Recognition;
2. Preservation, collection, and documentation;

3. Classification, comparison, and individualization;
4. Reconstruction.

The last two steps are the ones in which the evidence is analyzed. Casey points out that this is an evidence processing cycle, because the reconstruction can point to additional evidence that causes the cycle to begin again. The model is first presented in terms of standalone computer systems and then applied to the various network layers (from physical media up to the user applications layer, and including the network infrastructure) to describe investigations on computer networks. Casey's model is quite general and is successfully applied to both standalone systems and networked environments.

13.2.1.3 DFRWS

The First Digital Forensics Research Workshop (DFRWS) [4] produced a model that sets out the steps in a linear process for digital forensic analysis. The steps are as follows:

1. Identification;
2. Preservation;
3. Collection;
4. Examination;
5. Analysis;
6. Presentation;
7. Decision.

The model is not intended to be a final comprehensive one, but rather a basis for future work that will define a full model and a framework for future research. The DFRWS model is presented as linear, but the possibility of feedback from one step to previous ones is mentioned. The DFRWS report does not discuss the steps of the model in great detail, but for each step a number of relevant issues are listed (e.g., for preservation, the relevant issues are given as case management, imaging technologies, chain of custody, and time synchronization).

13.2.1.4 Reith, Carr, and Gunsch

Reith, Carr, and Gunsch [5] describe a model that is to some extent derived from the DFRWS model. The steps in their model are as follows:

1. Identification;
2. Preparation;
3. Approach strategy;
4. Preservation;
5. Collection;
6. Examination;
7. Analysis;

8. Presentation;
9. Returning evidence.

This model is notable in that it is explicitly intended to be an abstract model applicable to any technology or type of cybercrime. It is intended that the model can be used as the basis for developing more detailed methods for specific types of investigation (e.g., dealing with fixed hard drives or embedded nonvolatile memory), while identifying any commonality possible in procedures or tools.

13.2.2 The Extended Model

This model unifies and extends several existing models to provide a framework for the analysis of the requirements for technology supporting cybercrime investigations.

Given that a number of models already exist, what is the motivation for presenting yet another model? Based on research and analysis, it is viewed that existing models do not cover all aspects of cybercrime investigation and that, although valuable, they are not yet general enough to fully describe the investigative process in a way that will assist the development of new investigative tools and techniques as well as meeting the necessary security aspects of safeguarding the various information flows and hosting physical components under investigation, analysis, and ultimate presentation either in a court of law or in determining policy issues deemed necessary for an organization or business.

A comprehensive model can provide a common reference framework for discussion and for the development of terminology. It can support the development of tools, techniques, training, and the certification/accreditation of investigators and tools. It can also provide a unified structure for case studies/lessons learned materials to be shared among investigators, and for the development of standards, conformance testing, and investigative best practices.

The single largest gap in the existing models is that they do not explicitly identify the information flows in investigations (e.g., Reith et al. [5] have themselves noted the absence of any explicit mention of the chain of custody in their model). This is a major flaw when one considers the different laws, practices, languages, and so on that must be correctly dealt with in real investigations. It is important to identify and describe these information flows so that they can be protected and supported technologically (e.g., through the use of trusted public key infrastructures and *timestamping* to identify investigators and authenticate evidence).

A further issue with the existing models is that they have tended to concentrate on the middle part of the process of investigation (i.e., the capture and analysis of the evidence). However, the earlier and later stages must be taken into account if a comprehensive model is to be achieved and in particular if all the relevant information flows through an investigation are to be explicitly identified.

The activities in an investigation in a stepwise manner are as follows:

1. Awareness;
2. Authorization;
3. Planning;
4. Notification;

5. Search for and identification of evidence;
6. Capture of evidence;
7. Transport of evidence;
8. Storage of evidence;
9. Analysis of evidence;
10. Hypothesis;
11. Presentation of hypothesis;
12. Proof/defense of hypothesis;
13. Dissemination of information.

These activities are described next. In general, an investigation according to this model proceeds in a "waterfall" or "cascading" fashion, with activities following each other in a stepwise sequence. However, it is possible that an activity may require changes to the results of a previous activity or additional work in that activity, so the sequence of activities shown in the model allows backtracking. In fact, it is to be expected that there will be several iterations of some parts of the investigation. In particular, the analysis-hypothesis-presentation-proof/defense sequence of activities (steps 9–12) will usually be repeated a number of times, probably with increasingly complex hypotheses and stronger challenges to them at each iteration as the understanding of the evidence grows and becomes more objective in nature.

Together with the activities, the major information flows during the investigation are shown in Figure 13.2.

Information about the investigation flows from one activity to the next all the way through the investigation process. For example, the chain of custody is formed

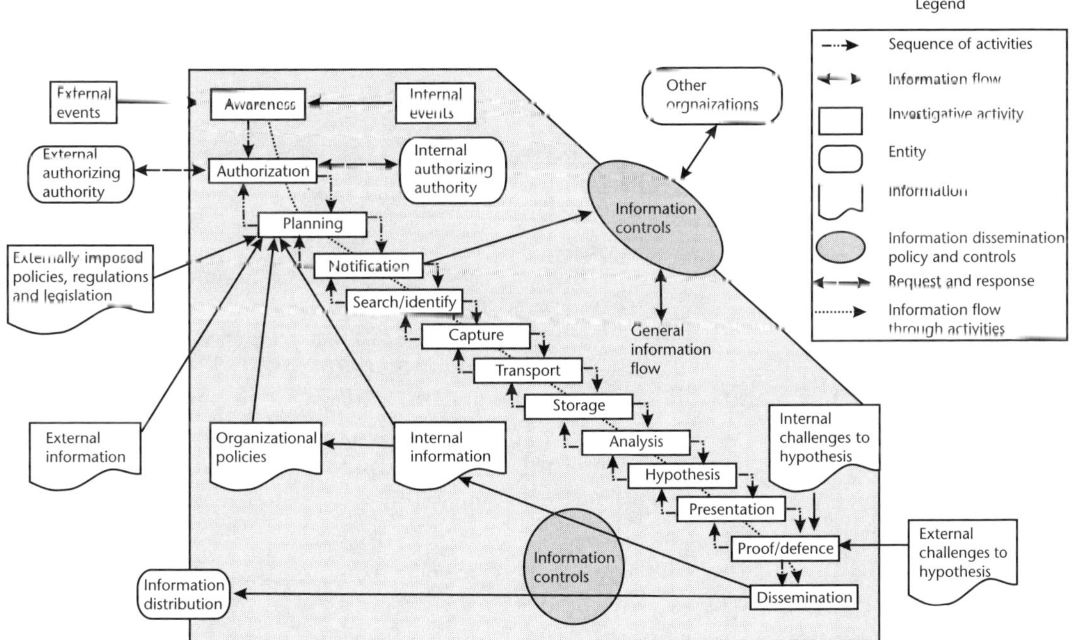

Figure 13.2 Extended comprehensive model for cybercrime investigations.

by the list of those who have handled a piece of evidence and must pass from one stage to the next with names being added at each step.

There are also flows to/from other parts of the organization and to/from external entities. We discuss the information flows in more detail in Section 13.2.2.14.

13.2.2.1 Awareness

The first step in an investigation is the creation of an awareness that investigation is needed. This awareness is typically created by events external to the organization (e.g., a crime is reported to the police or an auditor is requested to perform an audit).

It may also result from internal events (e.g., an intrusion detection system alerts a system administrator that a system's security has been compromised).

The awareness step is made explicit in this model because it allows the relationship with the events requiring investigation to be made clear. Earlier models do not explicitly show this step and so do not include a visible relationship to the causative events. This is a weakness of such models because the events causing the investigation may significantly influence the type of investigation required (e.g., an auditor can expect cooperation from a client, whereas a police investigator may not receive cooperation from suspects in an investigation). It is vital to take into account such differences to ensure that the correct approach is taken to an investigation in a particular context.

13.2.2.2 Authorization

After the need for an investigation is identified, the next step is to obtain authorization to carry it out. This may be very complex and require interaction with both external and internal entities to obtain the necessary authorization.

The level of formal structure associated with authorization varies considerably depending on the type of investigation. At one extreme, a system administrator may require only a simple verbal approval from company management to carry out a detailed investigation of the company's computer systems; at the other extreme, law enforcement agencies usually require formal legal authorization setting out in precise detail what is permitted in an investigation (e.g., court orders or warrants).

13.2.2.3 Planning

The planning activity is strongly influenced by information from both inside and outside the investigating organization. From outside, the plans will be influenced by regulations and legislation, which set the general context of the investigation and which are not under the control of the investigators. There will also be information collected by the investigators from other external sources. From within the organization, there will be the organization's own strategies, policies, and information about previous investigations.

The planning activity may give rise to a need to backtrack and obtain further authorization (e.g., when the scope of the investigation is found to be larger than the original information showed).

13.2.2.4 Notification

Notification in this model refers to informing the subject of an investigation or other concerned parties that the investigation is taking place. This step may not be appropriate in some investigations (e.g., where surprise is needed to prevent destruction of evidence). However, in other types it may be required, or there may be other organizations that must be made aware of the investigation.

13.2.2.5 Search for and Identification of Evidence

This activity deals with locating the evidence and identifying what it is for the next activity. In the simplest case, this may involve finding the computer used by a suspect and confirming that it is the one of interest to the investigators. However, in more complex environments, this activity may not be straightforward (e.g., it may require tracing computers through multiple ISPs and possibly in other countries based on knowledge of an IP address).

13.2.2.6 Capture of Evidence

Capture is the activity in which the investigating organization takes possession of the evidence in a form that can be preserved and analyzed (e.g., imaging of hard disks or seizure of entire computers).

This activity is the focus of most discussion in the literature because of its importance for the rest of the investigation. Errors or poor practices at this stage may render the evidence useless, particularly in investigations that are subject to strict legal requirements. It is also referred to as "acquisition" or "collection" of evidence in the literature.

13.2.2.7 Transport of Evidence

Following capture, evidence must be transported to a suitable location for later analysis. This could simply be the physical transfer of seized computers to a safe location; however, it could also be the transmission of data through networks. It is important to ensure during transport that the evidence remains valid for later use (i.e., that the means of transport used does not affect the integrity of the evidence).

13.2.2.8 Storage of Evidence

The captured evidence will in most cases need to be stored because analysis cannot take place immediately. Storage must take into account the need to preserve the integrity of the evidence.

13.2.2.9 Analysis of Evidence

Analysis of the evidence will involve the use of a potentially large number of techniques to find and interpret significant data. It may require repair of damaged data in ways that preserve its integrity. Depending on the outcomes of the search/

identification and capture activities, there may be very large volumes of data to be analyzed, so automated techniques to support the investigator are required.

13.2.2.10 Hypothesis

Based on the analysis of the evidence, the investigators must construct a hypothesis of what occurred. The degree of formality of this hypothesis depends on the type of investigation (e.g., a police investigation will result in the preparation of a detailed hypothesis with carefully documented supporting material from the analysis, suitable for use in court). An internal investigation by a company's systems administrators will result in a less formal report to management.

Backtracking from this activity to the analysis activity is to be expected as the investigators develop a greater understanding of the events that led to the investigation in the first place.

13.2.2.11 Presentation of Hypothesis

The hypothesis must be presented to persons other than the investigators (e.g., for a police investigation, the hypothesis will be placed before a jury, while an internal company investigation will place the hypothesis before management for a decision on action to be taken).

13.2.2.12 Proof/Defense of Hypothesis

In general the hypothesis will not go unchallenged: a contrary hypothesis and supporting evidence will be placed before a jury, for example. The investigators will have to prove the validity of their hypothesis and defend it against criticism and challenge.

Successful challenges will probably result in backtracking to the earlier stages to obtain and analyze more evidence and construct a better hypothesis.

13.2.2.13 Dissemination of Information

The final activity in the model is the dissemination of information from the investigation. Some information may be made available only within the investigating organization, while other information may be more widely disseminated. Policies and procedures that determine the details will normally be in place. The information will influence future investigations and may also influence the policies and procedures. The collection and maintenance of this information is therefore a key aspect of supporting the work of investigators and is likely to be a fruitful area for the development of advanced applications incorporating techniques such as data mining and expert systems.

An example of the dissemination activity is described by Hauck et al. [6]. They describe a system called *Coplink*, which provides real-time support for law enforcement investigators in the form of an analysis tool based on a large collection of information from previous investigations.

A further example is described by Harrison et al. [7]. Their prototype system is not real time but instead provides an archival function for the experience and knowledge of investigators.

13.2.2.14 Information Flows in the Model

A number of information flows are shown in the model. First, there is a flow of information within the investigating organization from one step to the next. This may be within a single group of investigators or between different groups (e.g., when evidence is passed to a specialist forensic laboratory for analysis). This flow of information is the most important in the course of the investigation but may not be formalized because it is within the organization, probably mostly within a single investigating team. However, there are benefits to be obtained by considering this information explicitly because by doing so we can provide support for it in the form of automated procedures and tools (e.g., case management tools).

However, before the investigation can begin, there is a need for information to come to the investigators, creating the awareness that an investigation is needed. We model this as being from either internal (e.g., an intrusion detection system alerting a system administrator to an attack) or external (e.g., a complaint being made to police) sources.

Obtaining authorization for the investigation involves further information flows to and from the appropriate authorities (e.g., obtaining legal authorization for a search or obtaining approval from company management to commit resources to investigating an attack).

The planning activity involves several information flows to the investigating team. From outside the organization, there will be policies, regulations, and legislation that govern how the investigation can proceed. Similarly, there will be the investigating organization's internal policies that must be followed by the investigators. Other information will be drawn in by the investigators to support their work (e.g., technical data on the environment in which they will be working).

If appropriate to the type of investigation, the notification activity will result in a flow of information to the subject of the investigation (e.g., in civil legal proceedings there will be requests for the disclosure of documents). This information will be subject to controls such as the policies of the investigating organization.

When the hypothesis based on the evidence must be justified and defended in the proof/defense activity, information will flow into the investigating team from within the organization and especially from outside (e.g., challenges to evidence presented in court).

When the investigation concludes (whether or not the outcome is successful from the investigators' point of view), there will be information flows as the results are disseminated. These flows are again subject to controls (e.g., names may have to be withheld or certain technical details may not be made known immediately to allow solutions to problems to be implemented). The information produced by the investigators may influence internal policies of the organization as well as becoming inputs to future investigations. It may also be passed through an organization's information distribution function to become available to other investigators outside the organization (e.g., in the form of a published case study used for training investigators or as a security advisory to system administrators).

At all times during the investigation, information may flow in and out of the organization in response to the needs of the investigators. These general information flows are subject to the information controls put in place by the investigating organization. In an abstract model it is not possible to clearly identify all the possible flows, and therefore further research is needed to refine this aspect of the model in particular contexts.

13.2.3 Comparison with Existing Models

Table 13.1 gives a comparison of the steps in our proposed model with those described earlier. It may be seen that there are a number of activities in this model that are not made explicit in the others. Information flows are not explicitly addressed by other models. The correspondence between the activities is not always one to one, but the overall process is similar.

It is believed that the additional aspects of an investigation captured in this model are important to the satisfactory modeling of the investigative process and that this model is a good basis on which to build a comprehensive reference framework.

13.2.4 Advantages and Disadvantages of the Model

This model has the advantages obtained from previous models, but extends their scope and offers some further benefits. A reference framework is essential to the development of cybercrime investigation because it allows for standardization, consistency of terminology, and the identification of areas in which research and development are needed. It can also provide a pedagogical tool and a basis for explaining the work of investigators to nonspecialists, whether they are jurors or company management.

Table 13.1 Comparison of the Cybercrime Investigations Models

	Existing Models			
Activities in New Model	*Interpol*	*Casey*	*DFRWS*	*Reith et al.*
Awareness				✓
Authorization				
Planning	✓			✓
Notification	✓			
Search/identification	✓	✓	✓	✓
Capture	✓	✓	✓	✓
Transport				
Storage				
Analysis	✓	✓	✓	✓
Hypothesis			✓	✓
Presentation			✓	✓
Proof/defense			✓	
Dissemination				

The most important advantage of this model in comparison to others is the explicit identification of information flows in the investigative process. This will allow us to specify and develop tools to support the investigator, dealing with case management, analysis of evidence, and the controlled dissemination of information. The model can also help us to capture the expertise and experience of investigators with a view to the development of advanced tools incorporating techniques such as data mining and expert systems.

Inevitably, the generality of the model presents some difficulties. It must be applied in the context of an organization before it will be possible to make clear the details of the process. For example, the model shows an information flow between activities that includes the recording of the chain of custody, but the procedures for this can only be specified in detail when the organizational and legal context of the investigators is known. We feel, however, that the benefits to be obtained from a general reference framework outweigh the extra work needed to apply it in particular contexts.

13.2.5 Application of the Model

The model described previously can be used to define requirements for supporting investigations (e.g., for tools to support the information flows identified in the model). The application of the model should be studied in different types of investigations in order to verify its viability and applicability as a general reference framework. The contexts of the different types of investigations that are of interest include the following:

- Police (criminal) investigations;
- Auditors;
- Civil litigation;
- Investigations by system administrators;
- Judicial inquiries.

We need to capture the characteristics of different investigation types (e.g., the applicable evidence standards), and add detail for different types of investigation. This is a general model that can be refined and extended in particular contexts. There is also a need to identify the actors in investigations and their roles more clearly in each context.

Although we have already made some efforts in this direction [8], there is a need to develop a more general and comprehensive model of how this type of information can be handled to the best advantage while still meeting the complex constraints imposed by considerations such as privacy and the protection of sensitive data.

13.3 Protecting the Evidence

An essential part in any cybercrime investigation is protecting the collected or acquired evidence from data and information sources. In order for this evidence to be valid and useful, we must be able to prove its integrity. Expert personnel must be trained in handling all gathered evidence in a secure manner, even when attacks are taking place. The damage the data could suffer by being handled improperly must be avoided, and the damage the data might suffer while stored in the system must be equally avoided by protecting it. The damage, for example, could be caused by an attacker trying to destroy evidence of his crime, or an employee trying to erase incriminating data from the log files or other digital media. Some essential methods to achieve this are briefly described next for the sake of completeness of this chapter, but see Chapter 7 for a more thorough explanation and details.

13.3.1 Password Protected

The evidence gathered should be protected at least with a password. However, password protection alone may not be enough to guarantee the security and integrity of the data. Passwords can be broken using password cracker software, so they are not very reliable. It is preferable that we use encryption as described next.

13.3.2 Encryption

In general, encryption is the most effective way to achieve data security with a high level of confidence. To read an encrypted file, you must have access to a secret key or password that enables you to decrypt it into its original form. Potential evidence such as log files, IDS output, and the data indexes should be encrypted to avoid either accidental or incidental tampering. It is important to ensure that before encrypted data or files are allowed into the evidence-gathering and analysis process, authorization is sought from a higher-level authority in order to obtain the keys from either a bona fide or known source to decrypt such data or files. It is imperative that during this process no new or extenuating information is introduced that will invalidate the data source and contaminate the related evidence. There is another reason that it is essential to have this policy of authorization in place—if an encrypted file contains information or evidence that we cannot access without a key, and specially if the file belongs to the attacker, then the only way to decrypt it will be to use brute-force techniques and methods to retrieve the data. We can try to break the encryption, but this process could take years, so it is better not to have to worry about encryption during a forensic investigation. This would be one case in which having encrypted files will actually help the computer forensics expert because a key to decrypt the files would be available, and the process guarantees the security, and therefore, the integrity of the evidence.

13.3.3 User Authentication

No unauthorized access to the system should be allowed. Whenever a user tries to connect to the system, we have to make sure that it is a valid user. Allowing anyone to enter the system without question could cause major disasters. That person may be trying to log in to erase all the files in the system, maybe to cover up something he did. In addition, user authentication also helps the last step of the forensic investigation, which is to identify the attacker(s). Knowing who is in the system at all times makes this step much easier.

13.3.4 Access Control

Authorization is establishing permission to access a resource, such as a file. One way to establish authorization is through access control. Access control is a mechanism for limiting use of resources to authorized users. This process establishes a relationship between users and files. We can establish the permissions on each file, specifying which users have access to the file, or we can establish the permissions on the users, specifying which files each user can access. This policy provides a starting point for the investigation, since we know that if an attacker modified a file, it had to be through one of the people that had permission to access that file. Instead of checking every employee to find a start point, we search the employees that have access to the particular file.

13.3.5 Integrity Check

Just protecting the data is not enough. We might need to provide proof that the evidence has not been corrupted, so there is a need to do periodic integrity checks on the data collected. There are different techniques to do integrity checks (e.g., there is a technique that uses hash functions). A hash, also called a message digest, is a one-way function that always produces the same unique output for each unique input. This type of function is called one-way because it is very impractical, and it is sometimes impossible to compute an inverse for it. So, how do we use this to provide integrity check? First we feed the data collected to a hash function and we store the hash result along with the data. At the time the integrity check is needed, we feed the same data to the same hash function, and if the hash output is exactly the same as the one we had stored, then the data has not been changed. However, if the output of the hash is not the same, then we know this data has been changed and it may no longer be useful.

13.4 Conclusions

As people get more and more comfortable with computers, and technology advances, society becomes more computer dependent. In an era where everything ranging from utilities such as electricity power distribution, telecommunications, and Internet services to stock markets to air traffic control and e-governance are managed through computers, security, trust, privacy, and safety become a survival issue. In today's society with *e-everything*, cybercrime is a serious problem. Simple preventive measures are not enough anymore;we must find a way to catch and prosecute cyber-criminals, and cybercrime investigations covering everything from digital forms to computers and network forensics is the fundamental gateway to archive it.

However, we should not leave everything to cybercrime forensics experts. If we are ever going to find a solution to the cybercrime problem, it has to be through a collaborative effort. Everyone from law enforcement agencies, legislators, and individual users to public and private institutions and business owners have to realize their responsibilities and play their part. Analyzing the risks, establishing appropriate policies, and adopting them to their specific needs enhances computer, data, and network security by ensuring that cybercrime investigations with detailed forensics will help the experts in the field do their jobs faster and more efficiently. It will not only reduce the security risks but also increase the possibilities for prosecuting cybercriminals. A new model of cybercrime investigations was described. The inclusion of information flows in this model, as well as the investigative steps, makes it more comprehensive than previous models. It provides a basis for the development of techniques and especially tools to support the work of investigators. The viability and applicability of the model now needs to be tested in different organizational contexts and environments.

References

[1] Judd, R. "An Explanation of Computer Forensics." February 2008, http://computerforensics.net/forensics.htm.

[2] Interpol (International Criminal Police Organization), "Interpol Computer Crime Manual," Lyon, France, 2001.

[3] Casey, E., *Digital Evidence and Computer Crime*, New York: Academic Press, 2000.

[4] Palmer, G. (ed.), "A Road Map for Digital Forensic Research: Report from the First Digital Forensic Workshop, 7–8 August 2001," DFRWS Technical Report DTR-T001-01, November 6, 2001, http://www.dfrws.org/.

[5] Reith M., C. Carr, and G. Gunsch. "An Examination of Digital Forensic Models," *International Journal of Digital Evidence*, Vol. 1, No. 3, Fall 2002, http://www.ijde.org/.

[6] Hauck, R. V., et al., "Using Coplink to Analyze Criminal-Justice Data," *IEEE Computer*, Vol. 35, No. 3, 2002, pp. 30–37.

[7] Harrison, W., et al., "A Lessons Learned Repository for Computer Forensics," International Journal of Digital Evidence, Vol. 1, No. 3, Fall 2002, http://www.ijde.org/.

[8] Patel, A., and S. Ó Ciardhuáin, "The Impact of Forensic Computing on Telecommunications," *IEEE Communications*, Vol. 38, No. 11, 2000, pp. 64–67.

CHAPTER 14

Systemic-Holistic Approach to ICT Security

Christopher C. Wills, Louise Yngström

14.1 Aims and Objectives

There are many problems associated with how to understand the concept of security in relation to computer-based information and communication systems. Typically, there may be cultural problems, such as that of language, or of understanding different views relating to differing definitions and understandings of security. Often, there are problems of a technical nature that need to be resolved, such as those associated with understanding the nature and functionality of cryptology. We need to decide if we are addressing information security or data security and whether we have understood and addressed the three main facets of security: confidentiality, integrity, and availability. What can ICT security evaluation criteria do to enhance our understanding, and what metrics and measurements of security can be used? The approaches are not presented in any predefined or preordered manner. Certainly, they could have been presented differently, or ordered in another fashion—that is exactly the point. Many perceive ICT security–related issues as a mess of ill-defined and often conflicting opinions and ideas, built on specific pools of knowledge, that are somehow connected to each other—although perhaps not obviously so.

To deal with these problems, this chapter presents an approach called the systemic holistic approach (SHA). It can be used to gain a basic understanding of problem domains such as that of security and control, by enabling the analysis and structuring of issues relating to ICT security. The aim of this chapter is to present the theoretical background to SHA, describe its model and framework, and present some examples. The objectives are to give a systemic-holistic view of the ICT security area, which makes it possible to grasp and deal with details as well as overviews in a subjective way based on objective knowledge.

14.2 Theoretical Background to the Systemic-Holistic Model

The model relies on three main building blocks:

1. General systems theory (GST) including cybernetics [11–18];
2. Sociotechnical design and soft system methodology [19–23];
3. General living systems theory (GLST) [24].

Furthermore, explanations and further comments are facilitated by Laufer [25, 26].

GST had its origin in observations of similar phenomena existing in many different sciences. In order to study these interdisciplinary phenomena, Bertalanffy [11] chose the concept of *system*. He used system as an epistemological device to describe organisms as *wholes*, and showed that the approach could be generalized and applied to wholes of any kind. Checkland [19] developed this further in discussions on the confusion between what exists (the ontological entity) and what is an abstraction (the epistemological entity).

Mumford's [20] view is that systems involving human inputs (human activity systems) are the sum of their parts—they are sociotechnical systems and need to be understood and designed as such. Thus her sociotechnical approach to understanding and designing computer-based human activity information systems involved the application of participative design methods, the idea that the users need to be engaged and facilitated in the design or redesign of systems. This view stems from the seminal work of Trist and Bamforth [21], who looked at production in the UK coal industry and who examined the counter-intuitive phenomena of falling output in the face of the increasing mechanization of coal getting. Checkland develops the thread of user involvement in analysis and design and suggests that people can best perceive reality through a methodology that uses abstract concepts. While perceiving a part of reality, humans are able to reflect on their findings—and in doing so, they are able to test and change their concepts in order to fit them better to the perceived reality. In this actual process of testing and changing, there is a multicreating relationship between the perceived reality and the intellectual concepts, which in fact constitutes a learning process.

Thus, in efforts to control, humans may choose to assume that the reality *is* a system rather than something that can be looked on *as* a system through the learning process. Checkland labels the control method used in the first case engineering or hard systems thinking, the second one, systemic or soft systems thinking. The main underlined differences between the methods are that in hard systems thinking, perceived realities are treated as existing systems and their problems solved by ordinary analytic *systematic* methods, while in soft systems thinking perceived realities are treated as problems and solved by subjectively analyzing and defining the problem space—*systemic* methods. Through soft systems thinking, humans can learn how the concept of a system *reflects* the real world and may represent one (and possibly a changing understanding) of the world. Checkland does not refrain from hard systems thinking and engineering; rather, he underlines that soft systems and hard systems thinking are *complementary* to each other. But the decision when to change from one to the other is a human subjective one.

The confusion that arises between "what *seems* to exist" and "what exists" was labeled by Checkland as "the confusion between the images of the systems and the systems image" [19]. Laufer [25] described it as the confusion between the science of nature and the science of culture; what is neither nature nor culture is artificial. And the science of the artificial is the science of systems (i.e., cybernetics).

Laufer offers one more explanation of importance to the security area: the main reason for the confusion between what is nature and what is culture is that the ultimate locus of control is undecided. This generates an ongoing crisis with two distinct states. Either the problem is very simplistic and implies a great number of similar events; in that case a manager can predict future states of the system and is confronted with the relatively safe risk of controlling the probable. Or (and more often), assumptions cannot be made about the similarity of future events, or about their independence, and management is confronted with the problem of *controlling the improbable*. The results of trying to control and cope with the improbable is to control it symbolically; for instance, through laws that authorize, commissions to deal with abuses or prevention, ad hoc commissions to deal with any new emerging problems, security norms produced by suitably composed commissions, or public opinion through opinion polls [26].

Checkland and Laufer, following Bertalanffy and GST, thus gives grounds for studying the concept of system as an epistemology for viewing and understanding perceived realities. The actual choice of when to change over to hard systems thinking becomes subjective, but is done consciously, and becomes a part of the conceptual model and the pedagogic used.

GLST [24] forms the third building block to the concept of systems, since it deals with systems that really exist—the ontological entity. It offers a concrete understanding of how physical realities restrict theoretical models and is so frequently used within ICT security that we tend to believe that the models are the reality.

GLST deals with living, concrete, open, homeostasis-aiming, systems composed of matter and energy and controlled by information. Matter and energy are considered in their physical form, and information is defined as physical markers carrying information. Thus a living system is composed of physical entities. Moreover, living systems exist on seven levels: cell, organ, organism, group, organization, nation, and supranational; each level needs 19 critical subsystems for its survival. Each subsystem is described through its structure and process and through measurable representative variables. The model is recursive on each level. General living systems theory offers knowledge and insights on how to link reality to theoretical models; through understanding of physical realities, restrictions of the domains of different theories can be understood.

14.3 The Systemic-Holistic Model and Approach

The systemic-holistic model is a conceptual model that will facilitate individuals to understand ICT security problems as related to originally existing physical entities, on specific abstract levels, in specific contexts. It consists of an epistemology and a framework. Taken together, they are called the systemic-holistic model; while in use, they are called the systemic-holistic approach (SHA)—see Figures 14.1 and 14.2.

The systemic module is the epistemology that defines security by using system concepts and directs the learning about facts in the framework. It presents general systems theory, cybernetics, and GLST as foundations for survival structures and mechanisms usable for control. Through this, the concept of security is defined as the attributes needed for a system to be in control of itself and survive—by means of being in control of its inflows, through-flows (processes), and outflows of matter, energy, data, and information, over a specific period of time.

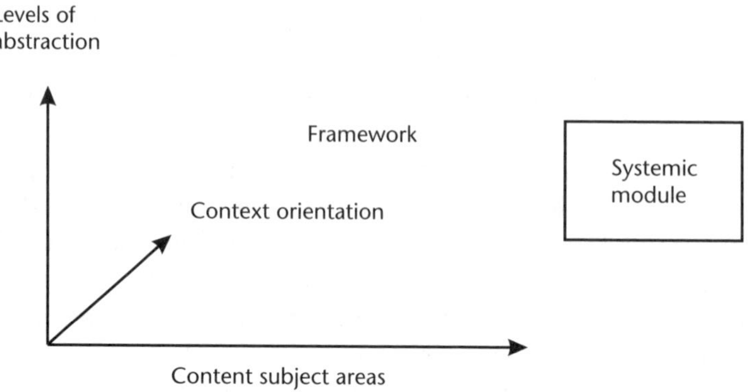

Figure 14.1 Overview of the framework and the methodology—the systemic-holistic model [27].

In the framework, each knowledge/subject area rests on one of three dimensions of the framework, the first level being some kind of basic reality or physical construction. Also, specific theoretical models are applicable for particular subject areas, and particular designs rely on these models. The second dimension of the framework is thus a level of abstraction. The third dimension is context, which adds meaning to a specific level of a subject area. All three dimensions form the framework for an understanding of security informatics. The systemic module may be viewed as an epistemology or a meta-science, but it also sets out criteria for control.

The model has been used for educational programs in ICT security in which the systemic module and a particular intersection of the dimensions of the framework are combined [27, 28]. This starts by generalizations—the abstract concepts that once were perceived by someone in interaction with reality. Through presenting examples and inviting participants to give or test their own examples, there is a shift from the ontological approach toward an epistemological approach. The results looked for—knowledgeable attitudes and conducts—will foster awareness and assurance that these generalizations are valid for each participant's own perceived realities of security.

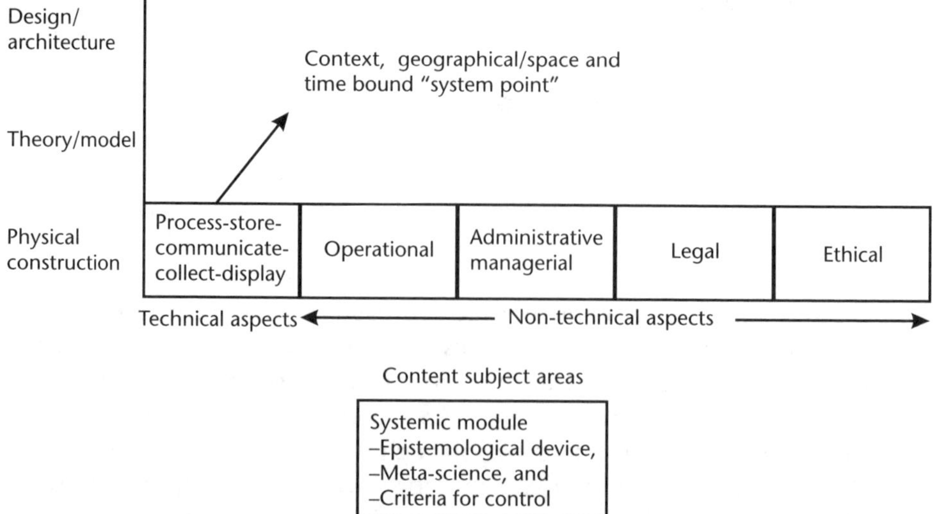

Figure 14.2 Details of the framework and the methodology—the systemic-holistic model [27].

SHA helps us understand and also express different points of view, backgrounds, weights, risks, and so on, through systems theories supplementing the ordinary analytic methods with the systems approach (SA).

The SA is different from traditional analytic methods. The main reason for complementing the analytic approach (AA) with an SA, is the increase in complexity to be dealt with in current systems. The differences between the two approaches can be characterized as follows:

- AA emphasizes parts, while SA emphasizes wholes.
- AA studies closed systems, while SA studies open systems.
- AA does not define explicit environments, while SA does.
- AA implies entropy since AA deals with closed systems, while SA deals with open systems that can compensate for entropy.
- AA considers fixed goals, while SA considers changing and learning new goals.
- AA considers few hierarchies, while SA considers many.
- AA is based on stable system states, while SA considers adaptive and changing systems states [29].

There is, however, not one single systems approach, but many variations of the same or similar "thinking." They can be considered as two main trends: general systems theory and particularized systems approaches. The latter may be divided into operations research, systems analysis, cybernetics, and systems engineering [29].

Ludwig von Bertalanffy founded the Society for General Systems Research because he regarded analytic methods as insufficient to describe current research problems. To him, different sciences appeared to try to research the same metaquestions, and current scientific paradigms and accepted methods were not adequate [11]:

> There is, however, another remarkable aspect. If we survey the evolution of modern science, as compared to science a few decades ago, we are impressed by the fact that similar general viewpoints of organization, of wholeness, of dynamic interaction, are urgent in modern physics, chemistry, physical chemistry and technology. In biology, problems of an organismic sort are everywhere encountered: it is necessary to study not only isolated parts and processes, but the essential problems are the organizing relations that result from dynamic interactions and make the behavior of parts different when studied in isolation than within the whole.

GST investigates the concept of system integrity: what makes a system a system. It studies how a specific system can be kept separate from its environment and what actions efficiently will control the system. It provides useful and general definitions. No separate discipline had established generalizable criteria of survival of a system separated from its outer or inner conditions. In other words, GST makes it possible to generalize specific phenomena in a way that they can be investigated from different angles, presumptions, techniques, fields of knowledge, and so forth. It is a metascience used for communication between sciences.

GST starts with unorganized complex structures and studies how chaos, haphazardness, and disorder can be steered into order, safety, and prediction. It uses control theory as set out in cybernetics—the study of control and communication in animal and machine [12]—and is built on 5 postulates [13, 29] and 10 hallmarks [14, 29]. The postulates are as follows:

1. Order, regularity, and nonrandomness are preferable to lack of order or to irregularity (chaos) and to randomness.
2. Orderliness in the empirical world makes the world good, interesting, and attractive to the systems theorist.
3. There is order in the orderliness of the external or empirical world (order to the second degree)—a law about laws.
4. The establishment of order, quantification, and mathematization are highly valuable aids.
5. The search for order and law necessarily involves the quest for those realities that embody these abstract laws and order—their empirical referents.

The postulates bring forward order, structure, and regularities as a base for what can be understood and thus controlled by laws. The foundation for laws lies in the real world that can be studied empirically.

The 10 hallmarks are as follows:

1. In all systems there exists interrelationships and interdependences of objects and their attributes. This facilitates delimiting the system from its environment and analyzing important chains of relations and dependencies. It also points out that unrelated and independent elements do not belong to the system and should not be granted access and authorization privileges to the system.
2. All systems have a gestalt—a wholeness that cannot be found through breaking up the system into parts. Analytic tools are necessary but not sufficient to make a system secure. It is, for instance, not enough to implement a technical security device and believe it will function as supposed. For the same reason, certified software might not perform as expected when implemented in a new environment.
3. All systems are goal seeking—a fact understood by all organizations. The tasks of those responsible for security and ICT security in organizations are to deduce, find, and specify security goals identified from the goals of the organization. Having done so, measurable security objectives must be established, as must strategies and policies to achieve these objectives. Efficient controls of these goals are attempted to be achieved using feedback from the system.
4. All systems are open systems, dependent on inflow to produce their goals, their outflow. Wrong, false, or untimely inflow can disturb production and result in inefficient or false outflow. The system must control, for instance, the type, sort, and frequency of its inflow and check on its production of outflow.
5. All systems exist to transform inputs to outputs. Inflow not used for production is useless to the system or can cause it harm—an obvious example is malicious software.

6. All systems have a degree of structural order or disorder—entropy. Natural systems left uncontrolled always seek disorder. Open systems can compensate disorder by adding extra matter, energy, and/or information. This way the system is controlled, which is really the aim of the security system. Entropy is also a central concept in information theory and cryptography.
7. All systems need to be managed in order to reach their goals. Management implies planning and controlling by feeding back information to check that plans and policies are kept.
8. There exist natural hierarchies within systems; that is to say that systems comprise subsystems, which in turn comprise further subsystems, and so on. The structure of a system is essential for its management and control—ill-structured systems are very difficult to control and thus make secure. Functional (application-oriented) and nonfunctional (security and control related) criteria must be coordinated.
9. In complex systems, specialized units perform specialized functions. This way, the total system can adapt more quickly and thereby more efficiently and effectively to changes in the environment or within the system itself. Security management and operation need to be such a specialized unit.
10. Open systems can reach their goals in many different ways—equifinality. They are not constrained by the simple cause-and-effect relationships found in physical systems, but can attain their objectives with varying inputs and transformational processes (i.e., there exist different valid ways to reach the same goals).

The hallmarks make a good and simple checklist and facilitate the understanding of the overall picture of a security organization. Security (control) is a part of management; its goal is to make the system survive; it needs planning and a structure that mirrors that of the organization—on all levels.

For our further studies we need a definition of the concept *system*. We will use the one from Schoderbek, [29], which is general enough to cover systems made of matter, energy, or information:

"A set of objects together with relationships between the objects and between their attributes related to each other and to their environment so as to form a whole" [29].

This definition specifically mentions the environment—and the fact that a system is looked on as one whole unit. Other definitions may also include a time reference.

From GST and this definition follows that we may perceive a system as comprising of chains of input-process-output, in which some input originates from the environment and some output is produced for the environment, while system internal chain of input-process-output form intermediate entities. Therefore, a system cannot usually control its environment; on the contrary, it is dependent on or controlled by its environment. However, the system may influence its environment through its output—but then at the earliest in the next cycle of the system. This highlights that the system is open; it exchanges matter, energy, and information with its environment. Many models of systems are closed—and they are not in tune with reality; all systems of importance to us are open. The question is only how open.

One main idea of GST is to be able to use analogies between systems. This way, one known system acts as a template for another system, and we may be able to use knowledge about one system's behavior to foresee the behavior of another system. Analogies are by no means absolute, but the following criteria should be considered: (1) the number of entities between which the analogies are said to hold, (2) the number of attributes in which the entities involved are said to be analogous, (3) the strength of the conclusion relative to the premises, (4) the number of disanalogies, (5) the dissimilarities in the attributes of the entities involved, and (6) and the relevance.

In using the analogy form, we may also interpret this as an attempt to explain different phenomena through a classification. System theories use many different classifications for the very same reason that humans have always tried to classify things (such as plants, animals, blood groups, weather patterns, diseases, and malware) in order to understand, explain, and eventually predict and control. Remember that every classification scheme is drawn up with some particular purpose in mind.

The analytic approach presupposes that all systems can be broken down into parts, which in turn can be broken down into parts, and so on—that there exists a hierarchy of parts and that each and all these eventually can be broken down into their *smallest* parts. But in reality there exist parts of a system that are dependent of each other and, if broken down further, these interactions are lost. These systems are *nondecomposable* systems. A middle form, in which the loss is only important in the long run, is *nearly nondecomposable* systems. These properties are important to security.

Another important property of systems is to understand adaptability. All systems adapt and change continuously based on their perception of their performance in relation to their environments. We as humans (along with all living systems) do this all the time without thinking too much about it, but for groups and organizations it has to be thought about. Essentially, there are two forms of adaptability: a functional appearing in the short run, and a structural appearing in the long run. The functional adaptation may be induced internally or externally of the system and concerns "what you do." Structural adaptation may imply either changing the system's structure or the structure of the environment; it concerns "what you are." Functional adaptation is easier to do, and changes may be changed back again, while once a structural adaptation has been performed, there is no easy return to an earlier structure.

14.4 Security and Control Versus Risk—Cybernetics

With the aim to investigate and discuss what general demands of control various systems have, Beer classified systems by two criteria: complexity (simple, complex, or exceedingly complex) and predictability (deterministic or probabilistic) [15, 16, 29].

For all deterministic systems, there exists a direct, predetermined relation between input and output; thus, it is enough to control only inflow. This holds for simple as well as for complex or exceedingly complex systems. If input is correct, output will be correct.

For probabilistic systems, this predetermined relation does not exist; the output will vary as a function of input and process, even if the system itself is classified as simple. Therefore, probabilistic systems need to be controlled through their output. The methods of output control may vary, depending on the system's degree of complexity. For simple probabilistic systems, like the performance of a machine, its efficiency can be measured toward expected behavior as stated by metrics such as run time and mean time between failures. For complex probabilistic systems, such as inventories, their behavior can be measured toward expected behaviors as stated by operational research or linear programming, whereas the behaviors of exceedingly complex probabilistic systems, such as firms or economies, need more complicated behavior patterns to be measured against—Beer simply calls this feedback or cybernetics, which we shall explain in detail later. In fact, the output of all these probabilistic systems will be measured toward some expected behavior, the differences between expected and actual behavior will be analyzed, and a signal to influence the behavior of the process in the next cycle will be fed back as new input. The differences of the control methods are concerned with how the expected behavior is expressed and what kind of analysis is performed [29].

The reason we place emphasis on these aspects is that few if any of the systems that we are dealing with within security and ICT security can be classified as deterministic. Thus, it is not enough to control input; we need to add control of output. Access control systems practice control of input; they need to be extended to check on the result of the access. Typical examples on program level are sandboxing and proof-carrying codes; another is in communication in which feedback is practiced through store-and-forward processes or by carrying redundant codes for checking on the results of a transmission. Budget control is another example of control of output in probabilistic systems.

The cybernetic principles are feed back and feed forward.

Cybernetic systems are classified on three levels called first-, second-, and third-order feedback systems. The least complex is the first-order system, which is also called a *thermostat*. Its behavior is measured toward predefined behaviors, called goals, and, if under or over functioning, a signal to increase or decrease its behavior is fed back as input into the next cycle. The thermostat cannot change its goals by itself, but it can control its own behaviors based on the goals and stay within the boundaries of these specified goals. This is called negative feedback.

The most complex cybernetic system is the third-order feedback system, also called the *reflective goal changer*. As the name indicates, these systems may themselves reflect on their goals and also change them. Basically, a third-order feedback system interacts between negative feedbacks with fixed goals as the thermostat uses, and positive feedbacks, when there are no goals to measure toward but some envisaged outcomes are reflected on vis-à-vis some higher stated missions or future states of the system. So the system is, by anticipation of the outcomes of an uncontrolled performance, able to reflect on which new goals to control its behavior toward and thus go back to use negative feedbacks based on the new goals.

A middle form of these two is the second-order feedback system, also called the *automatic goal changer*. As compared to the thermostat, it will use negative feedbacks for controlling its performance, but it is also able to choose which actual feedback to use in a specific instant. This is performed through a memory that has stored the control history—the outcomes of previous controlled cycles under specific conditions. This way, the system may increase its efficiency in reaching its goals.

All three feedback systems are thus controlled from within themselves through the cybernetic principles; first-order using negative feedback, second-order using negative feedback in combination with a history recording memory, and third-order using a combination of negative and positive feedbacks. Negative feedback thus is deviation minimizing, while positive feedback is deviation amplifying. Both forms exist within real systems—one for control purposes (feedback) and one for growth purposes (feed forward); in interaction with each other, they serve the purpose to see that the system survives and keeps its integrity in a particular environment.

But what if we instead want to find out externally how the system controls itself from within? For this we can also use feedback; by observing the external behavior of the system, hypothesize how it behaves, measure its performance by some observable external entity, and test our hypothesis through feedback.

This exemplifies Ashby's Law of Requisite Variety [30]; only through knowing all the possible states a system may take, and matching these with specific control signals, can we control a system. Uncertainty is reduced through information when the control mechanism exhibits the same amount of variety as the system to be controlled exhibits: "variety kills variety" is another popular formulation of this law.

Interpreting Ashby, this means that the control system must have the same degree of freedom as reality, contain or generate as large a variation as reality, and be able to identify and generate a particular control signal.

Complex systems thus may be studied from the outside through the black box technique, which involves manipulating input, classifying output, creating a hypothesis about the internal structure of the system, and concluding through repeated experiments to make a many-to-one transformation.

Ashby warns us against oversimplifications in the study of systems: if the complexity of a system is decreased, the possibilities to control its performance are also diminished. By complexity, we understand the number of elements comprising the system, the attributes of the specified elements of the system, the number of interactions among the specified elements of the system, and the degree of organization inherent in the system [29].

Returning to the demands for a control system, we may now conclude that a control system needs to contain functions that can discover and measure behavior, compare these behaviors to the prestated goal behavior, and decide which control signal to send; all this without simplifying a system's complexity. Thus, the main activities of a security organization (a control system) are to specify the control objects and their acceptable values together with setting up an efficient organization of the control (security) system.

The control system contains the three subsystems detector, comparator, and effector. The detector is designed to detect specified output variables of the chosen control objects and send their values on to the comparator. The comparator compares the actual values to accepted variances, and when not acceptable, sends signals about deviations to the effector. The effector decides which action is to be taken and feeds this back to the system to be controlled, which in turn takes these signals as new inputs. This way the system to be controlled is checked on its output and fed back control information to change its behavior for the next cycle.

The efficiency and effectiveness of the control system itself is crucial to the system to be controlled. Information on deviant output needs to be quickly analyzed and decided on for action—the time for feeding back a signal and the relevance of this signal are the most important features of the security system.

Stressing needs for effectiveness as well as efficiency of the control system gives us the occasion to mention the three principles of control in cybernetic systems [15]:

- *Principle I*: Implicit controllers depend for their success as much on continuous and automatic comparisons of expected and performed behavior as on continuous and automatic feedback of corrective actions.
- *Principle II*: In implicit controllers, control is synonymous with communication; to be in control is to communicate.
- *Principle III*: In implicit controllers, variables are brought back into control in the act of and by the act of going out of control.

This is to say that control (security) in the ideal case should be an inbuilt part of the system—or, practically, very close to the system so that security can perform all its checks, analyses, and corrections continuously and automatically. Security needs to be in constant contact and communication with the objects of control, but it is not until something extraordinary happens that security is activated. Also, a good control system may increase its own efficiency through recording and structuring its own memory for future control purposes and preparing for new situations through the use of feed-forward mechanisms. Looked on as the focal system and its control system in this way, control (security) is essential for the survival and growth of any system—the control system enables its focal system to survive through keeping the integrity of the system.

Now, we may think this sounds too simplistic and yet too difficult. It may sound too simplistic because we will never be able to foresee all possible events in complex systems; thus, we are in this case trying to simplify complexity and will thereby lose control. It is too difficult because in order to instantaneously monitor the focal system, the complexity of the control system will be of the same degree as the one of the focal system. This is exactly why we cannot expect to ever build totally proactive control systems; on the contrary, the control system, as well as the focal system, will only survive in its changing environment if it can combine the two cybernetic principles of feedback and feed forward also for its own behavior. The control system will have to be designed also to react to new and unforeseen events. Also, as we shall see later, living systems exhibit a variety of solutions for efficient control systems and mechanisms, all built on cybernetic principles and most of them structured along existing inherent natural hierarchies.

In conclusion, this section discusses the fundamental requirements of a subsystem that will protect the integrity of the total system. This is called the control system. It is shown that the sole use of input controls is inefficient. Instead, controls built on cybernetic principles are needed. The principles of feedback and feed forward are discussed, and it is concluded that the organization and functioning of the control system itself is crucial for the survival of the total system.

14.5 Example of System Theories As Control Methods

14.5.1 Soft System Methodology

Soft systems methodology (SSM) was developed by Peter Checkland [22, 23]. The approach seeks to deal with complex, fuzzy, poorly defined problem domains, particularly those problem domains in which there is the potential for high social drama. We have already discussed Ashby's Law of Requisite Variety and the issue of complexity, which we defined as the number of elements comprising the system. However, complexity is also a function of the relationships that exist between elements. Thus, we may have a system comprised of many elements, such as a business accounting system, in which the relationships between elements are well defined (bounded), well understood, and are therefore easily modeled. Such systems are known as being of *deterministic complexity*, are capable of being broken down and understood by applying reductionist techniques, and are referred to as *hard systems*.

Alternatively, we may have a system comprised of only a few elements, such as the party political system in the UK, in which the relationship between elements is not well defined (unbounded) and is highly complex, dynamic, and difficult to model. Such systems are viewed as being of *nondeterministic complexity*, do not respond well to the application of reductionist techniques, and are referred to as *soft systems*. We have previously discussed that systems survive only if they exert control both internally and externally in response to stimuli. The ability to respond is predicated on both the ability to detect change (i.e., the ability to measure) and the ability to effect change (i.e., the ability to apply effectors). However, that which we can't understand we can't measure, and that which we can't measure, we can't control or manage.

It is these types of soft system problem domains in which SSM can be used to express, understand, and if necessary effect change in the problem situation. Figures 14.3 and 14.4 depict an overview of the approach.

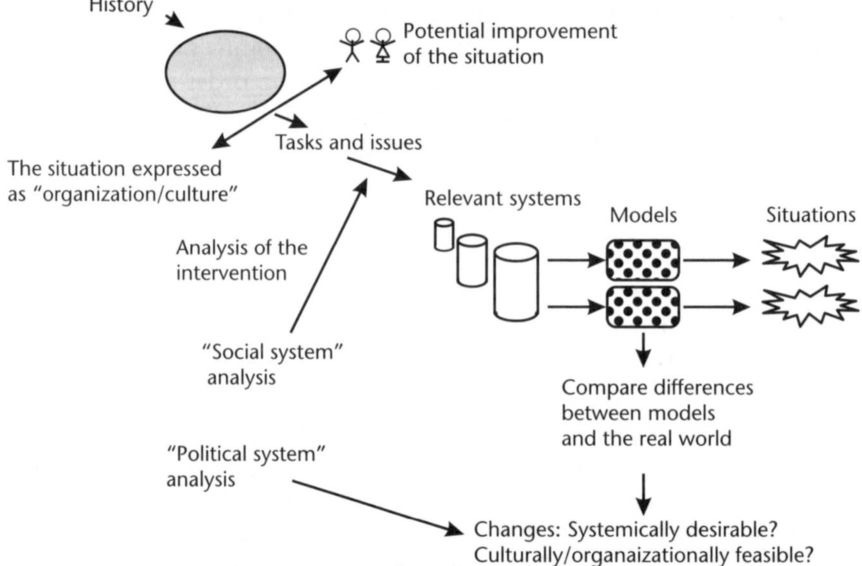

Figure 14.3 An overview of the SSM.

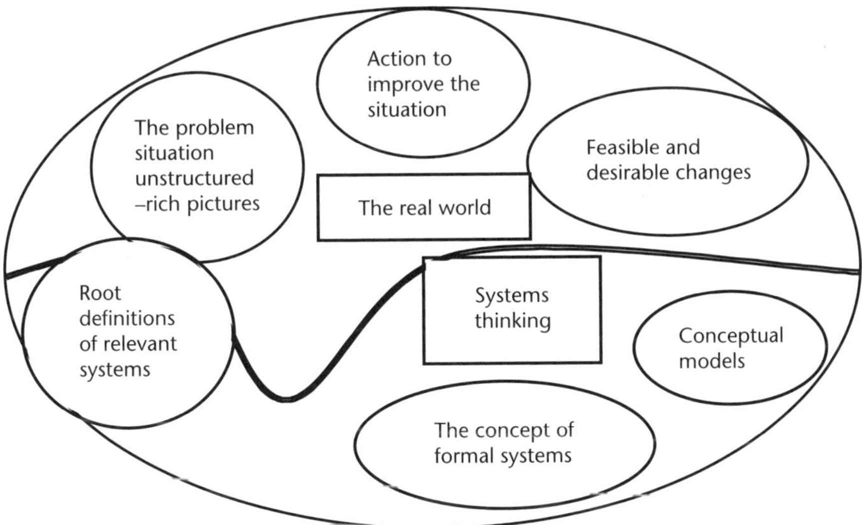

Figure 14.4 The real world and systems thinking.

The idea underpinning SSM is that of creating a comparison between what can be seen in the real world and what can be examined using systems thinking.

The approach consists of seven stages.

Stage 1

This stage consists of exploring the problem situation. Is there a problem? What does the system do? How does it work? What are the processes involved? Who are the main stakeholders? Figure 14.5 represents the problem situation unstructured.

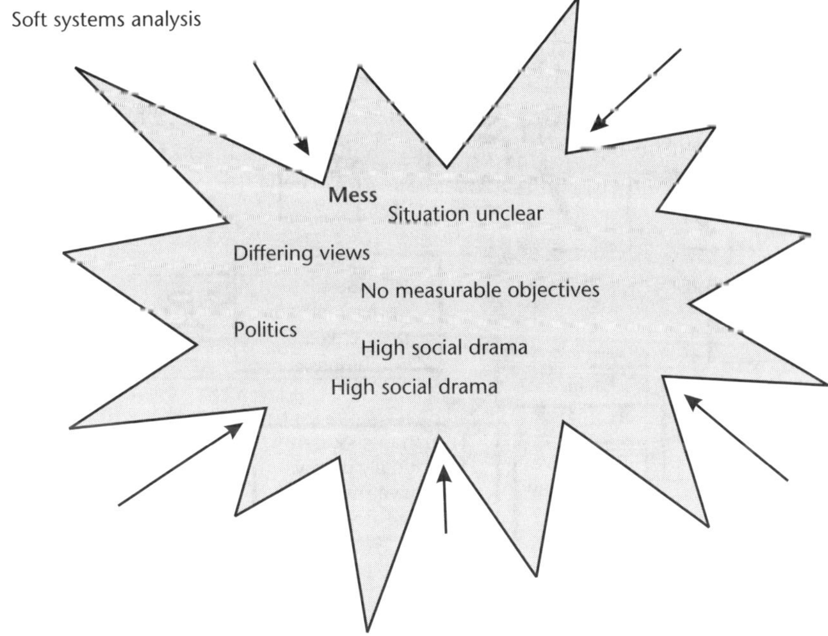

Figure 14.5 The problem situation unstructured.

Stage 2

The problem situation is expressed—the analyst(s), ideally with the active participation of the stakeholders, construct a diagrammatic representation of the problem domain. This representation, known as a *rich picture* (Figure 14.6), depicts both the physical (logical) and the political (cultural) elements that they have identified in the problem domain. The involvement of stakeholders at this stage of the method is not essential, but is highly desirable, as they will have a far deeper understanding of the problem situation than does the analyst(s). The analyst(s) attempts to capture both the physical infrastructure and the processes, as well as the relationships between the stakeholders, and depict these in the rich pictures in order to gain an understanding and appreciation of the problem situation. The picture should contain hard information and factual data, as well as soft information (i.e., subjective interpretations of situations, including aspects of conflict and emotions). The rich picture should give a holistic impression!

Stage 3

Once the rich picture has been constructed, relevant subsystems are identified and defined using *root definitions*, which encapsulate the essence of these systems—the "whats" (what does it do?) rather than the "hows" (how does it do it?). The root definitions describe the core transformation activities and processes of the system—the conversion of inputs into outputs. Checkland developed the use of a mnemonic "CATWOE" to enable the analyst(s) to ensure that the root definitions are complete:

- *Customers*: those who benefit in some form from the system;
- *Actors*: the people involved;

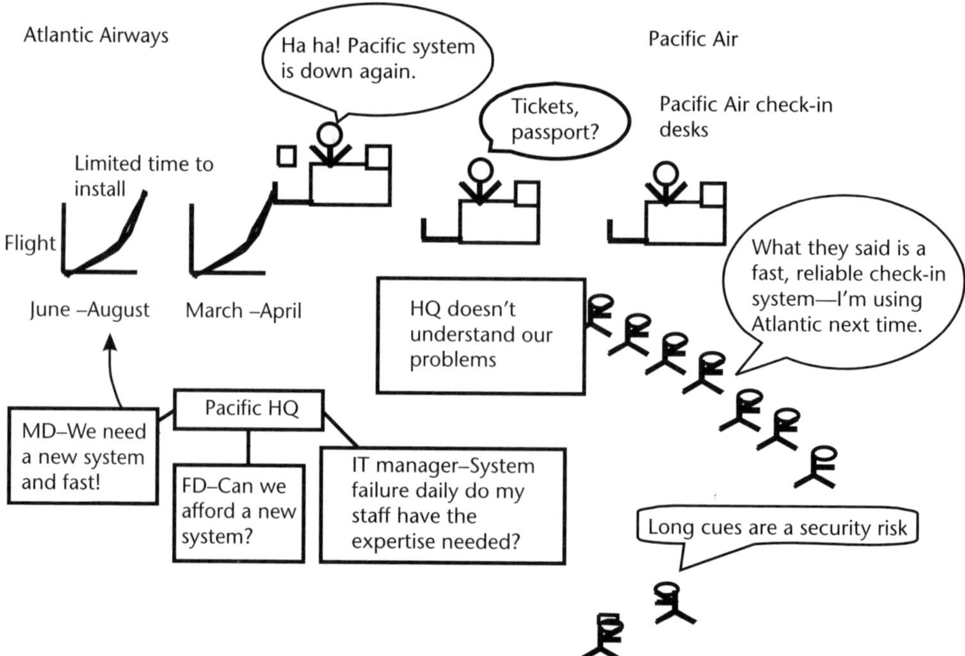

Figure 14.6 An example of a rich picture.

- *Transformation*: the development of outputs from inputs;
- *Weltanschauung*: the "world view," a holistic overview of both the transformation processes and the problem situation;
- *Owner*: the person(s) with control;
- *Environmental constraints*: physical boundaries, political, economic, ethical, or legal issues.

In order to construct the root definition(s), areas of clustered tasks and associated issues are selected, and some systems are named that could achieve those tasks and address the issues identified. Root definitions are written to describe the *activities* that must take place in order for that system to operate. These are the whats, not the hows.

Using the example of our rich picture in Figure 14.6, we might construct a root definition along the following lines (see Figure 14.7):

"An airline-owned passenger check-in system that enables passengers to check in their baggage and enables airline staff to issue passengers with boarding passes in a manner that is consistent with the safe and timely operation of the airline's departure schedules."

- *Customers*: the passengers;
- *Actors*: the airline staff;
- *Transformations*: unchecked baggage becomes checked; passengers' tickets are supplemented with boarding passes;
- *Weltanschauung*: the efficacious, effective, and efficient operation of the airport and the airline (it works with minimum waste and meets the expectations of the passengers, the airline, and the airport);
- *Owner*: the airline;
- *Environmental constraints*: time, safety, and security effectiveness (passengers and baggage departing to the same correct destination).

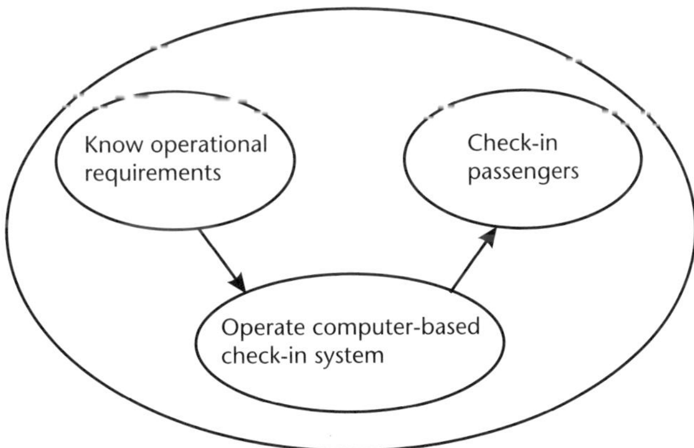

Figure 14.7 The check-in process—a top-level primary activity conceptual model.

Stage 4

Once the root definition(s) have been constructed, compared with the rich picture, and checked against CATWOE, conceptual models can be constructed. The conceptual models are formed from the actions stated or implied in the root definition(s). Of course each rich picture may be interpreted from quite differing world viewpoints. The example of the following conceptual model is from the perspective of the check-in staff. Other perspectives, such as those of Pacific Air's ICT manager, would be quite different. When applied, in practice several conceptual models would be developed from different viewpoints.

Having developed a top-level, primary activity conceptual model, each of the activities identified are modeled in a second-level conceptual model.

Stage 5

The conceptual model(s) are checked against both the root definition(s) and the rich picture. A good way of performing these checks is to ask three questions: Do the activities exist? Who does them? Why do it that way? The analyst(s) must imagine that the conceptual model is actually operating in the real world. They can identify a real process from the rich picture, follow its sequence in the conceptual model, and compare how the sequence would operate in reality (see Figure 14.8). This process can be represented using a chart.

Stage 6

Having ascertained that the conceptual models are in accord with the root definition and rich picture (Table 14.1), a discussion between the owner and the actors need to be facilitated by the analyst(s). The objective of the discussion is to identify changes that are both culturally and organizationally desirable and economically and technologically feasible. In terms of our simple example of an airline check-in system, it is obvious that the key aspect of the problem situation is the unreliability of the check-in system. However, as is clear from the rich picture, there are a number of competing pressures at Pacific Air's HQ. The soft systems method enables the analyst(s) to understand the technical problems within the wider context of the problem domain.

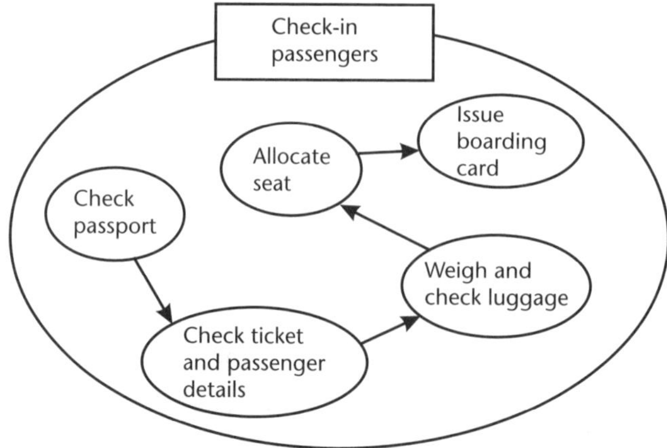

Figure 14.8 A second-level secondary activity conceptual model.

Table 14.1 Comparison Between Conceptual Model, Root Definition, and Rich Picture

Activity in conceptual model	Present in real world situation (rich picture)	Comments	Include on agenda
Check passport	Yes	Process tasks place independent of check-in computer system	No
Check passenger and ticket details	Yes	Dependent upon operation of check-in computer system	Yes
Weigh and check luggage	Yes	Process tasks place independent of check-in computer system	No
Issue boarding card	Yes	Dependent on operation of check-in computer system	Yes
Allocate seat	Yes	Dependent on operation of check-in computer system	Yes

Stage 7

The output from stage 6 is applied to the problem situation and may result in changes to procedures, policy, stakeholder attitudes, and technology.

14.5.2 General Living Systems Theory

The theory for living systems is a part of GST. While GST deals with all sorts of different systems, the theory for living systems deals with concrete, open, homeostasis-aiming systems that are complex, include DNA and organic material, have a decider, are composed of 19 critical subsystems, are actively self-regulating, and have aims and goals, as well as the need for a specific environment. The founder of the theory is James Grier Miller [24], a medical scholar, which in part explains his view of the world, sometimes called general living systems theory. According to Miller himself [24]:

> My analysis of living systems uses concepts of thermodynamics, information theory, cybernetics, and systems engineering, as well as the classical concepts appropriate to each level. The purpose is to produce a description of living structure and process in terms of input and output, flows through systems, steady state, and feedbacks, which will clarify and unify the facts of life.

Thus, with GLST he presents, describes, shows, and argues that there are analogies to be seen—and used—between living systems and the things around us we call systems. The presentations, descriptions, and arguments will use specific as well as generally known and accepted scientific methods, models, and concepts as proofs.

Miller's particular bias is to view the world as systems processing flows. In order to survive, the system needs to get the correct inflows and protect itself from incorrect ones. It must organize its internal work such that it can make use of inflows and can, if needed, get rid of or internally protect itself against unusable or incorrect flows. Living systems, in comparison to nonliving systems, are predefined through its DNA content for certain actions, processes, structures, and developments. Living systems are "blueprinted" through its DNA. In some circumstances, sometimes known, sometimes not, DNA may change, and then the blueprint has changed. All other changes are local adaptations within the range of DNA and the specific environment in time. Adaptation processes are controlled within the system itself—it is self regulated and strives to keep its integrity. All living systems strive to reach a steady state in which somehow the system is in balance with regard to handling flows. The actual physical place of the steady state may change over time, always aiming at keeping the total system in balance. Interestingly, there is no specific security system among the 19 critical subsystems that Miller defines. Control is internalized into all processes, and the information processing subsystems in particular are co-coordinating the control. If the control within in total cannot cope effectively enough with changes, the system will eventually go into decay and finally disintegrate.

One very obvious reason for studying living systems within security is that nature throughout millions of years has been able to design systems with the potential for survival, widely surmounting the technical designs of man. Many times nature can suggest how to build control systems or control functions. Within the ICT security area, it has also been suggested that networks, in particular, share many characteristics with biological organisms. Computer viruses were named viruses exactly for their similarities with living organisms.

We are going to use the theory for living systems first to understand how nature designs survivable structures, and second as a wealth of suggestions for how to solve security problems of different kinds.

Miller describes all living systems as consisting of 19 critical subsystems, which all process matter, energy, or information. All existing living systems are divided into 7 levels: cell, organ, organism, group, organization, society, and supranational. The complexity within each of the critical subsystems increases from one level to the other, but the purpose of the subsystems is still the same.

The critical subsystems are strictly ordered: two process matter/energy and information (reproducer and boundary), eight process only matter/energy (investor, distributor, converter, producer, matter-energy storage, extruder, motor, and supporter), and nine process only information (input transducer, internal transducer, channel and net, decoder, associator, memory, decider, encoder, and output transducer).

The matter/energy processing and information processing subsystems function similarly together; the matter/energy processing subsystems treat a matter/energy flow structurally the same way as the information processing subsystems treat the flow of information. The reason for eight and nine subsystems is exactly the physical treatment of the flows. In addition, the information processing subsystems control the matter/energy processing subsystems; together, they form the control system for the physical flows.

Each subsystem is described by specific structures and processes. In addition, possible control variables and their measurements are given for each process. This is described generally in Chapter 3 of Miller's book, and specifically for each level in his Chapters 6–12. Examples from different sciences are given as evidences; this way, all but seven of the particular subsystems on the seven levels are identified and examples are given.

The three main flows are defined as follows:

- Matter has mass and occupies physical space.
- Energy is the potential to do work. Matter and energy are exchangeable in the sense that they can transform into each other, and living systems can only process special types of matter and energy.
- Information is signals or markers.

Information in Miller's sense is "only data"—signals carry no meaning, but "meaning is the significance of information to a system which processes it" [24]. Signals are materialized in the form of matter or energy.

Thus, the three flows are actually of only two types, and they are all physical, concrete flows. As such, known laws for physical flows and processes can be applied.

Some examples of how GLST may be used are as follows.

Example 1

In Chapter 3, the structures and processes of all critical subsystems are outlined, including measurable representative variables of the structure of the subsystem For instance, for the subsystem matter-energy storage, the *structure* is described as storing matter/energy in all different consumable forms from directly consumable to parts from which all forms of consumables needed can be made. *Processes* are store, maintain, and take out components from the storage. *Representative measurable variables* are sort, storage capacity, percentage of matter/energy stored, place, changes over time and with different circumstances, rate, lag, costs of storage, rate of breakdown of organization of matter/energy storage, sorts retrieved, percentage retrieved, order of retrieval, rate of retrieval, costs of retrieval.

The equivalent subsystem for information is memory. *Structure* and *processes* are described similarly. *Measurable variables* are meaning of information stored in memory, sorts of information stored in memory, total memory capacity, percentage of information input stored, changes in remembering over time and with different circumstances, rate of reading into memory, lag in reading into memory, costs of reading into memory, rate of distortion of information during storage, the time information is in storage before being forgotten, sorts of information retrieved, percentage of stored information retrieved from memory, order of retrieval from memory, rate of retrieval from memory, lag in retrieval from memory, costs of retrieval from memory.

Each level, cell to supranational, can then be studied in detail for *structure*, *process*, and *representative measurable variables*.

Example 2

Chapter 4 lists hypothesis for different subsystems. Many of these have been proven valid for many subsystems and/or levels such as, "[t]wo-way channels which permit feedback improve performance by facilitating processes that reduce errors . . . As the noise in a channel increases, a system encodes with increasing redundancy in order to reduce errors in the transmission . . . Decisions overtly altering major values of a system are finalized only at the highest echelon . . . The signature identifying the transmitter of any message is an important determinant of the probability of the receiver complying with it . . . Under equal stress, functions developed later in history of a given type of system break down before more primitive functions do . . ."

Example 3

Chapter 5 presents research on which capacities living systems handle flows of information in steady states—and what happens in the systems when this is blocked: Cells can at a maximum process 4,000 bps, organs, 55; organisms, 4.75–5.75; groups, 3.44–4.60; and organizations, 2.89–4.55. After that, adjustment processes to simplify the flows start. Planned adjustment processes are filtering, queuing, multiple channels, and abstracting. Unplanned adjustment processes are omissions, errors, and escapes.

GLST presents important knowledge and insights into how reality can be linked to different logical descriptions of systems. Through understanding how a certain function is physically realized, limits of possible models to be used can be found. The theory makes it possible to generalize and use already existing physical laws. This way, GLST may function as a system of reference or a high-level model for creating survivable structures.

14.5.3 Beer's Viable Systems Model

While GLST focuses on essentials for the survival of living structures, Stafford Beer [31] investigates essentials for the survival of organizations. Both rely on cybernetics for control—Miller uses the overall concept of survivability, Beer uses viability. Both deal with problems such as complexity, rate of change, and interdependency between parts of a system—and identify what is minimally needed for a system:

- To create an organization that will meet the demands of the environments;
- To have internal structures that can deal with learning and adaptation;
- To have communications for connecting and transmitting information.

Miller sees, 19 (or 20) subsystems on seven levels; Beer, focusing on control (security), defines five different control functions; system 1–system 5, whose main task together, is to deal with complexity (variety) through amplifying and filtering variety. He calls this *variety engineering*. System 1 is the operational elements, those that do the actual job. System 2 coordinates the operational elements, so that they do not interfere with or destroy each other. System 3 ensures that operational elements together—daily and efficiently—produce what they are supposed to produce. System 4 sees that the organization, also in the future, will produce what the environment needs in the correct way. System 5 mediates the actions of systems 3 and 4 toward the identity of the organization as a whole.

Both Miller and Beer point to recursiveness; functionally the very same structures and processes should be found on all levels of living systems or organizations: Miller's critical subsystems and Beer's systems 1–5 all need to be present in order for the system to survive as an identifiable system. The differences between the systems of Miller and Beer are primarily the types of systems they approach—living systems versus organizations.

In the viable systems model (VSM), the production (of the entire system—the organization) goes on at the lowest level. This is called system 1 and can be pictured as consisting of four main elements: the operations, their environment, the management of the operations, and the models guiding the management. There are interactions between the environment and the operation, between operations and management, and between the management and the models used for management. At the same time, there is a scale of complexity; the environment is more complex than the operations, which are more complex than management, and which in turn is more complex than the used models.

In fact, in an organization system 1, at the lowest level, is made up of many operations, each seen as embedded in an environment, having a local management steered by this management's models. Each part is trying its best to optimize production through the local management and its models. But taken together, they all interact individually and in performance between each other. This creates problems. First, the environments of the operations are not the same, resulting in a problem of defining a boundary between the system and its environment, which will in total not be distinct. Second, there will be contacts directly between the separate operations, which in turn will affect the individual operational performance—and this may be different from what was intended by its local management. Third, there will be interactions between all control elements, management, and models.

This calls for more than one level of control and some further control functions. Beer calls these control functions the internal and now management (system 3), the external and future management (system 4), the closure and identity management (system 5), and the co-coordinating function (system 2). The four control levels in total will act on the meta-level of system 1, but the four functions themselves will only be strictly hierarchically organized in a particular instant; together, they strive toward keeping the organization functioning effectively (stay in homeostasis) over a longer period of time. Therefore, they interact in a pattern ultimately directed by the closure and identity management, which is system 5. System 5 is constructed to balance the total system between the needs of today and the needs of the future (i.e., between stability and rate of change, judging this toward the identity of the total system). Once the balance between the higher-level managements of today (system 3) and the future (system 4) has been decided, management of today can direct coordination needed (system 2) between lower level performances (system 1).

It is usually not possible to take an organization and immediately apply the VSM to its organizational chart—if you do that, you will find many inconsistencies. Much of that is owing to the history of the specific organization. In particular, you will most probably find only small fractions of system 4s and maybe unclear system 5s. You might also find inconsistencies on the lowest levels, although as Clemson [32] points out, it is more usual to concentrate on here and now operations, and thus to have systems 1–3 functioning, than to plan for the future.

14.6 Can Theory and Practice Unite?

When it comes to practice, security is physical and context oriented. It is implemented in a particular system, in a particular environment, and with a particular purpose.

Presented system theories offer possibilities to generalize control knowledge for application within our own environments.

Managements view security as risks to be handled and controlled often in economically feasible ways. This implies using effective and efficient control measures. The work will be directed through policies and plans, and work results will be fed back for control. Actual control measures or variables to choose are decided as a result of risk analyses.

Knowledge and understanding in the organizations are acknowledged by practitioners as fundamental principles to make security measures work in organizations, because, as systems theory expresses it: people close the nonevident and unstructured control cycles in open systems by their mere abilities to generate variety when needed. In case they cannot, and the control system has not foreseen this, there will be failures in the systems. Understanding, education, and training thus fills this function—and is always underlined as essential prerequisites for making any system work.

In general, the managers' and technicians' views of security are quite different, even though the structures are similar. That is what the whole SHA is about; the overall goal for the control of an organization will be to make effective and efficient use of assets, while controlling in a way that will make the organization stay in business. All detailed safeguards, protections, and preventions will have to be adjusted to the assets themselves, as well as to the external and internal environments, and to the purpose and identification of the organization. The general definitions, concepts, models, and methods that have been discussed make it possible to see both similarities and differences, and hopefully also choose adequate ones for our own organizations.

Usually when presenting system theories—and in particular the concepts of cybernetics and feedbacks—to various security and ICT security practitioners, there are two kinds of reactions:

- This is much too theoretical; security is a practical business.
- Yes, this is really the way it is.

The practitioners prefer security in the sense of what methodologies or safeguards to use; whether or not there should be an insurance, whether to start with a vulnerability analysis or a risk analysis, and how to perform these. Theoreticians such as Checkland, Miller, and Beer, on the other hand, view security in light of structuring the control systems correctly and what control functions to include. And, of course, both sides are right: there is a need to know what control functions are necessary *and* what methodologies and safeguards to use, and to implement these into one whole system.

So our own reaction is: Yes, both and simultaneously!

14.7 Conclusions

As we learn and understand more details about ICT security, we will be able to apply this general epistemology on control systems in various specific applications, levels, and contexts. However, we may not all have exactly the very same model in our heads when we work with preventive methods, analyses, architectures, security mechanisms, protection devices, and so on, but we are always able to refer back to the original systemic-holistic model, the systemic module, and the SHA—both for our own thoughts and to communicate to others, maybe even in other words. Thus, it will be possible for us to view and work with even minute details of a system, knowing well where they are of importance to the total system—and which this one is. It will be possible in each instant to decide for ourselves, as well as to explain to others, which actual restrictions, demands, and requirements we are regarding.

Using the SHA, we are also able to define, discuss, and question applications, architectures, environments, or paradigms, and we may also seek other or new solutions. There will, however, not be a fixed *Handbook of the Systemic-Holistic Method to ICT Security*—what you may acquire are personal attitudes and insights on how to handle and cope with ICT security issues: to embed the SHA into our minds in order to be guided by the principles of cybernetics adapting continuously in our own understanding, knowledge, and methods to the ever-changing inner and outer ICT security relevant environments.

References

[1] Frisinger, A., "A Generic Security Evaluation Method for Open Distributed Systems," PhD thesis, Department of Teleinformatics, Royal Institute of Technology, 2001.

[2] Tarimo, C. N., "ICT Security Readiness Checklist for Developing Countries. A Social-Technical Approach," PhD thesis, Department of Computer and Systems Sciences, Stockholm University, 2006.

[3] Chaula, J., "A Socio-Technical Analysis of Information Systems Security Assurance. A Case Study for Effective Assurance," PhD thesis, Department of Computer and Systems Sciences, Stockholm University, 2006.

[4] Bakari, J. K., "A Holistic Approach for Managing ICT Security for Non-Commercial Organizations. A Case Study in a Developing Country," PhD thesis, Department of Computer and Systems Sciences, Stockholm University, 2007.

[5] Björck, F. J., "Discovering Information Security Management," PhD thesis, Department of Computer and Systems Sciences, Stockholm University, 2005.

[6] Casmir, R., "A Dynamic and Adaptive Information Security Awareness (DAISA) Approach," PhD thesis, Department of Computer and Systems Sciences, Stockholm University, 2005.

[7] Näckros, K., "Visualising Security Through Computer Games. Investigating Game-Based Instruction in ICT Security: An Experimental Approach," PhD thesis, Department of Computer and Systems Sciences, Stockholm University, 2005.

[8] Kowalski, S., "IT Insecurity: A Multi-Disciplinary Inquiry," PhD thesis, Department of Computer and Systems Sciences, Royal Institute of Technology, 1994.

[9] Magnusson, C., "Hedging Shareholder Value in an IT Dependent Business Society—The Framework BRITS," PhD thesis, Department of Computer and Systems Sciences, Stockholm University, 1999.

[10] Zuccato, A., "Holistic Information Security Management Framework—Applied for Electronic Commerce," PhD thesis, Karlstad University Studies, 2005.

[11] von Bertalanffy, L., "Main Currents in Modern Thoughts," in *Yearbook of the Society for General Systems Research*, Vol. 1, 1956.

[12] Wiener, N., *Cybernetics or Control and Communication in the Animal and Machine*, New York: John Wiley & Sons, 1948.

[13] Boulding, K., "General Systems As a Point of View," in *Views on General Systems Theory*, M. D. Mesarovic (ed.), New York: John Wiley & Sons, 1964.

[14] Litterer, J. A., *Organizations: Systems, Control and Adaption, Vol, 2*, 2nd Ed., New York: John Wiley & Sons, 1969.

[15] Beer, S., *Cybernetics and Management*, New York: John Wiley & Sons, 1964.

[16] Beer, S., *Cybernetics and Management*, Science Edition, New York: John Wiley, 1964.

[17] Beer, S., *Brain of the Firm*, New York: John Wily and Sons, 1981.

[18] Ackoff, R. L., *Designing a National Scientific and Technological Communication System*, University of Pennsylvania Press, 1976.

[19] Checkland, P. B., "Images of Systems and the Systems Image," Presidential address to ISGSR, June 1987, *Journal of Applied Systems Analysis*, Vol. 15, 1988, pp. 37–42.

[20] Mumford, E., *Redesigning Human Systems*, Hershey, PA, and London: IGI Publishing, 2003.

[21] Trist, E. L, and K. W. Bamforth, "Some Social and Psychological Consequences of the Longwall Method of Coal Getting," *Human Relations*, Vol. 4, pp. 3–38.

[22] Checkland P. B., *Systems Thinking, Systems Practice*, New York: John Wiley & Sons, 1981.

[23] Checkland P., and J. Scholes, *Soft Systems Methodology in Action*, New York: John Wiley & Sons, 1990.

[24] Miller, J. G., *Living Systems*, New York: McGraw Hill, 1978.

[25] Laufer, R., "Cybernetics, Legitimacy and Society," in "Can Information Technology Result in Benevolent Bureaucracies?" *Proceedings of the IFIP TC9/WG9.2 Working Conference*, L. Yngström, et al. (eds.), Namur, Belgium, January 3–6, 1985, and North Holland, 1985, pp. 29–42.

[26] Laufer, R., "The Question of the Legitimacy of the Computer: An Epistemological Point of View," in *The Information Society: Evolving Landscapes*, J. Berleur, et al. (eds.), Springer Verlag & Captus University Publications, 1990, pp. 31–61.

[27] Yngström, L., "A Systemic-Holistic Approach to Academic Programmes in IT Security," PhD thesis, Report No. 96:021, Department of Computer and Systems Sciences, Stockholm University, Kista Sweden, 1996.

[28] Yngström, L., "Towards a Systemic Holistic Approach to Academic Programmes in the Area of IT Security," Licentate Thesis, Stockholm University, Sweden, 1992.

[29] Schoderbek, P., G. Schoderbek, and A. Kefalas, *Management Systems: Conceptual Considerations*, 4th Ed., Boston: Irwin, 1990.

[30] Ashby, R., *Introduction to Cybernetics*, New York: John Wiley & Sons, 1963.

[31] Beer, S., *The Heart of the Enterprise*, New York: John Wiley & Sons, 1979.

[32] Clemson, B., *Cybernetics: A New Management Tool*, Turnbridge Wells, Kent, Abacus Press, 1984. (UK)

[33] Drucker, P., *Management: Tasks, Responsibilities, Practices*, New York: Harper & Row, 1973.

CHAPTER 15

Electronic Voting Systems

Costas Lambrinoudakis, Emmanouil Magkos, and Vassilis Chrissikopoulos

For several years now the identification of the user requirements that an electronic voting system should satisfy has attracted the interest of both governments and research communities. The main difficulty of the requirements elicitation process seems to be the different perspective of each side: governments refer to requirements as the set of applicable laws pertaining a certain voting procedure, while researches don't go much further than simply providing a narrative description of system's nonfunctional characteristics related to security. Both sides seem to underestimate the fact that an electronic voting system is an information system with functional, as well as nonfunctional, requirements.

Functional requirements may vary from one system to the other, since they depend on the needs of the market segment that the system will serve. However, this is not the case for the vast majority of security requirements. They are similar to all e-voting systems since they aim to ensure compliance of the system with the election principles and the security and privacy issues dictated by the international legal frameworks. Security requirements are, to a large extent, fulfilled by the voting protocol adopted by the system.

The first part of this chapter includes a complete list of functional and nonfunctional requirements for an electronic voting system, taking into account the European Union legislation, the organizational details of currently applicable voting procedures, and the possibilities offered, as well as the constraints imposed by the latest technology. Following that, there is a detailed presentation of several generic and enhanced models, proposed in the cryptographic literature, for remote e-voting, as well as of a new class of cryptographic voting schemes for paper-based elections in polling stations.

15.1 Requirements for an Internet-Based E-Voting System

The decision to build an electronic voting system in order to conduct elections over public networks (i.e., the Internet) is neither an easy nor a straightforward one. The reason is that a long list of legal, societal, and technological requirements must be fulfilled [1, 2]. A further difficulty is that a vast majority of the system requirements has been produced by transforming abstract formulations (i.e., laws or principles like "preserve democracy") to a concrete set of functional and nonfunctional requirements, as clearly illustrated in Figure 15.1.

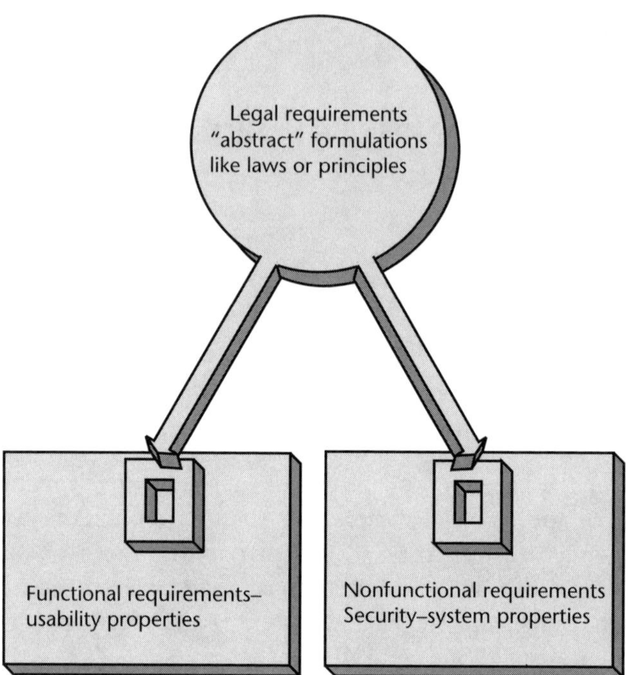

Figure 15.1 Requirements for an e-voting system.

The *functional requirements* of an e-voting system specify, in a well-structured way, the minimum set of services (tasks) that the system is expected to support, highlighting at the same time their desired sequence and all possible interdependencies. For instance, the number and type of elections processes (e.g., polls, referendums, internal elections, general elections) supported by an e-voting system are determined by its set of functional requirements. Furthermore, functional requirements are related to many of the usability properties of the system, dominating the properties and characteristics of its interaction model with the user. On the other hand, *nonfunctional requirements* are related to the underlying system structure. In principle, they are invisible to the user, and they normally have a severe impact on architectural decisions. Security requirements and several systemwide properties like flexibility, voter convenience, and efficiency are derived through the set of nonfunctional requirements.

Clearly, an Internet-based voting system is just a special case of an electronic voting system. Considering that the task of securing the voting process over a public network is significantly harder than over a private one, the security requirements for Internet voting can be assumed to be a superset of those for any other type of electronic voting system. For that specific reason, the requirement analysis that follows has been based on Internet voting.

15.1.1 Functional Requirements

In principle, functional requirements for e-voting systems may vary a lot, since each system is aiming to fulfill the specific requirements of the market segment that it is targeting. However, the most common objectives of an e-voting system are to do the following [2]:

1. Provide the entire set of required services for organizing and conducting a voting process;
2. Support, in accordance to a well-defined operational framework, all "actors" that have a need to interact with the system;
3. Support different "types" of voting processes like polls, plebiscites, interorganizational elections, and general elections;
4. Be customizable with respect to the geographical coverage of the voting process, the number of voting precincts, the number of voters, and other specific characteristics of the process, like starting date and time, number of candidates, and so on;
5. Ensure the following:
    - Only eligible persons can vote.
    - No person can vote more than once.
    - The vote is secret.
    - Each vote is counted in the final tally.
    - The voters trust that their vote is counted.

Assuming that the supported voting process is a *general election*, which, when compared to polls, internal elections, and so on, is the broadest and most complicated election process, the functionality that must be exhibited by an Internet-based system in order to meet the aforementioned objectives is listed next:

1. *Authorize actor*: This is the starting point for any interaction with the information system. It provides access to the system functions that a specific actor (organizer, user) is authorized to perform.
2. *Define election districts*: Define the districts and the corresponding number of candidates that will be elected to parliament.
3. *Define electors*: All persons over a certain age have the right, and in some countries the obligation, to participate in the election process. In countries where voting is obligatory, all eligible voters are included in the elector list, unless convicted to attainder or excluded by judicial judgment; in countries where voting is not obligatory, only the people wishing to vote are included in the elector list.
4. *Manage parties and candidates*: Notify the system about candidate parties and insert, modify, and delete a party's candidates for a specific election district.
5. *Create ballots*: Each participating party requires a ballot and a list of its representatives per election district.
6. *Provide authentication means*: Create and distribute authentication means to electors in order to allow them to identify themselves during the voting process.
7. *Cast vote*: The voter is allowed to cast her vote, provided that she has been successfully authenticated. The voter may be supplied with a receipt, confirming that she has voted.
8. *Tally votes*: Calculation of the number of votes each participating party has received, along with not valid votes. This process cannot be performed before the end of the election.

9. *Verify result integrity*: This process takes place in case a voter—or any other interested party—requests to verify that any of the aforementioned election procedures has been conducted properly.

15.1.1.1 Nonfunctional (Security) Requirements

The vast majority of security requirements are common to all e-voting systems, since they determine the required compliance of the system with the election principles (democracy) and the security and privacy issues dictated by the international legal frameworks. Security requirements are, to a large extent, fulfilled by the voting protocol adopted by the system (refer to Section 15.2). Specifically, as presented in [3], the security requirements of an Internet-based e-voting system can be identified in terms of the properties that a voting protocol must exhibit. A short description follows.

Accuracy

Accuracy, also referenced as correctness in [4], demands that the announced tally exactly matches the actual outcome of the election. This means that no one can change anyone else's vote (inalterability), all valid votes are included in the final tally (completeness), and no invalid vote is included in the final tally (soundness).

Democracy

A system is considered to be *democratic* if only eligible voters are allowed to vote (eligibility) and if each eligible voter can only cast a single vote (unreusability). An additional characteristic is that legitimate votes cannot be altered, duplicated, or removed without being detected.

Privacy

According to this requirement, no one should be able to link a voter's identity to her vote, after the latter has been cast (unlinkability). *Computational privacy* is a weak form of privacy ensuring that the relation between ballots and voters will remain secret for an extremely large period of time, assuming that computational power and techniques will continue to evolve in today's pace. *Information-theoretic privacy* is a stronger and, at the same time, harder to obtain form of privacy, ensuring that no ballot can be linked to a specific voter as long as information theory principles remain sound.

Robustness

This requirement guarantees that no reasonably sized coalition of voters or authorities (either benign or malicious) may disrupt the election. This includes *allowing abstention* of registered voters, without causing problems or allowing other entities to cast legitimate votes on their behalf, as well as *preventing misbehavior* of voters and authorities from invalidating the election outcome by claiming that some other actor of the system failed to properly execute its part. Robustness implies that security should also be provided against external threats and attacks (e.g., denial of service attacks).

Verifiability

Verifiability implies that there are mechanisms for auditing the election in order to ensure that it has been properly conducted. It can be provided in three different forms: (a) *universal or public verifiability* [5], meaning that anyone (voters, authorities, or even external auditors) can verify the election outcome after the announcement of the tally, (b) *individual verifiability with open objection to the tally* [6], which is a weaker requirement allowing every voter to verify that her vote has been properly taken into account and file a sound complaint, in case the vote has been miscounted, without revealing its contents, and (c) *individual verifiability*, which is an even weaker requirement, since it allows for individual voter verification but forces voters to reveal their ballots in order to file a complaint.

Uncoercibility

The concept of *receipt freeness*, introduced by Benaloh and Tuinstra [7], implies that no voter should be able to prove to others how he voted (even if he wants to). On the other hand, uncoercibility means that no party should be able to coerce a voter into revealing her vote. Clearly, the notion of receipt freeness is stronger than *uncoercibility*, and thus more difficult to achieve [8], especially in online (general) elections.

Fairness

This property ensures that no one can learn the outcome of the election before the announcement of the tally. Therefore, acts like influencing the decision of late voters by announcing an estimate or providing a significant but unequal advantage (being the first to know) to specific people or groups are prevented.

Verifiable Participation

This requirement, often referred as *declarability*, ensures that it is possible to find out whether a particular voter actually has participated in the election by casting a ballot or not. This requirement is necessary in cases where voter participation is compulsory by law (as in some countries, such as Australia, Belgium, and Greece) or social context (e.g., small or medium-scale elections for a distributed organization board) where abstention is considered a contemptuous behavior.

15.2 Cryptography and E-Voting Protocols

Cryptography is naturally used to secure transactions in complex systems in which the interests of the participating entities may be in conflict. Not surprisingly, cryptography is one of the most significant tools for securing online voting protocols. While in traditional elections, most ideal security goals such as *democracy*, *privacy*, *accuracy*, *fairness*, and *verifiability* are supposedly satisfied, given a well-known set of physical and administrative premises, this same task is quite difficult in online elections. For example, receipt-freeness and verifiability seem to be contradictory: when voting electronically, the very means that allow a voter to verify that her vote was counted properly (e.g., paper receipts, vote encrypting keys, user-selected randomness), may also allow a dishonest third party to force the voter to reveal her vote.

In Section 15.2.1 we highlight several well-known cryptographic models, proposed in the academic literature, for securing remote elections (e.g., Internet voting). In Section 15.2.2 we will discuss some recent cryptographic schemes for securing e-voting at the polling place.

15.2.1 Cryptographic Models for Remote E-Voting

Any scheme for remote e-voting must employ some kind of cryptographic transformation to establish secrecy and/or integrity for a set of crucial transactions. Since the first cryptographic protocols for electronic elections [9–11], several solutions have been described in academia to deal with the security problems in online voting. We consider how a variety of remote e-voting schemes in the literature apply some of the generic security requirements. We will use the *unlinkability* requirement to attempt a first categorization of the cryptographic schemes, and then we will consider how properties such as *fairness* and *robustness* are established. The notions of *verifiability* and *receipt-freeness* will be examined separately, due to their importance. Depending on the exact phase in which the *unlinkability* property is applied on the encrypted votes, the majority of e-voting schemes can be categorized as follows:

- *Unlinkability at the tallying stage*: Unlinkability is achieved at the tallying stage, by taking advantage of the algebraic properties of several public key encryption schemes. In what is known as the *homomorphic model* (e.g., [8, 12–15]), the originally submitted votes are combined, and a "sum" of encrypted votes is produced. The encrypted tally can later be decrypted by a set of election authorities. In the *mix-net model* (e.g., [16–19]), encrypted votes are shuffled (e.g., re-randomization and re-ordering of the list of votes) by a set of mix servers in a verifiable manner.
- *Unlinkability at the vote preparation stage*: the voter proves her eligibility to vote and then submits a "blinded" (i.e., randomized) version [9] of her encrypted vote to an election authority for validation. This "blinding" is later removed and the unblinded, validated vote is anonymously submitted to the election authorities. This model is also known as the *blind signature* model (e.g., [20, 21]).

A well-known technique to establish *fairness* in any critical system is to share power among several independent entities, hopefully with colluding interests. In the election paradigm, no single authority should violate the privacy of voters or the correctness of the final tally. An extra requirement would be to establish *robustness* against a (reasonably sized) set of entities who may wish to prevent the completion of the election. As a result, a majority of election authorities is usually enough to accomplish a task (e.g., decrypt the final tally). The notion of *threshold cryptography* [22], adapted for several public key encryption schemes, has been a building block for most cryptographic schemes for remote e-voting.

15.2.1.1 The Mix-Net Model

At a high level, each node in a mix-net *shuffles* and *re-randomizes* the input messages before passing them to the next node in the network. Rerandomization can be either *reencryption* [23] or *partial decryption* [9] of the input messages in order to increase the entropy of the system. For *verifiability*, each node must also construct a *zero knowledge* proof of correctness that it has accomplished its task without altering, removing, or adding false votes. Correctness can be verified among the mix servers or be universally verifiable [16]. In the universal scenario, each server would construct a *non-interactive* proof of correct transformations, to be later checked during system audit or by any external observer. In their more robust form, mix servers are mutually distrusted and privacy is ensured as long as at least one mix server refuses to divulge its random choices. There have also been proposals for removing misbehaving mix servers, without disrupting the mixing process [25].

In a generic scenario for mix net e-voting, voters sign and publish their encrypted votes on a public bulletin board: *unlinkability* is then established at the tallying level, where a set of mix servers sequentially perform mixing and prove correctness of their computations. By separating the mixing and tallying mechanisms, any interested party could perform the shuffling and provide proofs of correctness [26]. Finally, a sufficiently large set of election authorities cooperate to decrypt the individual encryptions and produce the result of the election.

Mix-nets naturally support write-in ballots and allow post-election auditing by preserving the complete list of submitted ballots. In comparison with homomorphic elections, the tallying process in mix-net–based systems is considerable slower. Late schemes have improved significantly the efficiency of mix nets (e.g., [27–31]).

15.2.1.2 The Homomorphic Model

The idea of combining the encrypted votes in an additive way to construct the final encrypted tally is due to [4, 32]. Later, a more practical scheme for large-scale elections was presented in [12], where an exponential version of the ElGamal cryptosystem was used to allow for homomorphic addition. In a generic homomorphic election, each voter signs and publishes an encryption of her vote on a bulletin board. Unlinkability is established during tallying, by "adding up" the encrypted votes without ever decrypting them. Later, a sufficiently large set of multiple authorities cooperate in decrypting the final tally, and the results are published on the bulletin board.

Baudron et al. [33] proposed an efficient variation of the model in [12] for multiple candidates and races. Damgard et al. [34] proposed a generalization of the Paillier cryptosystem to support very large tallies. An attempt to bring down the costs of such proofs of validity, especially in elections with multiple races and candidates, was made in [35].

Homomorphic elections naturally establish universal verifiability and are characterized by a very fast tallying process. Note that each vote must belong to a well-determined set of possible votes such as {+1, -1} for {"yes," "no"} votes. Moreover, each voter must provide a universally verifiable proof that her vote belongs to the predefined set of votes, or else it would be easy for a malicious voter to manipulate the final tally. Obviously, schemes based on this model seem unsuitable for running elections in which votes cannot be combined additively [36]. An example election based on the homomorphic model is shown in Figure 15.2.

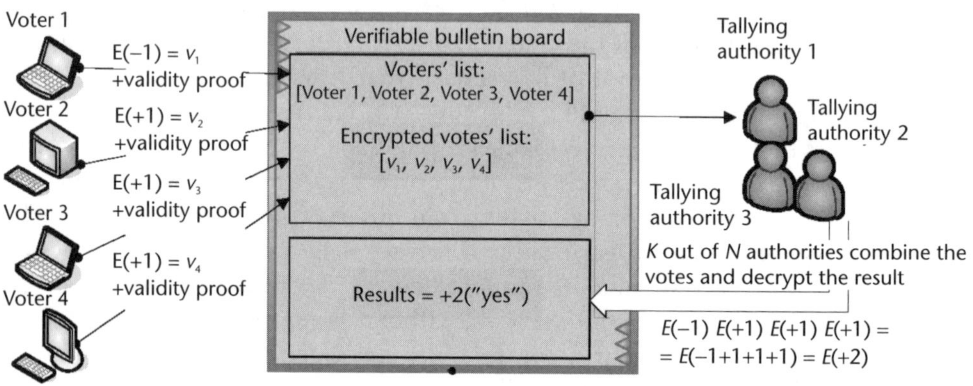

Figure 15.2 An example election based on the homomorphic model.

15.2.1.3 The "Blind Signature" Model

Election protocols of this category, introduced in [20], enable voters to get their vote validated by an election authority, while preserving the secrecy of their vote. *Blind signatures* [37] are the electronic equivalent of signing carbon paper–lined envelopes. In an online voting protocol, a voter encrypts, then blinds, the vote and presents it to an election authority who blindly signs it. Then, the voter removes the blinding factor and gets a validated and encrypted vote that cannot be correlated to the original blinded message. The voter then uses an anonymous channel to submit the validated vote to the election authorities. Later, the voter may even anonymously *object to the tally* [6], if her vote is missing.

Schemes according to this model usually result in a complex election setup phase. Due to the anonymity in the vote-casting phase, a series of known internal attacks, such as invalid vote submission by malicious election administrators, have made it difficult to establish universal verifiability. Much trust is placed on the election administrators and the anonymity network, concerning both voter privacy and tally correctness. In recent proposals (e.g., [38–41]), the power of administration is distributed among multiple authorities to augment the security of such schemes.

Observe that the random factor used in blinding as well as in the vote's encryption could also be used as a receipt in a coercion protocol. In the next section, we will discuss receipt-free elections. On the other hand, protocols within this model are simple, easily manageable, computationally efficient, and naturally support "write-in" ballots. The model is also easily adapted for elections in which the list of voters who actually voted is never published [21], which is a prerequisite against a specific class of coercion attacks (e.g., the *forced abstention attack* [17], also discussed in the next section).

15.2.1.4 Receipt-Freeness in Remote E-Voting

In every cryptographic election, where a vote is to be *encrypted* (or *shuffled*, as we have already seen, by a mix-net), with the help of a public key cryptosystem, the encryption operation needs to be *randomized*: the ciphertext will depend on both the plaintext vote and some random value. Otherwise, trivial chosen-plaintext attacks would be possible, given that usually there is a small set of possible votes in the system. Most generic models, discussed in the previous sections, use some randomness during the vote generation protocol to achieve this level of security. In a generic scheme based on blind signatures, for example, the voter randomly chooses a blind factor for her vote to be validated. Similarly, in homomorphic and mix-net generic schemes, voters are required to choose some randomness to encrypt their vote with a randomized encryption scheme. In the mix-net model, mix servers also use randomization for reencryption or partial decryption purposes. However, it has been shown that the randomness used in voting protocols could also be used to undermine the privacy of a voter. As noted in [8], if a scheme requires the voter to choose her own randomness, then this scheme cannot be receipt-free: the randomness may constitute a receipt in a *coercion* or *vote buying* protocol. The notion of *receipt-freeness* in e-voting was introduced by Benaloh [7] and independently by Niemi and Renvall [45].

Special Channels

In cryptographic research, for remote e-voting, most proposals for receipt-freeness involve some ad hoc physical assumptions and procedural constraints, such as *untappable channels* (e.g., [8, 21, 46]) or *voting booths* (e.g., [7, 15, 18, 19, 47, 48]). An untappable channel may require a physically separated and closed communication medium (e.g., a leased line inaccessible from outsiders). In [8] it was claimed that one-way untappable channels between voters and authorities constitute a minimal assumption for receipt-free elections.

Schemes in [49, 50] assume the existence of a *secondary communication channel* between the voter and election authorities: A vote buyer would have difficulties in tapping both channels (or doing that for a large population of voters). In [17], an untappable channel during the registration phase (e.g., postal mail) and an *anonymous channel* during vote-casting phase, were assumed. These assumptions are weaker than the assumption of [8], in that untappability involves an offline transaction between the voter and the authority, which may happen before the election day, thus being more practical. Furthermore, solving the *forced abstention* threat [17] would require establishing anonymity in the vote-casting phase [21, 36]. Intuitively, however, this could also undermine the public auditability of the final tally [51].

Special Proofs of Knowledge

Often, a voter needs to verify that a third party (which she does not trust) has performed a correct transformation concerning her vote (e.g., a correct re-encryption in a mix-net [8]). This "receipt" should be *privately verifiable* (i.e., not transferable to a vote buyer). The notion of *designated verifier proofs* [52] has often been used in receipt-free protocols to establish nontransferability of cryptographic assertions (e.g., [8, 13, 14]).

Any e-voting scheme in which the name of the voter who participated in the election is publicly announced is subject to a forced abstention attack [17], in which the coercer may simply demand a voter to abstain from voting. Furthermore, in elections in which write-in ballots are allowed, the decrypted ballot itself could also constitute a receipt for the vote-buyer. Furthermore, most remote e-voting schemes fail to provide protection, in a practical and affordable way, against an *identity theft* attack (also referred to as *simulation* attack in [17]), in which the coercer (or vote buyer) may collect part or all of the voter's secrets and credentials, and even cast the vote on the voter's behalf. Another difficulty in establishing receipt-freeness in remote e-voting is the *secure platform problem* [53]. In remote e-voting, the PC becomes the voting machine and loses the inherent physical and logical security of precinct systems. An adversary may have access to the client's computed and/or communicated data, either directly (e.g., physical presence, man in the middle attacks) or indirectly (e.g., Trojans, backdoors, spyware). In this way, the attacker may actually control all electronic communication channels between the voter and the election authorities. In another attack scenario not directly related to receipt-freeness, the client may become a *zombie* in a *botnet* and be used in distributed denial of service attacks against other voters or against the election servers.

Tamper-Resistant Hardware

The work in [54] transformed the assumption of an untappable channel between the voter and election authorities into the weaker assumption of an untappable channel between the voter and a tamper-resistant token. Later, Lee and Kim [13] employed *designated verifier* and *divertible zero-knowledge* proofs to correct a security flaw in [54]. Admittedly, voting with a personal election smartcard would be a costly alternative in the large-scale setting: according to a scenario, all eligible voters would be given, during registration, a voting card (and possibly, some necessary reading peripheral). Furthermore, unless additional access control mechanisms are imposed (e.g., fingerprint identification), the mere use of a smartcard cannot protect against identity theft attacks, in which a vote buyer may be in possession of all the credentials and secrets of a vote seller. On the other hand, such devices are expected to be applied to a wide range of applications in the near future, when everybody is expected to store their signing and cryptographic keys in their ID cards. It remains to be answered whether e-voting could become an extra application without any extra cost.

Admittedly, it seems hard to implement receipt-freeness in remote e-voting without any untappability assumptions. Intuitively, such a scheme would probably tweak the privacy/accuracy tradeoff against accuracy or would be too complex to implement for large-scale elections. This is the main reason why recent cryptographic voting schemes require voters to be physically isolated in a *voting booth* during vote casting. As we will see in Section 15.2.2, the voting booth may indeed guarantee privacy and establish verifiability in a nontransferable way.

15.2.1.5 Implementations of the Generic Models

No cryptographic protocol for remote e-voting was ever implemented in a large scale system. On the other hand, several protocols have actually been implemented in small-scale environments. The blind signature model has been implemented in several projects, mainly due to its simplicity and flexibility. The first implementations were the Sensus system [56] and the EVOX [57] system. The EVOX system was improved by EVOX Multiple Administrators [39], which in turn was succeeded by the REVS system [40] in an effort to eliminate single entities from disrupting the election. Improved implementations of the REVS system [41] increase the robustness of REVS. This is achieved with a scheme that prevents specific denial of service attacks against protocol participants from colluding malicious servers. The REVS system is fully implemented in Java and is publicly available [58].

A series of publications (e.g., [19, 28]) marketed by VoteHere.net have led to the implementation of several cryptographic assurances in a real system for polling place e-voting under the mix-net model.

15.2.2 Cryptographic Protocols for Polling-Place E-Voting

A fact in the current polling-place (DRE-based) e-voting infrastructure is that the integrity of an election is more or less dependent on the correctness of the vendor's software [53, 59]. Similarly, most cryptographic schemes for remote e-voting, as the ones described in the previous sections, assume a computationally capable voter, then consider verifiability only at the tallying stage, and more or less ignore the vote generation phase.

Recent proposals [15, 18, 19, 47, 48] have established the notion of a *voter-verifiable* election. These are actually hybrid paper/electronic systems for *polling place* voting. However, instead of verifying the voting equipment, an emphasis is given on the voter's verifying the election results in an *end-to-end* way [60]. In voter verifiable schemes, verifiability comes in three flavors: First, the voter needs to have confidence that her vote is *cast as intended* (also referred to as *casting assurance* [15]). In this context, it is important that the human voter will get casting assurance *without* or with *minimal* external help [15, 19]. Verification of correctness for the vote-generation stage is not always an all-or-nothing fact. *Cut-and-choose* techniques have been proposed by many recent schemes to establish correctness that can be verified by election officials [47] or by human voters [15, 18, 19, 48] before leaving the polling station. If enough voters perform the audit, then fraud and/or errors will be detected (and even corrected) with a non-negligible possibility. Second, the voter needs to have confidence that her vote was *tallied as cast*. During the interaction with the system, a receipt is printed that will permit the voter to verify that the final tally contains her vote. Third, for *public verifiability*, the voter must be sure that the final tally is not tampered with by anyone. A vital issue in all schemes discussed in this section is that whatever evidence about the voter taken from the system will not be transferable to a vote buyer or a coercer. Toward the direction of designing secure systems with relatively low complexity, the use of cryptographic primitives in conjunction with the *voting booth* assumption seems very promising.

Chaum [18] was the first to propose a cryptographic scheme for paper ballots. In [18], the voter is presented with two ballot halves, whose superposition yields the plaintext vote and establishes confidence that the vote represents the correct voter's choice. The voter then destroys one half and keeps the other as a receipt. This scheme was recently evolved into the *punchscan* system [62]. A variant of the Chaum's original scheme was also proposed in [47] under the name of *Prêt-a-Voter*. All these schemes use verifiable mix-nets to establish unlinkability at the tallying stage. Another scheme, the *scratch and vote* system [15], implements the homomorphic model in paper-based voting. Each ballot contains the candidate names on its left half in random order. On the right half, there are the optical-scan bubbles, a two-dimensional barcode that contains the probabilistic encryptions for each candidate choice, and a scratch surface that hides the random values for each encryption. Each voter can select a second ballot for audit purposes, and casting assurance is established with a cut-and-choose protocol: The voter selects one of the two ballots for auditing, scratches it off, and verifies that the ballot was formed correctly. Then, she goes into the booth with the second ballot, fills her choices, and discards the left half of the ballot into a receptacle. Out of the booth, an election official verifies that the scratch surface is intact and publicly discards it. The voter casts (what remained of) the ballot and keeps a copy as a receipt for later verification. All encrypted votes are posted on a bulletin board, and the final tally can be constructed by "adding up" the votes in a publicly verifiable way.

In the *threeballot* voting approach for polling place elections, recently proposed by Rivest [63], end-to-end verifiability is achieved *without* using any cryptography. Each voter in [63] gets a multiballot with three identical ballots (except for the unique ID number on each ballot). The voter fills in bubbles in rows corresponding to candidates, in a way that no two ballots will ever reveal the voter's choices. The voter chooses at random one ballot to be kept as a receipt and casts (optically scans) all three ballots. The protocol has been shown to be uncoercible but not receipt-free [63].

15.3 Conclusions

Historically, in physical elections, most verifiability checks were delegated to election officials at the precinct, during the voting and counting stages. Accordingly, in electronic communication protocols, when cryptography cannot guarantee by itself all properties of a secure electronic transaction, certain reliance must by placed on the behavior of a set of third parties. It is not clear whether it is realistic to expect that there can be several mutually distrustful, independent parties who can be trusted on crucial security properties in remote e-voting schemes [64, 65]. On the other hand, in polling place elections, such parties could be representatives from opposing political parties, or even helper organizations [15]. The goal of a cryptographic protocol is to establish security by trusting a third party as little as possible and on as few security properties as possible. Trust on third parties will never be eliminated, but part of it may be transferred on certain properties of mathematics and cryptography.

Another critical factor is security versus complexity. A secure but complex system is unlikely to be adopted for large-scale voting [66]. Schemes for remote e-voting are by default complex protocols: In the absence of a voting booth, high security must be provided mainly by cryptographic means. Any system built on a highly secure cryptographic remote e-voting protocol would probably suffer considerable usability issues. Recent schemes for paper-based voting (Section 15.2.2) take advantage of the voting booth primitive to protect voter privacy and add minimal cryptographic primitives to achieve end-to-end verifiability while maintaining privacy for the encrypted votes. However, all these schemes impose a few additional requirements whose purpose may not be clear to voters [26].

Until today, no cryptographic scheme for remote e-voting or for polling place e-voting has been implemented in a real election of significant scale. Transition to remote Internet voting cannot be a one-off step. Toward this transition, it seems natural to take the intermediate step of performing secure e-voting *at the precinct*: Recent cryptographic schemes for paper-based voting seem feasible for large-scale elections and easily implementable in the next future. An open question today is whether such advances will increase or decrease public confidence in the voting process.

References

[1] Mitrou, L., D. Gritzalis, and S. Katsikas, "Revisiting Legal and Regulatory Requirements for Secure E-Voting," *Proceedings, IFIP TC11 17th International Conference on Information Security (SEC2002)*, Cairo, Egypt, 2002, pp. 469–480.

[2] Ikonomopoulos, S., et al., "Functional Requirements for a Secure Electronic Voting System," *Proceedings, IFIP TC11 17th International Conference on Information Security (SEC2002)*, Cairo, Egypt, 2002, pp. 507-520.

[3] Lambrinoudakis, C., et al, "Secure e-Voting: The Current Landscape," *Secure Electronic Voting: Trends and Perspectives, Capabilities and Limitations*, Kluwer Academic Publishers, 2002.

[4] Cohen, J. D., and M. J. Fischer, "A Robust and Verifiable Cryptographically Secure Election Scheme," *Proceedings, 26th Annual Symposium on Foundations of Computer Science*, IEEE, 1985, pp. 372–382.

[5] Schoenmakers, B., "A Simple Publicly Verifiable Secret Sharing Scheme and Its Application to Electronic Voting," in *Advances in Cryptology—CRYPTO '99*, LNCS, Vol. 1666, Springer-Verlag, 1999, pp. 148–164.

[6] Riera, A., J. Borell, and J. Rifà, "An Uncoercible Verifiable Electronic Voting Protocol," *Proceedings, IFIP-SEC'98 Conference*, Vienna-Budapest, 1998, pp. 206–215.

[7] Benaloh, J., and D. Tuinstra, "Receipt-Free Secret-Ballot Elections," *Proceedings, 26th Annual ACM Symposium on Theory of Computing*, ACM, 1994, pp. 544–553.

[8] Hirt, M., and K. Sako, "Efficient Receipt-Free Voting Based on Homomorphic Encryption," *Proceedings, Eurocrypt 2000*, LNCS, Vol. 1807, Springer, 2000, pp. 539–556.

[9] Chaum, D., "Untraceable Electronic Mail, Return Addresses, and Digital Pseudonyms," *Communications of the ACM*, Vol. 24, No. 2, 1981, pp. 84–88.

[10] Demillo, R., N. Lynch, and M. Merritt, "Cryptographic Protocols," *Proceedings, 14th Annual ACM Symposium on Theory of Computing,* ACM, 1982, pp. 383–400.

[11] Benaloh, J., *Verifiable Secret Ballot Elections*, PhD thesis, Yale University, 1987.

[12] Cramer, R., R. Gennaro, and B. Schoenmakers, "A Secure and Optimally Efficient Multi-Authority Election Scheme," *European Transactions on Telecommunications*, Vol. 8, No. 5, 1997, pp. 481–490.

[13] Lee, B., and K. Kim, "Receipt-Free Electronic Voting Scheme with a Tamper-Resistant Randomizer," *Proceedings, ICISC'02,* LNCS, Vol. 2587, Springer-Verlag, 2002, pp. 389–406.

[14] Acquisti, A., "Receipt-Free Homomorphic Elections and Write-in Ballots," Technical Report 2004/105, CMU-ISRI-04-116, Carnegie Mellon University, 2004.

[15] Adida, B., and R. L. Rivest, "Scratch & Vote—Self-Contained Paper-Based Cryptographic Voting," *Proceedings, Workshop on Privacy in the Electronic Society—WPES '06*, 2006, to be published.

[16] Sako, K., and J. Kilian, "Receipt-Free Mix-Type Voting Scheme—A Practical Solution to the Implementation of a Voting Booth," *Proceedings, EUROCRYPT 95*, LNCS, Vol. 921, Springer, 1995, pp. 393–403.

[17] Juels, A., D. Catalano, and M. Jakobsson, "Coercion-Resistant Electronic Elections," *Cryptology ePrint Archive: Report 2002/165*, http://eprint.iacr.org/.

[18] Chaum, D., "Secret-Ballot Receipts: True Voter-Verifiable Elections," *IEEE Security and Privacy*, Vol. 2, No. 1, 2004, pp. 38–47.

[19] Neff, A., "Practical High Certainty Intent Verification for Encrypted Votes," 2004, http://votehere.com/vhti/documentation/vsv-2.0.3638.pdf.

[20] Fujioka, A., T. Okamoto, and K. Ohta, "A Practical Secret Voting Scheme for Large Scale Elections," *Proceedings, AUSCRYPT '92*, LNCS, Vol. 718, Springer-Verlag, 1993, pp. 244–251.

[21] Okamoto, T., "Receipt-Free Electronic Voting Schemes for Large Scale Elections," *Proceedings, 5th Security Protocols Workshop '97*, LNCS, Vol. 1163, Springer-Verlag, 1997, pp. 125–132.

[22] Desmedt, Y., "Threshold Cryptography," *European Transactions on Telecommunications*, Vol. 5, No. 4, 1994, pp. 449–457.

[23] Park, C., K. Itoh, and K. Kurosawa, "Efficient Anonymous Channel and All/Nothing Election Scheme," *Proceedings, EuroCrypt '94*, LNCS, Vol. 765, Springer, 1994, pp. 248–259.

[24] Goldreich, O., S. Micali, and A. Widgerson, "Proofs That Yield Nothing but Their Validity, or All Languages in NP Have Zero-Knowledge Proof Systems," *Journal.of the ACM*, Vol. 38, 1991, pp. 691–729.

[25] Ogata, W., et al., "Fault Tolerant Anonymous Channel," *Proceedings, 1st International Conference on Information and Communications Security—ICICS*, LNCS, Vol. 1334, Springer-Verlag, 1997, pp. 440–444.

[26] Benaloh, J., "Simple Verifiable Elections," *Proceedings, Workshop on Electronic Voting Technology*, Vancouver, BC, Canada, USENIX, August 2006.

[27] Abe, M., "Universally Verifiable Mix-Net with Verification Work Independent of the Number of Mix-Centers," *Proceedings of the Advances in Cryptology—EUROCRYPT 98*, LNCS, Vol. 1403, Springer-Verlag, 1998, pp. 437–447.

[28] Neff, A., "A Verifiable Secret Shuffle and Its Application to E-Voting," *Proceedings, 8th ACM Conference on Computer and Communications Security*, 2001.

[29] Groth, J., "A Verifiable Secret Shuffle of Homomorphic Encryptions," *Proceedings, Public Key Cryptography 2003*, LNCS, Vol. 2567, Springer-Verlag, 2003, pp. 145–160.

[30] Jakobsson, M., A. Juels, and R. L. Rivest, "Making Mix Nets Robust for Electronic Voting by Randomized Partial Checking," *Proceedings, USENIX Security Symposium*, 2002, pp. 339–353.

[31] Furukawa, J., "Efficient and Verifiable Shuffling and Shuffle-Decryption," *IEICE Trans. Fundamentals E88-A*, Vol. 1, 2005, pp. 172–189.

[32] Benaloh, J., and M. Yung, "Distributing the Power of Government to Enhance the Power of Voters," *Proceedings, Symposium on Principles of Distributed Computing*, ACM Press, 1986, pp. 52–62.

[33] Baudron, P., et al., "Practical Multi-Candidate Election System," *Proceedings, 20th ACM Symposium on Principles of Distributed Computing*, ACM Press, 2001, pp. 274–283.

[34] Damgard, I., M. Jurik, and J. Nielsen, "A Generalization of Paillier's Public-Key System with Applications to Electronic Voting," Manuscript, 2003, www.daimi.au.dk/~ivan/GenPaillier-finaljour.ps.

[35] Groth, J., "Noninteractive Zero-Knowledge Arguments for Voting," *Proceedings, ACNS 2005*, LNCS, Vol. 3531, 2005, pp. 467–482.

[36] Smith, W. D., "New Cryptographic Voting Scheme with Best-Known Theoretical Properties," *Proceedings, Workshop on Frontiers in Electronic Elections (FEE 2005)*, Milan, Italy, September 2005.

[37] Chaum, D., "Blind Signatures for Untraceable Payments," *Proceedings, Crypto '82*, Plenum Press, 1982, pp. 199–203.

[38] Ohkubo, M., et al., "An Improvement on a Practical Secret Voting Scheme," *Proceedings of the Information Security Conference—IS'99*, LNCS, Vol. 1729, Springer-Verlag, 1999, pp. 225–234.

[39] Durette, B. W., "Multiple Administrators for Electronic Voting," Bachelor's Thesis, Massachusetts Institute of Technology, May 1999.

[40] Joaquim, R., A. Zuquette, and P. Ferreira, "REVS—A Robust Electronic Voting System," *Proceedings, IADIS'03 International Conference of e-Society*, 2003, pp. 95–103.

[41] Lebre, R., et al., "Internet Voting: Improving Resistance to Malicious Servers in REVS," *Proceedings, International Conference on Applied Computing* (IADIS'2004), 2004.

[42] ElGamal, T., "A Public Key Cryptosystem and a Signature Scheme Based on Discrete Logarithms," *IEEE Trans. on Information Theory*, Vol. 30, No. 4, pp. 469–472.

[43] Paillier, P., "Public Key Cryptosystems Based on Discrete Logarithms Residues," in *Advances in Cryptology—EuroCrypt '99*, LNCS, Vol. 1592, Springer-Verlag, 1999, pp. 221–236.

[44] Canetti, R., et al., "Deniable Encryption," in *Advances in Cryptology—Crypto '97*, LNCS, Vol. 1294, Springer-Verlag, 1997, pp. 90–104.

[45] Niemi, V., and A. Renvall, "How to Prevent Buying of Votes in Computer Elections," *Proceedings, ASIACRYPT '94*, LNCS, Vol. 917, Springer-Verlag, 1994, pp. 164–170.

[46] Aditya, R., et al., "An Efficient Mixnet-Based Voting Scheme Providing Receipt-Freeness," *Proceedings, 1st Trustbus 2004*, LNCS, Vol. 3184, Springer-Verlag, 2004, pp. 152–161.

[47] Chaum, D., P. Y. A. Ryan, and S. Schneider, "A Practical Voter-Verifiable Election Scheme," *Proceedings, ESORICS '05*, LNCS, Vol. 3679, Springer-Verlag, 2005, pp. 118–139.

[48] Ryan, P. Y. A., and S. A. Schneider, "Prêt a Voter with Re-Encryption Mixes," *Proceedings, ESORICS 2006*, LNCS, Vol. 4189, Springer, 2006, pp. 313–326.

[49] Damgard, I., and M. J. Jurik, "Client/Server Tradeoffs for Online Elections," *Proceedings, PKC '02*, LNCS, Vol. 2274, 2002, pp. 125–140.

[50] Groth, J., and G. Salomonsen, "Strong Privacy Protection in Electronic Voting," BRICS Report Series—RS-04-13-2004, BRICS, 2004.

[51] Kiayias, A., and M. Yung, "The Vector-Ballot E-Voting Approach," *FC 2004*, LNCS, Vol. 3110, Springer-Verlag, pp. 72–89, 2004.

[52] Jakobsson, M., K. Sako, and R. Impagliazzo, "Designated Verifier Proofs and Their Applications," *Advances in Cryptology—Eurocrypt '96*, LNCS, Vol. 1070, Springer-Verlag, 1996, pp. 143–154.

[53] Rubin, A., "Security Considerations for Remote Electronic Voting over the Internet," Technical Report, AT&T Labs, 2002, http://avirubin.com/evoting.security.html.

[54] Magkos, E., M. Burmester, and V. Chrissikopoulos, "Receipt-Freeness in Large-Scale Elections Without Untappable Channels," *Proceedings, 1st IFIP Conference on E-Commerce/E-business/E-Government*, Kluwer, 2001, pp. 683–693.

[55] Kelsey, J., B. Schneier, and D. Wagner, "Protocol Interactions and The Chosen Protocol Attack," *Proceedings, Security Protocols International Workshop*, Springer LNCS, Vol. 1361, 1997, pp. 91–104.

[56] Cranor, L., and R. Cytron, "Sensus: A Security-Conscious Electronic Polling System for the Internet," *Proceedings, Hawaii International Conference on System Sciences*, Wailea, Hawaii, 1997.

[57] Herschberg, M., "Secure Electronic Voting Using the World Wide Web," Master's Thesis, MIT, June 1997, http://theory.lcs.mit.edu/~cis/theses/ herschberg-masters.pdf.

[58] REVS—Robust Electronic Voting System, http://www.gsd.inesc-id.pt/~revs.

[59] Shamos, M., *Paper v. Electronic Voting Records—An Assessment,* Mimeo, Carnegie Mellon University, 2004, http://euro.ecom.cmu.edu/people/faculty/mshamos/paper.htm.

[60] Rivest, R. L., "Remarks on the Technologies of Electronic Voting, Harvard University's Kennedy School of Government Digital Voting Symposium," http://theory.lcs.mit.edu/ ~rivest/2004-06-01%20Harvard%20KSG% 20Symposium%20Evoting%20remarks.txt.

[61] Naor, M., and A. Shamir, "Visual Cryptography," in *Advances in Cryptology: EUROCRYPT '94*, LNCS, Vol. 950, Springer, 1995, pp. 1–12.

[62] Fisher, K., R. Carback, and A. Sherman, "Punchscan: Introduction and System Definition of a High-Integrity Election System," *Proceedings, IAVoSS Workshop On Trustworthy Elections (WOTE'06)*, Cambridge UK, June 2006.

[63] Rivest, R. L., "The ThreeBallot Voting System," 2006, http://theory.lcs.mit.edu/rivest/ Rivest-TheThreeBallotVotingSystem.pdf.

[64] Karlof, C., N. Sastry, and D. Wagner, "Cryptographic Voting Protocols: A Systems Perspective," *Proceedings, USENIX Security Symposium*, pp. 33–50, 2005.

[65] Kubiak, P., M. Kutylowski, and F. Zagórski, "Kleptographic Attacks on a Cascade of Mix Servers," *Proceedings, ASIACCS'07*, Singapore, March 20–22, 2007, to be published.

[66] Hoffman, L. J., K. L. Jenkins, and J. Blum, "Trust Beyond Security: An Expanded Trust Model," *Communications of the ACM*, Vol. 49, No. 7, 2006, pp. 94–101.

CHAPTER 16

On Mobile Wiki Systems Security

Georgios Kambourakis

A wiki, also known with the term *wiki wiki*, is a software component that allows multiple authors to edit web page content. Ward Cunningham originally described a wiki as "the simplest online database that could possibly work" [1]. One of the most well-known wikis is Wikipedia (www.wikipedia.org). A wiki enables documents to be processed collaboratively (i.e., asynchronously or synchronously by multiple users using standard web browsers). A single page in a wiki is referred to as a *wiki page*, while the entire range of pages, which are typically densely interconnected via hyperlinks, is *the wiki* itself. To put it simply, a wiki is essentially a database for creating, browsing, altering, and searching information. Wiki popularity mainly stems from its easy access, at least from anywhere an Internet or intranet connection exists, and simple editing capabilities.

Naturally, the aforementioned characteristics make wikis very popular for collaboration projects, either open-source or research-oriented [2–4]. Wikis encourage contribution and spontaneity in a way that most other tools do not; thus, they are generally considered to be used in the context of e-learning to facilitate a plethora of collaboration activities [5–9]. Wikis are also easy to install, customize, and maintain. Generally, they have small hardware requirements and run on Linux-based systems. These properties make them an attractive solution for organizations with limited funds. This simplicity has also led to many extensions transforming many wikis into content management systems (CMSs) and groupware.

The key characteristic of wiki technology is the straightforwardness in which pages can be created, modified, and maintained. In other words, the true spirit of the "wiki way" is to allow for the online collaboration of documents for visitors or contributors to be able to create their own pages or to edit existing pages. This of course works well for open source or public projects, but it has several problems with closed ones. For example, consider an industry organization that is currently working on a new project. A good collaboration practice would be the employment of a wiki among the developers so that they could add and exchange their contributions and comments. But should all employees have the same status of access and administration privileges?

As a result, several reviewers of open-source wiki systems have stressed that these systems could be easily tampered with and in some cases even vandalized. That is, allowing anyone to edit wiki content does not ensure by any means that all contributors have virtuous intentions. On the other hand, as explained in the following, wiki supporters argue that the community of legitimate users is able to detect malicious or offensive content and correct it (e.g., due to automated backups). Lars

Aronsson, a data systems specialist, summarizes this debate as follows: "Most people, when they first learn about the wiki concept, assume that a website that can be edited by anybody would soon be rendered useless by destructive in-put. It sounds like offering free spray cans next to a grey concrete wall. The only likely outcome would be ugly graffiti and simple tagging, and many artistic efforts would not be long lived. Still, it seems to work very well" [10].

Usually, there is no review phase before modifications made by a wiki user are verified and accepted permanently. Many wiki webs are even open to everyone without requiring from potential users to previously register any user account. In some cases however, session login is mandated to acquire a wiki-signature cookie for auto-signing edits. At the perils, in the majority of wikis, editing can be made in real time and become visible almost immediately online. Thus, the absence of any access policy can lead to abuse of the system. In contrast, private wiki servers require user authentication to edit, sometimes even to read, pages.

Apart from small scale wiki content abuse, known sometimes as *trolling*, vandalism can be a major problem. Especially, in larger wiki sites, vandalism may not be detected even after a considerable period of time. Currently, most wikis deal with vandalism by adopting a soft security [11] approach. In practice, this means that they cope with the particular problem by making damage easy to roll back rather than attempting to proactively prevent damage. Often, larger wikis utilize more advanced mechanisms. These include *bots* which automatically identify and revert vandalism and JavaScript add-ons that track how many characters have been added in each edit. Using these countermeasures in parallel, vandalism attempts can be greatly confined; the characters added or deleted are in a number that bots do not identify, and users are not annoyed much.

The greater the openness of the wiki, the larger the amount of vandalism or abuse it may suffer. Some wikis, as in the case of Wikipedia, allows nonregistered users (known only by their IP addresses) to edit content, while others do not. Once a user shows malicious behavior, his IP becomes blacklisted. However, this option is rather simplistic; the aggressor is able to use another machine with a different IP address. Often, IP identified–only users afford a restricted set of editing rights, while normal registered ones enjoy some extra editing functionality. However, note that on most wikis, becoming a registered user is a very simple and quick task. Still, newly registered users may be held up for some time before they can actually use new editing functions. For instance, in the English Wikipedia, registered users must wait for three days after creating an account in order to gain access to the new tools. Similar to that, in the Portuguese Wikipedia users must perform at least 15 constructive edits before authorization to use the added tools become active. By this scheme the wiki engine attests users' trustworthiness and usefulness on the system. Beyond doubt, closed or more restrictive wikis are more secure and reliable but are not so popular so far, while open wikis have penetrated the market a great deal but result in being an easy target for various aggressors. Also note that the majority of existing wikis can be configured and used as public or as private, corporate/enterprise, education intranets [12].

In our opinion the ideal wiki should be accessible on an anywhere, anytime, basis and optionally secure to its legitimate users. The former attribute or requirement makes wiki accessible from virtually any mobile device with a web browser, while the latter ensures that when (optionally) extra security is needed, the wiki will be able to support it. In this chapter we shall investigate both aforesaid requirements discussing ways that can be realizable. Furthermore, a novel wiki multiplatform prototype implementation will be presented, and its major components will be analyzed.

16.1 Blending Wiki and Mobile Technology

Nowadays, mobile devices such as PDAs, handhelds PCs, and cell phones have dominated the market, becoming very popular. Nevertheless, until now, very few works try to intertwine wiki webs with security [13–15]. Moreover, to the best of our knowledge, none of them explicitly focus on mobile wikis (i.e., wikis that can be accessed from low-end mobile devices). As we already mentioned, the main aim of a wiki should be the facilitation of the user for accessing and altering information—preferably, to produce an anytime, anywhere wiki experience. Hence, as with all m-learning services, accessing a wiki through mobile devices, is becoming a very challenging issue to explore. Nevertheless, there are several obstacles that must be surmounted before wiki webs can be really functional on mobile devices.

At first, the vast majority of wikis today have been designed for access from desktop systems. They usually utilize standards such as HTML and technologies such as JavaScript for enabling users to access and alter their pages. But, while the aforesaid technologies and standards have long proved their value on desktop systems, the majority of the mobile devices do not fully support them. Specifically, certain incompatibilities exist (e.g., a wiki site that is based on HTML probably will not be displayed at all on mobile devices that incorporate browsers based on wireless markup language). Even worse, most wikis have evolved from simple script-based implementations written in Perl [16], thus making them difficult to extend. Thus, the first challenge is the implementation of a truly portable wiki.

On the other hand, the possibility of equipping a mobile wiki with robust security services is neither quite easy nor straightforward. When speaking of closed wikis in which security matters, the protection of the wireless link is of top priority. Generally, roaming handheld devices that transfer data over public networks are vulnerable to a seemingly limitless number of security compromises. This is invigorated from (a) the uncontrolled or semiuncontrolled wireless terrain (i.e., the wireless link itself), (b) the diversity of wireless access technologies (i.e., universal mobile telecommunications system, IEEE 802.11, 802.16), and the inherent lower layer security vulnerabilities or weaknesses such as wired equivalent privacy (WEP) and standards they rely on, and (c) the existence of various service providers with different security policies and/or access technologies.

Besides that, one has to consider the fact that most mobile devices incorporate very limited resources (i.e., processing power, memory, small screen, battery reserves to mention just a few). Also, mobile applications have limitations of data that can be kept in memory. It is obvious that if security is desirable then it is certain that we will have to execute some sort of cryptographic operations that may be quite demanding in computational power. Thus, the second challenge here is the implementation of a mobile wiki system that will provide security with as few requirements in computational power as possible.

In a nutshell, our contribution is twofold: as far as the portability characteristic is concerned, we have tried to make possible that the only true requirement from the mobile device to run a wiki will be to have some sort of access on the network and perhaps a Java virtual machine (JVM) installed (as today is the case with virtually every mobile device). Second, our goal is to minimize cryptographic operations and the associated protocol demands without sacrificing security much at the same time.

16.2 Background Information

As already mentioned in the introduction, in wikis that control access to web pages, a user registers a username and password with the wiki. After that, whenever the user attempts to log in, the wiki retrieves user credentials and uses them to decide whether to permit the user to log in. In practice, there are several authentication methods that can be used:

1. *No authentication*: Anyone that can connect to the wiki is able to edit its pages.
2. *A wiki-supplied HTML form*: The wiki software can then use the input provided to perform multiple authentication methods, including, but not limited to: lightweight directory access protocol (LDAP), Windows NT challenge/response authentication protocol (NTLM), and custom database.
3. *Web server authentication* (in which the web server passes the authentication information to the wiki software): The web server software can then use the input provided to perform multiple authentication methods as before.

Many private, corporate, or intranet wikis utilize access control lists (ACLs) for granting or revoking read and write access to wiki pages (see [12]). An ACL for a wiki page can be made up of individual users, or groups of users, or a combination of both. Groups are created with different access rights. Note that access control is only available on a subscription basis. When a particular user tries to view or edit a particular page, the wiki application determines whether the page is restricted, and, if it is, whether the user is allowed to perform the requested action on the page. In large wikis, which may accept thousands of simultaneous user queries in a given moment, this ACL approach must be carefully considered in terms of both administrative costs and wiki performance.

Server access is usually controlled by means external to the wiki software itself (e.g., the access restriction of IP addresses by placing the wiki behind a firewall). Numerous software applications can run a wiki in a variety of programming languages including Perl (CGI), active server pages (ASP) check, and PHP. Applications may or may not be integrated with a database that stores all wiki topics. The databases employed can also vary. Commonly, a PHP wiki is integrated with a MySQL database. Some wikis use not a database but a flat file system based on standard or proprietary text files. The reader can also consult the wiki comparison page, which is available on [12].

As depicted in Figure 16.1, today most wikis are based on a centralized, three-tier architecture, which includes presentation, logic or application, and data tiers. Actually, this architecture is well known in the context of web development. For example, Wikipedia operates according to the following model: (a) it accepts wiki requests generated from the presentation tier in the clients, (b) requests are processed by a set of PHP scripts in the logic tier, which in turn (c) issues the appropriate SQL request toward the data tier. Finally, the corresponding reply may return to the client through the logic layer (e.g., some HTML pages are generated by the server and transferred to the client's browser). As said before, wikis that follow this particular centralized approach employ the SSL/TLS protocol [17, 18] to enforce confidentiality.

On the downside, the disadvantages of such a server-client approach are significant. First, the data transferred between the client and the server is in HTML format. As known, an HTML file consists of two parts: data and markup information. The latter determines the exact way the actual data must be displayed in the client's browser. While HTML is a de facto standard for desktop machine browsers, the overwhelming majority of mobile devices today support only a subset of it or other standards solely. As a result, some of the information (markup) being transferred to the mobile device is useless, while at the same time the data is not displayed properly. Second, utilizing SSL means that all HTML files being transferred are scrambled without exceptions. In practice, however, only some sensitive or private portions of the actual data must be cryptographically protected—not all. This unfortunately means that the mobile device side, where processing power matters, must decrypt/encrypt more information (markup, nonsensitive data) than it actually needs to.

Another wiki architectural approach based on peer-to-peer (P2P) philosophy has been lately emerged [19–21]. A generic representation of a P2P wiki is depicted in Figure 16.2. This decentralized option tries to mitigate the problem of a single or small number of wiki servers storing and administering huge numbers of data. For example, there are currently close to 1.8 million articles in the English Wikipedia alone, and around 2,000 new articles emerge every day [22]. Obviously, this means high administration costs, diminished performance, single point of failure, and so on. As it is well known, a P2P computer network relies primarily on the computing

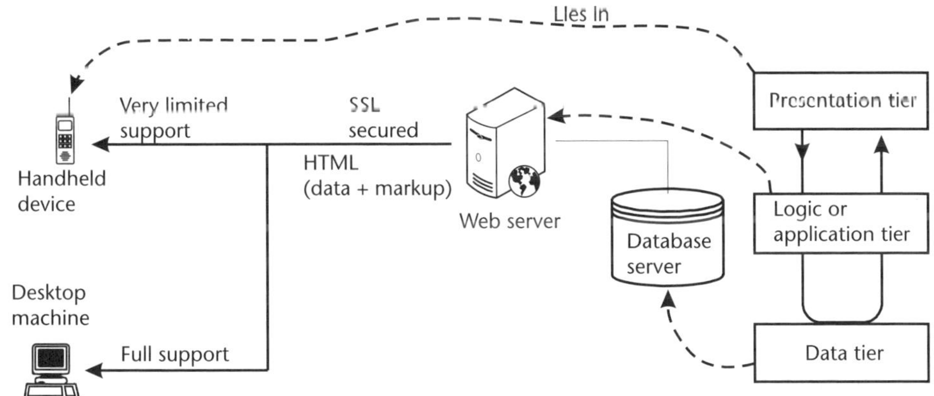

Figure 16.1 Wiki centralized approach (used by Wikipedia).

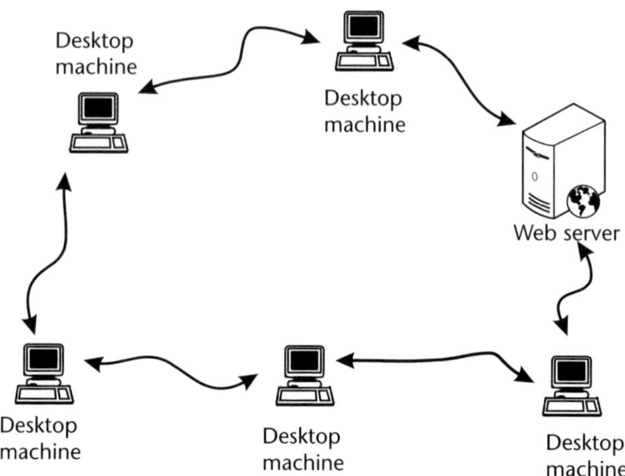

Figure 16.2 Abstract view of a P2P wiki architecture.

power and bandwidth of the nodes connected to the network. In such a network all clients provide resources, including bandwidth, computing power, and storage space. Thus, the more nodes that are connected, the greater the total capacity of the system becomes.

P2P wiki is a serverless system that lets wiki sites to be shared between participants. It is based on a P2P version control system, which is responsible for sharing, transmitting the updates, and storing the history of wiki pages. Rather than using an Internet server, a P2P wiki site (or at least a portion of it) is stored directly on each user's computer as a collection of wiki files. In addition, a P2P may enable wiki machines to actively interchange wiki pages. This happens either to increase reliability by creating redundant copies or for better performance by moving resources closer to the location where they are being frequently used. The first P2P wiki system was designed and implemented by Reliable Software [23]; more information can be found in [20].

Today, P2P wiki technology is still in its infancy. Although a secure implementation of a P2P wiki does not exist, it is obvious that the SSL option does not fit well. More specifically, SSL provides hop-by-hop, not end-to-end, fashioned security. This means that all the intermediate peers must be trusted, as they have access to the data being transferred.

16.3 The Proposed Solution

In this section we shall analyze our solution toward a secure multiplatform mobile wiki. Regarding security services, our goal is to support them on demand. Therefore, our model can support the following three models according to users' or organizations' specific needs: (a) classic wiki transactions, in which anyone can read and write a page, (b) traditional web transactions, in which everybody can read a page but only a group of them can edit it, and (c) closed project transactions, in which only project contributors can read or write pages.

16.3.1 General Issues

The architecture in which our solution is based on is in fact a variation of the multitier client-server model presented in the previous section. Our contribution lies in enhanced portability and security, making it a good choice for low-end mobile devices and closed environments in which security is at stake. Moreover, as explained later, the theoretical model in which our solution is based fits well to P2P wiki architectures, too.

To deal with the portability issue, we employed the extended markup language (XML) standard [24, 25]. One the other hand, to satisfy security needs we used XML security [26], as well as a lightweight custom authentication protocol described in the following. XML is a general-purpose markup language. Its extensible character stems from the fact that it allows its users to define their own custom tags. Its primary purpose is to assist the sharing of data across heterogeneous information systems or platforms. Due to these properties, custom languages based on XML do not carry any information related to the presentation of the data, as with HTML, but rather only for the data themselves. This attribute makes XML the ideal choice in our case, when compared to HTML as explained previously in Section 16.2.

In our case it is adequate to define a very simple XML-based language that is able to describe wiki data. On receiving a wiki request, the server will query for the matching data stored in the database. After that, it will transform the data back to XML form (which the client can also understand) and forward it to the client. The exact way the data should be presented on the client's screen is up to the client itself. On the other hand, the client must also understand this custom-tailored XML-based language and fabricate its queries accordingly (i.e., building the appropriate XML file before transmitting it to the server).

As far as the confidentiality and integrity of the data in transit are concerned, we utilize the XML encryption and XML signature specifications [27, 28]. XML encryption is recommended by the W3C specification that defines the exact way to encrypt the content of an XML element. XML encryption covers the encryption of arbitrary data, including an XML document, an XML element, or XML element content. When encrypting an XML element or element content, the encrypted data element replaces the element or content in the encrypted version of the XML document, respectively.

XML encryption provides end-to-end security for applications that require secure exchange of structured data. For instance, in contrast to XML encryption, the following are two important areas not addressed by SSL: (a) encrypting part of the data being exchanged, and (b) secure sessions between more than two parties. As already said, if the application requires that the whole communication be secure, then SSL is the proper choice. On the other hand, XML encryption is an excellent choice if the application requires a combination of secure and insecure communication; this means that some of the data will be securely exchanged and the rest will be exchanged as plaintext. This feature fits best in our case, ensuring maximum portability and improved performance for low-end mobile devices. Moreover, it is fully compatible with the P2P wiki model.

XML signature, also known with the terms XMLDsig, XML-DSig, XML-Sig, is a W3C recommendation that defines an XML syntax for digital signatures. XML signature has many similarities with PKCS #7 [29] but is far more extensible for signing XML documents. It is also used by various web technologies such as simple object access protocol (SOAP) [30, 31] and security assertion markup language (SAML) [32].

16.3.2 Architecture

The overall architecture of our solution is presented in Figure 16.3. On registration to the wiki system, a new user is created to the MySQL database using the credentials *{username, password}* of her choice. Particular rights are assigned to her by utilizing a role-based system. Currently, there are three access modes, each one assigned to the corresponding role:

- Read-only access to the wiki topics (the user is able to execute only the select command toward the database);
- Read/write access (both select and update commands toward the database are allowed);
- Full access (for advanced only).

Of course, several other roles may be created according to the specific needs. Moreover, each wiki topic in the database contains both sensitive (when in transit, it must be cryptographically protected) and nonsensitive fields. Those fields can be read and altered by advanced users only. When creating a new page, an advanced user may designate by clicking on a check box which particular fields are sensitive.

Another option is to associate each page with certain permissions (e.g., none, read only, read/write, full for sensitive data) mapped to individual users. This way every page is associated with an ACL, while each newly created page will inherit its parent's permissions. This means that a page's permissions are totally independent of other pages, except if it inherits its parent's page rights. Using this principle, any page tree may include subpages that have different accessibility than their root pages. Anyone who has granted read/write access to a page can read and modify all fields except sensitive ones. In contrast, everybody that has full access to a page can read and alter the sensitive data it contains as well as its access permissions. To ease

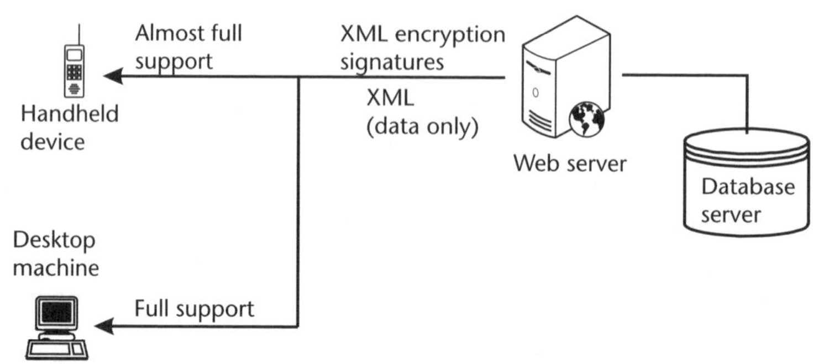

Figure 16.3 The proposed architecture.

administration tasks and compact ACLs, groups of users may be defined. Hence, every authenticated user may create a group, administer a page he has access to, and add new users. This option is also well coupled with wiki philosophy.

Normally, a closed project will involve a group of users, and that group will have read/write or full access to project pages. The group may permit some of the project page tree to be read by others. So, once a top-level page has been created, all other subpages to the leaf can simply inherit its permissions. Using the sensitive field option, users with advanced status may or may not restrict access even to certain fields from certain pages.

As previously mentioned, in the logic tier lies our application, which is in charge of serving clients' requests. When receiving a request from a client in the form of an XML file, it parses it and undertakes the mission to serve it. If the request is a query, the application retrieves wiki data from the database, converts it to an XML file, and transmit it toward the client. On the downside, if the request concerns an update or insert, the application executes the corresponding command to the data base. Whenever some fields or even an entire topic must be scrambled before transmission, the application enciphers it using a symmetric session key and places it into the XML file as designated by the XML encryption specs.

In the client side resides a wiki application, too. The client application receives XML files from the server and parses it to isolate the data from the tags. In case the XML file carries some cryptographically protected sensitive fields, then the application deciphers them by utilizing the session key. After that, it displays the information according to the GUI implemented. The inverse procedure is followed when the client must send some sensitive data to the server. As can be easily noticed, ciphering/deciphering procedures on the client is based on a symmetric session key, and it is flexible by the means that it can be optionally applied only to certain sensitive fields.

16.3.3 Authentication and Key Agreement Protocol Description

In this section we describe a simple lightweight authentication and key agreement (AKA) protocol that enables a user to securely authenticate himself to the wiki server. The protocol also produces a 256-bit-length key to serve as the session key. The authentication process illustrated in Figure 16.4 is one way; the client is authenticated to the server, not the opposite. Actually, mutual authentication is not necessary here, since fake or rogue wiki servers cannot harm clients by any means, except causing DoS. That is, the client repeatedly attempts to authenticate to a fake server unsuccessfully. The server is not in a position to eavesdrop on any valuable information transmitted, so this attack is considered harmless. Of course, DoS can be achieved by various and far more profitable ways in a wiki system; thus, this attack is not even attractive to the attackers themselves.

In a nutshell, the wiki AKA protocol utilizes both symmetric and asymmetric cryptography, depending on the case or its execution phase. The well-known advanced encryption standard (AES) [33] with a key length of 256 bits is used for symmetric ciphering/deciphering, while the Rivest Shamir Adleman (RSA) algorithm [34] with a key length of 1,024 bits is used for public key operations. It is also assumed that all clients hold a copy of the server's public key in the form of a base-64 encoded X.509 certificate [35, 36] issued by a CA. Note that a base-64 encoded

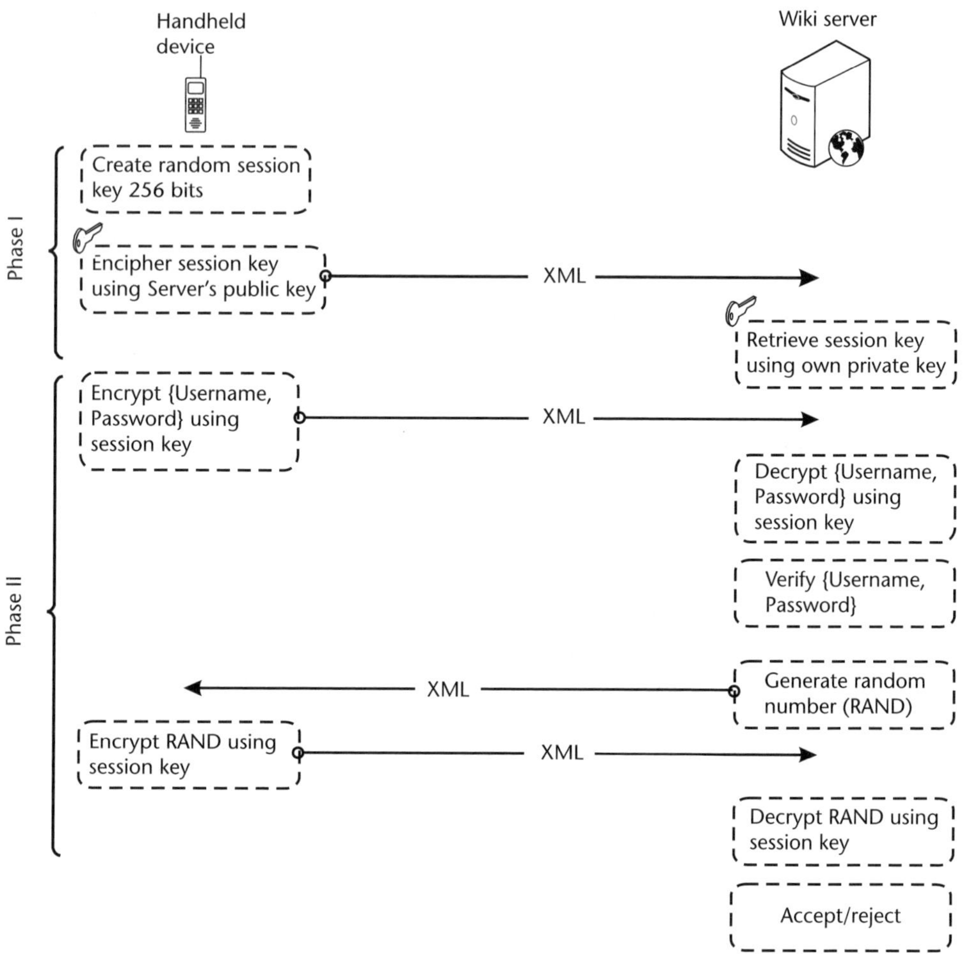

Figure 16.4 Authentication and key agreement protocol message flow.

certificate is very easy to manage and transfer to virtually every mobile device, as it is in plain text. For instance, a wiki server's certificate can be distributed by email to all wiki clients.

The protocol is executed in two distinct phases. During the first phase:

1. The client generates a random 256-bit session key, encrypts it using the public key of the server, and sends it to the server.
2. On receipt, the server retrieves the session key using its private key.
3. The client encrypts its credentials {username, password} employing the session key and sends them to the server.
4. The server retrieves user's credentials using the same session key. Until that point, the client has been successfully authenticated to the server but the process itself is vulnerable to joint man-in-the-middle/replay attacks. More specifically, assuming that the attacker eavesdrops on the link between the

server and the client, she can record all the messages transmitted. After that, at a later time, she can replay them toward the server and become successfully authenticated (still, she does not possess the session key and cannot read sensitive data). Therefore, some further actions must be taken in order to guarantee that the client attempting to authenticate is the legitimate one (i.e., does have the proper session key). For this reason we additionally perform a challenge-response procedure consisting of the following steps:

- The server generates a random number (RAND) and sends it toward the client.
- On receipt, the client encrypts RAND using the session key and returns the result to the server.
- After decryption, the server verifies if the original RAND matches with the one returned by the client.

Measurements showed that the wiki AKA service timespan was between 0.5 and 1.0 seconds. Tests were conducted using as client device a Fujitsu-Siemens Loox N560 Pocket PC, which incorporates a 624-MHz Intel X-scale PXA270 processor, and an IEEE 802.11b/g wireless connection. The operating system running on the device was Windows Mobile in version 5.0. Wiki prototype applications for both the client and the server were written in C# [37]. A Java-based client is also under consideration to support mobile clients running other operating systems like Linux, Symbian, or Palm OS.

16.3.4 Confidentiality & Integrity of Communication

As soon as the user has been successfully authenticated, confidentiality of sensitive data is based on XML encryption, which utilizes the established session key known to both parties. The sender encrypts only the sensitive fields of the wiki data, inserts them to the XML file as appropriate, and delivers them to the other part. The other part will follow the opposite procedure to acquire the original content. It should also be remembered that XML encryption is not proposed here to replace or supersede SSL/TLS. Rather, it provides a mechanism for security requirements (lightness) that are not covered by SSL, as explained in Section 16.3.1. Requests issued by normal (nonauthorized) users to read data that includes sensitive fields return only nonsensitive (nonprotected data), if any.

The integrity of the XML messages in transit can also be supported. Every time the client wishes to protect the integrity of the contents of an XML file prior to transmitting it: (a) the client computes a hash over the XML file, (b) the hash is digitally signed (i.e., encrypted using the session key), and (c) the signature is placed on the XML file as a special tag according to the XML signature specifications. The other part verifies the acquired signature by recalculating the hash of the file and comparing it with the received one after decryption. It is stressed that confidentiality and integrity services can be applied either to the XML file as a whole or to some sensitive parts of it.

16.4 Conclusions

During the last few years, wikis have emerged as one of the most popular tool shells. It is not an exaggeration to say that wikis dominate in every context that calls for effective collaboration and knowledge sharing at a low cost. Wikipedia has certainly boosted their popularity, but they also keep a significant share in intranet-based applications such as defect tracking, requirements management, test-case management, and project portals.

In this chapter, we analyzed the requirements for a novel multiplatform secure wiki implementation. It is made clear that existing wiki systems cannot fully support mobile clients due to several incompatibilities that exist. Moreover, an effective secure mobile wiki system must be lightweight enough to support low-end mobile devices having several limitations. Consequently, XML encryption and signature specifications were employed to realize end-to-end confidentiality and integrity services. Our scheme can be applied selectively and only to sensitive wiki content, thus diminishing by far computational resources needed at both ends. To address authentication of wiki clients, a simple one-way authentication and session key agreement protocol was also introduced. This protocol is also of low demands, requiring only one public operation on the client side. On top of everything else, the proposed solution can be easily applied to both centralized and forthcoming P2P wiki implementations.

Acknowledgments
I would like to thank Mr. Stefanos Demertzis and Mr. Costantinos Kollias for their valuable contribution to this work.

References

[1] Cunningham, W., “Original Wiki,” http://c2.com/cgi-bin/wiki?WikiWikiWeb, 2004.

[2] Louridas, P., “Using Wikis in Software Development,” *IEEE Software*, Vol. 23, No. 2, 2006, pp. 88–91.

[3] Fichter, D., “Intranets, Wikis, Blikis, and Collaborative Working,” *Online*, Vol. 29, No. 5, 2005, pp. 47–50.

[4] Leuf, B., and W. Cunningham, *The Wiki Way: Quick Collaboration on the Web*, Upper Saddle River, NJ: Addison Wesley, 2001.

[5] Raitman, R., and N. Augar, “Constructing Wikis As a Platform for Online Collaboration in an E-Learning Environment,” *Proceedings of International Conference on Computers on Education*, Australia, 2004.

[6] Kim, E. E., “The Future of Wikis,” *Proceedings of WikiSym’06—2006 International Symposium on Wikis 2006*, p. 17.

[7] Beldarrain, Y., “Distance Education Trends: Integrating New Technologies to Foster Student Interaction and Collaboration,” *Distance Education*, Vol. 27, No. 2, 2006, pp. 139–153.

[8] Volkel, M., S. Schaffert, and E. Pasaru-Bontas, “Wiki-Based Knowledge Engineering,” *Proceedings of WikiSym’06—2006 International Symposium on Wikis*, p. 133.

[9] Rafaeli, S., “Wiki Uses in Teaching and Learning,” *Proceedings of WikiSym’06 International Symposium on Wikis 2006*, pp. 15–16.

[10] Ebersbach, A., M. Glaser, and R. Heigl, "The Wiki Concept," in *Wiki*, Heidelberg, Berlin: Springer, 2006, pp. 9–30.

[11] Meatball Wiki, "Soft Security,"http://www.usemod.com/cgi-bin/mb.pl?SoftSecurity.

[12] Wikipedia, "Comparison of Wiki Software," http://en.wikipedia.org/wiki/Comparison_of_wiki_software.

[13] Mason, R., and P. Roe, "RikWik: An Extensible XML Based Wiki," *Proceedings of the 2005 International Symposium on Collaborative Technologies and Systems*, pp. 267–273.

[14] Dondio, P., et al., "Extracting Trust from Domain Analysis: A Case Study on the Wikipedia Project," *Lecture Notes in Computer Science*, LNCS 4158, 2006, pp. 362–373.

[15] Raitman, R., et al., "Security in the Online E-Learning Environment," *Proc. 5th IEEE International Conference on Advanced Learning Technologies*, ICALT 2005, pp. 702–706.

[16] Perl, "The Perl Language Directory," http://www.perl.org/.

[17] Frier A., P. Karlton, and P. Kocher, "The SSL 3.0 Protocol Version 3.0," http://home.netscape.com/eng/ssl3/draft302.txt.

[18] Dierks, T., and C. Allen, "The TLS Protocol Version 1.0," IETF RFC 2246, January 1999.

[19] Morris, J., and C. Lüer, "DistriWiki: A Distributed Peer-to-Peer Wiki," 2007, submitted for publication, http://www.cs.bsu.edu/homepages/chl/P2PWiki/.

[20] Urdaneta, G., G. Pierre, and M. Van Steen, "A Decentralized Wiki Engine for Collaborative Wikipedia Hosting," *Proceedings of the 3rd International Conference on Web Information Systems and Technology (WEBIST)*, March 2007.

[21] Zhang, G., and Q. Jin, "Scalable Information Sharing Uilizing Decentralized P2P Networking Integrated with Centralized Personal and Group Media Tools," *Proceedings of IEEE International Conference on Advanced Information Networking and Applications (AINA)*, 2006, pp. 707–711.

[22] Wikipedia, "Wikipedia: Size of Wikipedia," http://en.wikipedia.org/wiki/Wikipedia:Size_of_Wikipedia.

[23] Code Co-op, "Peer-to-Peer Version Control for Distributed Development," Reliable Software, http://www.relisoft.com/co_op/.

[24] W3C architecture domain, "Extensible Markup Language (XML)," http://www.w3.org/XML/.

[25] W3C, "Extensible Markup Language (XML) 1.1 (Second Edition)," http://www.w3.org/TR/xml11/, 2006.

[26] XML.org, "XML Security," http://www.xml.org/xml/resources_focus_security.shtml.

[27] W3C, "XML Encryption Syntax and Processing," W3C Recommendation, December 2002, http://www.w3.org/TR/xmlenc-core/.

[28] W3C, "XML-Signature Syntax and Processing," W3C Recommendation, February 2002, http://www.w3.org/TR/xmldsig-core/.

[29] Kaliski, B., "PKCS #7: Cryptographic Message Syntax," Version 1.5, RFC 2315 RSA Laboratories, March 1998.

[30] W3C, "SOAP Version 1.2 Part 0: Primer (Second Edition)," W3C Recommendation, April 27, 2007, http://www.w3.org/TR/soap12-part0/.

[31] W3C, "SOAP Version 1.2 Part 1: Messaging Framework (Second Edition)," W3C Recommendation, April 27, 2007, http://www.w3.org/TR/soap12-part1/.

[32] Ragouzis, N., et al., "Security Assertion Markup Language (SAML) V2.0 Technical Overview," OASIS Draft, February 2007. Document ID sstc-saml-tech-overview-2.0-draft-13, http://www.oasis-open.org/committees/download.php/22553/sstc-saml-tech-overview-2%200-draft-13.pdf.

[33] NIST, "Announcing the Advanced Encryption Standard (AES)," Federal Information Processing Standards Publication 197 November 2001, http://www.csrc.nist.gov/publications/fips/fips197/fips-197.pdf.

[34] RSA Laboratories, "PKCS #1 v2.1: RSA Cryptography Standard," June 14, 2002, ftp://ftp.rsasecurity.com/pub/pkcs/pkcs-1/pkcs-1v2-1.pdf.

[35] Housley, R., et al., "Internet X.509 Public Key Infrastructure: Certificate and CRL Profile," RFC 3280, April 2002, http://tools.ietf.org/html/rfc3280.

[36] Josefsson, S., "The Base16, Base32, and Base64 Data Encodings," RFC4648, October 2006, http://tools.ietf.org/html/rfc4648.

[37] ISO/IEC, "Languages—C#," International Standard 23270, Second edition, November 2006, http://standards.iso.org/ittf/PubliclyAvailableStandards/c042926_ISO_IEC_23270_2006(E).zip.

About the Authors

Manfred Aigner graduated from a technical college for communication and electronics, after which he studied Telematics at the Graz University of Technology with special emphasis on integrated circuits. Since 2001, he has been responsible for the activities of the VLSI group at the IAIK. He was responsible for initiation and coordination of the European FP-6 project SCARD and the FIT-IT projects ART and SNAP. His special interest is in hardware implementation of cryptographic algorithms and side-channel analysis.

Juhani Anttila graduated from Helsinki Technical University in 1967 (MSc-E Eng.) and completed the General Management Programme for Specialists at Cranfield School of Management, UK, 1997. He has been International Academician for Quality (Member of the International Academy for Quality) since 1995. He has been professionally involved more than 40 years with different quality-related tasks and positions, and during that period worked 35 years for the leading Finnish telecommunications operator, Sonera Corporation. He has been broadly involved with national and international standardization. He has been an expert of quality in many national and international projects including developing countries. During 1990–1994 he was the chairman of the criteria committee of the Finnish National Quality Award, and in 1993 he was Assessor of the European Quality Award (EFQM). He served for many years from the 1970s to the 1990s as board member in the Finnish Society for Quality; in 1984–1987 he was the President of the Finnish Society for Quality and from 1998 he was an Honorary Member of the Society. For several years he was the expert responsible for international contacts of the Finnish Society for Quality, including EOQ General Assembly and bilateral scientific-technical cooperation between Finland and other countries. In 1994–1996 he was Vice President of European Organization for Quality. Mr. Anttila has many publications, including contributions in professional periodicals, conferences, and seminars, in the fields of telecommunications, quality/reliability, information security, and business crises. He has been a lecturer in several educational institutes and universities, including the University of Lapland, Finland, and the University of Fribourg, Switzerland. After retiring in 2003 from Sonera Corporation and from the position of vice president of quality integration, Mr. Anttila has been an independent expert—Venture Knowledgist, Quality Integration.

Vassilis Chrissikopoulos is a professor of informatics in the Department of Informatics at the Ionian University. He received his BSc from the University of Thessaloniki, Greece (1976), and his MSc and PhD from the University of London (1979, 1983). During the period 1985–2000 he was a member of staff (assistant professor, associate professor, and professor) in the Department of Informatics at the University of Piraeus. During the period 2000–2007 he was affiliated as a professor in the Department of Archives and Library Science at the Ionian University. His research interests include Information Security and Cryptography, e-commerce, mobile agents, e-voting, and digital libraries. Prof. Chrissikopoulos has participated in several research projects funded by Greece or European Community as a coordinator or as a member. He also was member of several technical committees and working groups on subjects relating to informatics and information security. He is a member of the Greek Mathematical Society and of the Greek Computer Society.

Nathan Clarke is a lecturer in information systems security within the Information Security and Network Research Group at the University of Plymouth. His research interests reside in the area of biometrics, mobility, and wireless security, having published 31 papers in international journals and conferences. Dr. Clarke is a chartered engineer, a member of the British Computing Society (BCS) and the Institute of Engineering Technology (IET), and a UK representative in the International Federation of Information Processing (IFIP) working groups relating to information management and identity management.

Paul Dowland, MBCS, is a senior lecturer in the School of Computing, Communications and Electronics, at the University of Plymouth, UK, and is a member of the Information Security and Network Research Group, a post-graduate and post-doctoral team encompassing 22 researchers and 18 staff. His current research interests include information systems security and online learning. Research within the Network Research Group encompasses a range of industrial and European projects, and details can be found at http://www.network-research-group.org. Dr. Dowland is a professional member of the British Computer Society, member and secretary of IFIP Working Group 11.1 (Information Security Management), and was appointed as an honorary fellow of the Sir Alister Hardy Foundation for Ocean Science (www.sahfos.ac.uk). He has authored/edited more than 40 publications and has organized 8 conferences since 2000 as well as reviewing for a further 20.

Simone Fischer-Hübner has been a full professor at the computer science department of Karlstad University since June 2000, where she is the head of the PriSec (Privacy and Security) research group. She received her Doctoral (1992) and Habilitation (1999) degrees in computer science from Hamburg University. Her research interests include technical and social aspects of IT security, privacy, and privacy-enhancing technologies. She was a research assistant/assistant professor at Hamburg University (1988–2000) and a guest professor at the Copenhagen Business School (1994–1995) and at Stockholm University/Royal Institute of Technologies (1998–1999). She is the vice chairperson of International Federation for Information Processing (IFIP) Working Group 11.6 on "Identity Management" and served as the chair of IFIP WG 9.6/11.7 on "IT Misuse and the Law" (1998–2005). She is a member of the External Advisory Board of the <http://www.research.ibm.com/privacy/> IBM Privacy Institute, board member of the IEEE-Sweden–Section Computer/Software Engineering Chapter, member of the NordSec (Nordic Workshop on Secure IT Systems) steering committee, coordinator of the Swedish IT Secure

Network for PhD students, and member of the International Editorial Review Board of the *International Journal of Information Security and Privacy* (IJISP). She is currently representing Karlstad University in the EU FP6 projects Privacy and Identity Management in Europe (PRIME) and Future of Identity in the Information Society (FIDIS).

Steven Furnell is the head of the Center for Information Security and Network Research at the University of Plymouth in the United Kingdom, and an adjunct professor with Edith Cowan University in Western Australia. He specializes in computer security and has been actively researching in the area for 15 years, with current areas of interest including security management, computer crime, user authentication, and security usability. Prof. Furnell is a Fellow and Branch Chair of the British Computer Society (BCS), a senior member of the Institute of Electrical and Electronics Engineers (IEEE), and a UK representative in International Federation for Information Processing (IFIP) working groups relating to information security management (of which he is the current chair), network security, and information security education. He is the author of more than 180 papers in refereed international journals and conference proceedings, as well as the books *Cybercrime: Vandalizing the Information Society* (2001), Addison-Wesley and *Computer Insecurity: Risking the System* (2005), Springer. Further details can be found at www.cisnr.org.

Stefanos Gritzalis (www.icsd.aegean.gr/sgritz) holds a BSc in physics, an MSc in electronic automation, and a PhD in informatics all from the University of Athens, Greece. Currently he is an associate professor, the head of the Dept. of Information and Communication Systems Engineering, University of the Aegean, Greece, and the director of the Lab. of Information and Communication Systems Security (Info-Sec-Lab). He has been involved in several national and EU funded R&D projects in the areas of information and communication systems security. His published scientific work includes several books on information and communication technologies topics, and more than 150 journal and national and international conference papers. The focus of these publications is on information and communication systems security. He has served on program and organizing committees of national and international conferences on informatics and is an editorial advisory board member and reviewer for several scientific journals. He was a member of the board (secretary general, treasurer) of the Greek Computer Society. He is a member of the ACM and the IEEE. Since 2006 he is a member of the "IEEE Communications and Information Security Technical Committee" of the IEEE Communications Society and of the IFIP WG 11.6 "Identity Management."

Norleyza Jailani is currently a lecturer at the National University of Malaysia (UKM), Department of Computer Science. She graduated from the National University of Malaysia with a BSc in computer science in 1992, and holds an MSc degree in computer science from University College, Dublin, Ireland, in 1996. She is currently pursuing her PhD at the University of Malaysia (UM) in Kuala Lumpur. Her research interests are in distributed computing and computer networks, agent-mediated electronic commerce, agent-based auction systems, security in mobile agent-based systems, and forensic investigation of transactions in the marketplace for mobile users, location-based services in public transportation systems, and information visualization in transportation network.

Jorma Kajava holds an MSc degree in control and automation engineering and a Lic.Tech degree in telecommunication from the University of Oulu, Finland, and currently is research professor in information security at the University of Lapland in Rovaniemi, Finland. He first joined the University of Oulu in 1974 with the Control and Systems Engineering laboratory as a teacher and with the wireless telecommunications group as a researcher. Since 1978 he taught at the Department of Information Processing Science of the University of Oulu as a head of Systemeering laboratory, lecturer, and professor until the end of the year 2005. His international experience ranges from the participation of international EU-based research projects and teaching in Sweden, Greece, Austria, Spain, and Russia. His major research focus is on information security management, end-user security and security education. He has been the head of several industry-related domestic projects during 27 years. Since January 2006 he is research professor in information security with the Department of Research Methodology of the University of Lapland.

Georgios Kambourakis was born in Samos, Greece, in 1970. He received the Diploma in Applied Informatics from the Athens University of Economics and Business (AUEB) in 1993 and the PhD in information and communication systems engineering from the department of Information and Communications Systems Engineering of the University of the Aegean (UoA). He also holds a MEd from the Hellenic Open University. Currently, Dr. Kambourakis is a lecturer at the Department of Information and Communication Systems Engineering of the University of the Aegean, Greece. His research interests are in the fields of mobile and ad hoc networks security, VoIP security, security protocols, public key infrastructure and m-learning, and he has more than 35 publications in these areas. He has been involved in several national and EU-funded R&D projects in the areas of information and communication systems security. He is a reviewer of several IEEE and other international journals and has served as a technical program committee member in several conferences.

Sokratis K. Katsikas was born in Athens, Greece, in 1960. He received the Diploma in Electrical Engineering from the University of Patras, Patras, Greece, in 1982, the Master of Science in Electrical and Computer Engineering degree from the University of Massachusetts at Amherst in 1984 and the PhD in Computer Engineering and Informatics from the University of Patras, Patras, Greece, in 1987. Currently he is a professor with the Dept. of Technology Education and Digital Systems of the University of Piraeus, Greece. From 1990 to 2007, he was with the University of the Aegean, Greece, where he served as rector, vice-rector, department head, professor of the Department of Information and Communication Systems Engineering and director of the Information and Communication Systems Security Lab. His research interests lie in the areas of information and communication systems security and of estimation theory and its applications. He has authored or co-authored more than 150 journal publications, book chapters, and conference proceedings publications in these areas. He is serving on the editorial board of several scientific journals, he has authored/edited 20 books, and has served on/chaired the technical program committee of numerous international conferences.

Dogan Kesdogan has been senior researcher at the Computer Science Department (Informatik IV) of the Aachen University of Technology (RWTH) since 2000. His current research interests and education fields include security in networks, privacy techniques, foundations and modeling of mixes, security in mobile communication systems, and distributed systems. Dr. Kesdogan has spent his sabbatical year in the United States at the IBM T. J. Watson Research Center, in New York from 2001 to 2002. He was security expert at the Mannesmann o.tel.o Communications GmbH & Co. from 1998 to 2000. He received his doctoral degree from the Aachen University of Technology in 1999. Currently he is a guest professor at the Norwegian University of Science and Technology (NTNU).

Costas Lambrinoudakis was born in Greece in 1963. He holds a BSc (Electrical and Electronic Engineering) degree from the University of Salford (UK), and MSc (Control Systems) and PhD (Computer Science) degrees from the University of London (UK). Currently he is an assistant professor at the Department of Information and Communication Systems of the University of the Aegean Greece and the associate director of the Laboratory of Information and Communication Systems Security (Info-Sec-Lab). His current research interests include information systems security, privacy enhancing technologies, and smart cards. He is an author of several book chapters and refereed papers in international scientific journals and conference proceedings. He has participated in many national and EU-funded R&D Projects. He has served on program and organizing committees of many national and international conferences on Informatics, and he is a reviewer for several scientific journals. He is a member of the ACM and the IEEE.

Javier Lopez received his MS in Computer Science and PhD in Computer Engineering in 1992 and 2000, respectively, from University of Malaga. After a period of four years as system analyst in the industrial sector, he joined the Computer Science Department at the University of Malaga, completing during that period several research stays at University of Wisconsin–Milwaukee (US), Yale University (US), Queensland University of Technology (Australia), and the University of Tsukuba (Japan). He has been actively involved in research projects of the V, VI, and VII European Framework Programmes and has participated in more than 80 international program committees of security and cryptography events. He is also co-editor in chief of the *International Journal of Information Security* (IJIS), and member of the editorial boards of the *Information Management and Computer Security Journal* (IMCS), *Security and Communications Network Journal* (SCN), and *International Journal of Internet Technology and Secured Transactions* (IJITST). Additionally, he is Spanish representative of the IFIP TC-11 WG (Security and Protection in Information Systems) and member of the Spanish Mirror Committee JTC1 of ISO.

Emmanouil Magkos received his first degree in Computer Science from the University of Piraeus, Greece, in 1997. In April 2003 he received a PhD in Security and Cryptography from the University of Piraeus, Greece. His PhD research was related to cryptographic protocols and techniques for securing electronic transactions between mistrusted parties that communicate over open networks such as the Internet. Since 2003 he has been teaching computer science at the Archives and Library Sciences Department, Ionian University, Corfu, Greece. Currently he is affiliated with the Department of Informatics, Ionian University, Corfu, Greece, where he holds the position of lecturer in computer security and cryptography. His current research interests include key management in wireless ad hoc networks, access control in logging systems, monitoring infrastructures for worm (malware) propagation, and cryptographic security of Internet voting systems.

Leonardo A. Martucci is a researcher at the Department of Computer Science at Karlstad University, Sweden, in the field of computer security and privacy. He has been involved in education, research, and deployment in the field of wireless network security and privacy since 2001 in academic and industrial projects. His research focuses in privacy problems in dynamic and distributed environments, such as mobile ad hoc networks. He holds a Licentiate in Engineering Degree from Karlstad University (2006), a Masters in Electrical Engineering (2002), and an Electrical Engineer Degree (2000) from the University of Sao Paulo, Brazil.

Natalia Miloslavskaya graduated from the Moscow State Engineering Physical Institute (Technical University) on an "engineer-mathematician" speciality. Since that time all her working activities have been connected with the MEPhI. At first she was a research associate, then a post-graduate, scientific associate, and a head of some scientific groups. She received the PhD (1988) degree from the MEPhI in Technical Sciences. She is currently associate professor, vice dean of the Information Security Faculty of the MEPhI and deputy head of the Information Security of Banking Systems Department. Her research interests lie in network security of different types of systems. She also studies distance learning and testing methods that she implements in the distance learning and progress testing system that has been created for her original educational course "Vulnerability and Protection Methods in Networks" and "Information Security of Open Systems." She does research on security solutions, services, and policies for electronic commerce applications (especially on secure protocols). She is a lecturer at the MEPhI and at the retraining courses for specialists from the Russian financial and banking sector. More than 6,500 trainees have taken her educational course. She acts as a supervisor and consultant of graduates and post-graduate students. She has written or co-authored 140 papers and 10 textbooks.

Lilian Mitrou is assistant professor at the Department of Information and Communication Systems Engineering, University if the Aegean, Greece. She holds a PhD in Data Protection (Law School of Johann Wolfgang Goethe Universitaet Frankfurt). L. Mitrou teaches electronic democracy, electronic governance, information law, and data protection at the University of Athens (Dept. of Law, Dept. of Political Sciences—Postgraduate Studies) as well as information law and electronic governance at the National Academy of Public Administration. She has served as a member of the Hellenic Data Protection Authority (1999–2003) and as Adviser to the Prime Minister in the sectors of information society and public administration (1996–2004). From 1998 until 2004 she was the national representative in the EC Committee on the Protection of Individuals with Regard to the Processing of Personal Data. She served as member of many committees working on law proposals in the fields of data protection (transposition of Directives 95/46/EC, 97/66/EC and 02/58/EC), electronic government, electronic commerce, and so on. Her professional experience includes senior consulting and researcher positions in a number of private and public institutions. She has published 12 books or chapters in books (in Greek, German, and English) and several journals and national and international conference papers.

Ahmed Patel received his MSc and PhD degrees in Computer Science from Trinity College Dublin, specializing in the design, implementation, and performance analysis of packet-switched networks. He is a lecturer and consultant in ICT and computer science. He is a visiting professor at Kingston University in the UK and currently lecturing at University Kebangsaan Malaysia. His research interests span topics concerning high-speed computer networking and application standards, network security, forensic computing, autonomic computing, heterogeneous distributed computer systems, and distributed search engines and systems for the Web. He has published well over 170 technical and scientific papers and co-authored two books on computer network security and one book on group communications. He also co-edited a book on distributed search systems for the Internet. He is a member of the editorial advisory board of the following international journals: (a) *Computer Communications*, (b) *Computer Standards and Interfaces*, (c) *Digital Investigations*, (d) *Cyber Criminology*, and (e) *Forensic Computer Science*. He lecturers on cybercrime investigations and digital forensics on the IPICS school courses.

Günther Pernul received both the diploma and the doctorate degrees (with honors) from the University of Vienna, Austria. Currently he is chair and full professor at the Department of Information Systems at the University of Regensburg, Germany. His current major research interests are information systems security, individual privacy and data protection, identity management, authorization, and access control. Additionally, he has developed interests in web-based information systems, advanced data-centric applications, and ICT for user groups at risk of becoming excluded from the information society. Dr. Pernul is co-author of a text book, has edited or co-edited 10 books, published more than 100 papers in scientific journals and conference proceedings on various information systems topics, and has actively participated in nationally and internationally funded research. He has been involved in several European projects (e.g., in the FP7 project SPIKE, which he is also leading as a project coordinator). Dr. Pernul served as member in editorial boards and in program committees of more than 50 international conferences, is founding editor of the conference series Electronic Commerce and Web Technologies (EC-Web, since 2000), Trust, Security and Privacy in Digital Business (Trustbus, since 2004), and associate editor of the *International Journal of Information System Security*.

Karl C. Posch is professor at the Institute for Applied Information Processing and Communications, Graz University of Technology, Austria. He holds a Master's Diploma in Electrical Engineering (1979) and a PhD in Computer Science (1988). He has been working in the industry, first leading a hardware design group in 1983–84, then as guest professor at Denver University (1988) in Colorado, and at the Memorial University of St. John's (1990–91) in Newfoundland, Canada. His research interests are microchip design and information security. In particular he is interested in smartcards and contactless technology.

Torsten Priebe holds a Diploma Degree in Information Systems from the University of Essen, Germany (2000), and a PhD in economics from the University of Regensburg (2005). He has been working for the Department of Information Systems at the University of Essen since 1999 and was involved in various international research projects of the group. Together with Prof. Günther Pernul, he moved to the University of Regensburg in 2002. Since 2006 he has been working as a consultant for Capgemini in Vienna, Austria. Since 2007 he has also been teaching business intelligence and data warehousing at the University of Regensburg. Dr. Priebe is coauthor of three books, has published in international conference proceedings and journals, and gives talks on business intelligence and data warehousing, knowledge management, and information security. Dr. Priebe is member of ACM, IEEE, AIS, and GI, member of IFIP WG 11.3, serves on program committees of international conferences such as the International Conference on Trust, Privacy and Security in Digital Business (TrustBus), and as a reviewer for international journals such as the *Journal of Systems and Software* (Elsevier).

Gerald Quirchmayr holds doctorate degrees in computer science and law from Johannes Kepler University in Linz (Austria) and currently is a professor at the Institute of Distributed and Multimedia Systems at the University of Vienna. In 2001–2002 he held a chair in Computer and Information Systems at the University of South Australia. He first joined the University of Vienna in 1993 from the Institute of Computer Science at Johannes Kepler University in Linz (Austria), where he had previously been teaching. In 1989–1990 he taught at the University of Hamburg (Germany). His wide international experience ranges from the participation in international teaching and research projects, very often UN- and EU-based, several research stays at universities and research centers in the United States and EU member states to extensive teaching in EU staff exchange programs in the United Kingdom, Sweden, Finland, Germany, Spain, and Greece, as well as teaching stays in the Czech Republic and Poland. International teaching and specialist missions include UN-coordinated activities in Egypt, Russia, and the Republic of Korea. His major research focus is on information systems in business and government with a special interest in security, applications, formal representations of decision making, and legal issues. In July 2002 he was appointed adjunct professor at the School of Computer and Information Science of the University of South Australia. Since January 2005 he heads the Department of Distributed and Multimedia Systems, Faculty of Computer Science, at the University of Vienna.

Chris Wills is the Director of the Centre for Applied Research in Information Systems at Kingston University in London. He was educated in the UK, at the universities of Oxford and Brunel, and has worked for a number of years as a management consultant. His research interests include the analysis of complex human activity systems, and he worked in this field for a number of years for the Royal Navy. Recently, he has undertaken research in software reliability and software process models on behalf of some of the world's largest rail and metro companies. Chris is a member of the City of London's Worshipful Company of Information Technologists and is a Freeman of the City of London.

Louise Yngström is a professor in computer and systems sciences with specialization in security informatics in the joint department of Computer and Systems Sciences at Stockholm University and the Royal Institute of Technology. Her research base is systems science, and, since 1985, she has applied this within the area of ICT security, forming holistic approaches. Her research focuses on various aspects of how ICT security can be understood and thus managed by people in organizations, but also generally on criteria for control. She has been engaged in various activities of the International Federation of Information Processing (IFIP) since 1973; the Technical Committee 3(TC3) with an educational scope, the TC9 with focus on social accountabilities of ICT structures, and the TC11 with focus on ICT security. She founded the biannual conference World Conference on Security Education (WISE) in 1999. She was also engaged in European ERASMUS networking for curricula developments within the ICT security area during the 1990s and involved in introducing ICT security in academic and business life in African countries through her research students who simultaneously with their research are academic teachers in their home countries. Over the years she has traveled and networked extensively with international peers. Presently she is the principal advisor of seven PhD students. Further information can be found on http://www.dsv.su.se/~louise.

Index

J

K

L

M

Q

R

Recent Titles in the Artech House Computer Security Series

Rolf Oppliger, Series Editor

Bluetooth Security, Christian Gehrmann, Joakim Persson and Ben Smeets

Computer Forensics and Privacy, Michael A. Caloyannides

Computer and Intrusion Forensics, George Mohay, et al.

Defense and Detection Strategies against Internet Worms, Jose Nazario

Demystifying the IPsec Puzzle, Sheila Frankel

Developing Secure Distributed Systems with CORBA, Ulrich Lang and Rudolf Schreiner

Electric Payment Systems for E-Commerce, Second Edition, Donal O'Mahony, Michael Peirce, and Hitesh Tewari

Evaluating Agile Software Development: Methods for Your Organization, Alan S. Koch

Implementing Electronic Card Payment Systems, Cristian Radu

Implementing the ISO/IEC 27001 Information Security Management System Standard, Edward Humphreys

Implementing Security for ATM Networks, Thomas Tarman and Edward Witzke

Information Hiding Techniques for Steganography and Digital Watermarking, Stefan Katzenbeisser and Fabien A. P. Petitcolas, editors

Internet and Intranet Security, Second Edition, Rolf Oppliger

Introduction to Identity-Based Encryption, Luther Martin

Java Card for E-Payment Applications, Vesna Hassler, Martin Manninger, Mikail Gordeev, and Christoph Müller

Multicast and Group Security, Thomas Hardjono and Lakshminath R. Dondeti

Non-repudiation in Electronic Commerce, Jianying Zhou

Outsourcing Information Security, C. Warren Axelrod

Privacy Protection and Computer Forensics, Second Edition, Michael A. Caloyannides

Role-Based Access Control, Second Edition, David F. Ferraiolo, D. Richard Kuhn, and Ramaswamy Chandramouli

Secure Messaging with PGP and S/MIME, Rolf Oppliger

Securing Information and Communications Systems: Principles, Technologies and Applications, Javier Lopez, Steven Furnell, Sokratis Katsikas, and Ahmed Patel

Security Fundamentals for E-Commerce, Vesna Hassler

Security Technologies for the World Wide Web, Second Edition, Rolf Oppliger

Techniques and Applications of Digital Watermarking and Content Protection, Michael Arnold, Martin Schmucker, and Stephen D. Wolthusen

User's Guide to Cryptography and Standards, Alexander W. Dent and Chris J. Mitchell

RECEIVED
JUL 23 2008